# Applications of Remote Sensing/ GIS in Water Resources and Flooding Risk Managements

Special Issue Editors

**Hongjie Xie**
**Xianwei Wang**

MDPI • Basel • Beijing • Wuhan • Barcelona • Belgrade

MDPI

*Special Issue Editors*
Hongjie Xie
University of Texas at San Antonio
USA

Xianwei Wang
Sun Yat-sen University
China

*Editorial Office*
MDPI
St. Alban-Anlage 66
Basel, Switzerland

This edition is a reprint of the Special Issue published online in the open access journal *Water* (ISSN 2073-4441) from 2016–2018 (available at: http://www.mdpi.com/journal/water/special_issues/ rs_gis_flooding).

For citation purposes, cite each article independently as indicated on the article page online and as indicated below:

Lastname, F.M.; Lastname, F.M. Article title. *Journal Name* **Year**, *Article number*, page range.

**First Editon 2018**

**ISBN 978-3-03842-982-1 (Pbk)**
**ISBN 978-3-03842-981-4 (PDF)**

# Table of Contents

# About the Special Issue Editors

**Hongjie Xie** is a professor in the Department of Geological Sciences and Director of Laboratory for Remote Sensing and Geoinformatics, University of Texas at San Antonio (UTSA), USA. He obtained a Ph.D. in Geological Sciences from the University of Texas at El Paso in 2002, and then worked as a Post Doc in New Mexico Tech before joining UTSA in 2004. He is interested in interdisciplinary research by integrating remote sensing, GIS, field measurements correlated with geology (the Earth and Mars), surface hydrology, hydrometeorology, cryosphere, terrestrial ecosystems, urban development, and environmental studies. His technical expertise allows him to approach many research questions regarding spatio-temporal patterns and processes through quantitative modeling. His research has been funded by U.S. NSF, NASA, NOAA, USGS, and the U.S. Department of Education. He has published several books/chapters and over 100 peer-reviewed journal articles.

**Xianwei Wang** is a professor at the School of Geography and Planning, Sun Yat-sen University (SYSU), Guangzhou, China. He obtained a Ph.D. in Environmental Science and Engineering from the University of Texas at San Antonio in 2008, and then worked as a Post Doc from 2008–2010 at the University of California, Irvine. He joined SYSU in 2011 and worked as a Deputy Director of the Guangdong Engineering and Technology Research Center for Public Security and natural Hazards (GETRC-PuSH). His recent research focus is on the fields of water resource mapping, hydraulic infrastructures survey and management, flooding modeling and risk management. His research has been funded by the National Science Foundation of China, High Technology Development Program, the Department of Science and Technology of China, Department of Hydraulics in Guangdong Provinces, among others. He has published several books/chapters and over 40 peer-reviewed journal articles.

# Preface to "Applications of Remote Sensing/GIS in Water Resources and Flooding Risk Managements"

Water, one of the most important natural resources, supports our daily life, maintains the ecosystems that we rely on, provides transportation, recreation, ecotourism, and much more. Pressures on water resources and disasters are rising primarily due to unequal distribution, urbanization, extreme and frequent drought and flooding, pollution, deforestation, and also partly due to poor knowledge about the distribution of water recourses and poor management of water resources and usage. Remote sensing provides critical data for mapping water resources and changes, while GIS provides the best tool for water resource and flood risk management, presentation, visualization and public education. This Special Issue collects the best practices, cutting-edge technologies and applications of remote sensing, GIS and hydrologic models for water resources mapping, satellite rainfall measurements, runoff simulation, urban water body and flood inundation mapping. We hope you will find this issue valuable and it helps to improve water and water-related risk research and management moving forward to a new era.

**Hongjie Xie, Xianwei Wang**
*Special Issue Editors*

*water*

MDPI

*Editorial*

# A Review on Applications of Remote Sensing and Geographic Information Systems (GIS) in Water Resources and Flood Risk Management

**Xianwei Wang [1,\*] and Hongjie Xie [2,\*]**

1   School of Geography and Planning, and Guangdong Key Laboratory for Urbanization and Geo-Simulation, Sun Yat-Sen University, Guangzhou 510275, China
2   Department of Geological Sciences, University of Texas at San Antonio, San Antonio, TX 78249, USA
\*   Correspondence: wangxw8@mail.sysu.edu.cn (X.W.); hongjie.xie@utsa.edu (H.X.);
    Tel.: +86-20-84114623 (X.W.); +1-210-458-5445 (H.X.)

Received: 4 April 2018; Accepted: 3 May 2018; Published: 7 May 2018

**Abstract:** Water is one of the most critical natural resources that maintain the ecosystem and support people's daily life. Pressures on water resources and disaster management are rising primarily due to the unequal spatial and temporal distribution of water resources and pollution, and also partially due to our poor knowledge about the distribution of water resources and poor management of their usage. Remote sensing provides critical data for mapping water resources, measuring hydrological fluxes, monitoring drought and flooding inundation, while geographic information systems (GIS) provide the best tools for water resources, drought and flood risk management. This special issue presents the best practices, cutting-edge technologies and applications of remote sensing, GIS and hydrological models for water resource mapping, satellite rainfall measurements, runoff simulation, water body and flood inundation mapping, and risk management. The latest technologies applied include 3D surface model analysis and visualization of glaciers, unmanned aerial vehicle (UAV) video image classification for turfgrass mapping and irrigation planning, ground penetration radar for soil moisture estimation, the Tropical Rainfall Measuring Mission (TRMM) and the Global Precipitation Measurement (GPM) satellite rainfall measurements, storm hyetography analysis, rainfall runoff and urban flooding simulation, and satellite radar and optical image classification for urban water bodies and flooding inundation. The application of those technologies is expected to greatly relieve the pressures on water resources and allow better mitigation of and adaptation to the disastrous impact of droughts and flooding.

**Keywords:** remote sensing; geographic information systems (GIS); glaciers; water body; soil moisture; groundwater; flooding; rainfall measurements; design storm; runoff simulation

## 1. Introduction

Human-accessible freshwater resources primarily include mountain glaciers, snow, surface water bodies (lakes, rivers and reservoirs), soil moisture and ground water. Since they have very unequal distribution spatially and temporally, pressures on water resources are increasing globally. On the other hand, people may not know where and how much water resources are available regionally, especially for the remote mountain glaciers/snow, and deep confined groundwater. In extreme cases, if there is too little or too much water within a certain period and area, severe drought and torrent flooding could occur, often resulting in catastrophic impacts and damages to the local and regional community. Therefore, it is of great significance to map and manage water resources, drought and flooding risk precisely by using the cutting-edge technologies of remote sensing, geographic information systems (GIS), geostatistics and hydrologic models (Table 1).

**Table 1.** Latest remote sensing technology and sensors used for water resources, hydrological fluxes, drought and flood mapping.

| Application Fields | Specific Contents | Examples of Sensors or Satellites |
|---|---|---|
| Water resources | Snow | AVHRR, Terra/Aqua MODIS, Landsat, SSM/I, AMSR-E, Cryosat etc. |
| | Glaciers | Landsat, ASTER, SPOT, ICESat, SRTM, etc. |
| | Soil moisture | SSM/I, AMSR-E, SMAP, SMOS, etc. |
| | Groundwater | GRACE |
| | Lakes, reservoirs, rivers, and wetlands | MODIS, Landsat, SPOT, ICESat, GRACE, SRTM etc. |
| Hydrological fluxes | Precipitation | NEXRAD, TRMM, GPM, etc. |
| | Evapotranspiration | MODIS, Landsat, GRACE, etc. |
| | River, reservoir or lake discharge | MODIS, ENVISAT, Landsat, SRTM, ICESat, etc. |
| Drought and flooding | Drought and flooding | MODIS, Landsat, GRACE, UAV, AMSR-E, SMAP, SMOS, ENVISAT, ASAR, Sentinel-1A/2A, etc. |

Remote sensing provides critical data for water resource mapping (Table 1). Satellite remote sensing techniques can make continuous and up-to-date measurements with global coverage depending on their orbital features, while they count on ground observations for algorithm development and validation [1]. For example, the Moderate Resolution Imaging Spectroradiometer (MODIS) on-board on Terra and Aqua satellites has provided daily global snow cover products since February 2000 [2]. They have been widely applied in different fields, such as hydrology, agriculture and climate studies [3,4]. Relative high-resolution images from the Landsat series could be used to recovery and monitor the global state of mountain glaciers, thus making it possible to update the global glacier inventories at high accuracy and confidence, such as the Global Land Ice Measurements from Space (GLIMS), and Glacier Area Mapping for Discharge from the Asian Mountains (GAMDAM), and the second Chinese Glacier Inventory [5–7]. Besides areas, glaciers surface elevation information extracted from satellite instruments of the National Aeronautics and Space Administration (NASA) Ice, Cloud and land Elevation Satellite (ICESat), Shuttle Radar Topographic Mission (SRTM), Advanced Spaceborne Thermal Emission and Reflection Radiometer (ASTER), SPOT 5 and even airborne stereo images were used to investigate glaciers' thickness and volume changes in the vast, high Asia Mountains [8–10]. MODIS images and ICESat elevation data were used together to map the lake water body areas and surface elevation changes in human-inaccessible regions in the Tibetan Plateaus for the first times on record [11,12]. The Advanced Microwave Scanning Radiometer for NASA's Earth Observing System (AMSR-E), NASA's Soil Moisture Active and Passive (SMAP) mission, and the European Space Agency (ESA) Soil Moisture Ocean Salinity (SMOS) mission all can provide global soil moisture mapping [1,13]. Groundwater is the most difficult to detect by satellite sensors, while the Gravity Recovery and Climate Experiment (GRACE) has been successfully used to measure groundwater depletion and the filling of the Three Gorges Reservoir [14,15]. Ground penetration radar can even obtain accurate estimations of glacier thickness, soil moisture, and groundwater [16,17].

Besides water resource mapping, remote sensing can also quantitatively measure hydrological fluxes, such as precipitation, evapotranspiration, river stages and discharges (Table 1). Ground-based radar such as the US Next Generation Weather Radar (NEXRAD) has been used to quantitatively measure precipitation on US territory since 1990s at relatively high accuracy, and has been widely applied to monitoring precipitation locally and regionally worldwide [18,19]. The Tropical Rainfall Measuring Mission (TRMM) was the first satellite to measure the global mid-latitude precipitation at unprecedented 0.25° and 3-h product since 1998, e.g., the Multi-satellite Precipitation Analysis (TMPA), and the Global Precipitation Measurement (GPM) products have even been able to provide global near-real time precipitation estimates of 0.1° and 30-min products since 2014, e.g., the Integrated Multi-satellitE Retrievals of GPM (IMERG) [20,21]. Those precipitation products greatly improve hydrological simulation and flood prediction due to their large coverage and relatively high spatial resolution [22]. Evapotranspiration (ET) can be estimated based on the Surface Energy Balance

Algorithms for Land (SEBAL) using the radiance detected by satellite sensors, such as Landsat and MODIS retrievals [23]. Now, more energy balance-based models have been developed to estimate the field actual ET in agricultural management [24]. Together with in situ lake water level observations, daily MODIS images were also used to map the Poyang Lake's water volume and lake bed topography changes by the elevation contours derived from the land-water boundary line [25]. River discharge estimation was traditionally done by in situ observations, and now can also be detected by the synthetic width/stage-discharge rating curves via measuring the river's effective wet width and water level using MODIS, the Environmental Satellite (ENVISAT), Landsat and other high-resolution images [26–28].

Remote sensing techniques are playing increasingly important roles in drought monitoring and flooding emergency response (Table 1). Many drought indices were developed using MODIS reflectance data under different climate and land cover conditions, such as the Normalized Difference of Vegetation Index (NDVI), Normalized Difference of Water Index (NDWI), Visible Atmospherically Resistant Index (VARI), Enhanced Vegetation Index (EVI), Normalized Difference Infrared Index Band 6 or Band7 (NDIIB6/7), and so on [29–31]. Emerging Unmanned Aerial Vehicles (UAV) provide a more flexible low-altitude platform to monitor vegetation growth, soil moisture conditions, flood inundation mapping and damage assessment [32]. Flood mapping from various data sources can greatly improve disaster response, e.g., for the widespread and sustained flood events in several river basins in Texas and Oklahoma of USA in late April and May 2015, a total 27,174 space- and airborne images were applied to monitor the daily variations of flood inundation extents [33].

GIS is very versatile, especially in spatial analysis, modeling, visualization, data processing and management. At most times e.g., in this special issue, GIS operates heroically behind the scenes. Almost every paper published in this special issue uses GIS for data preprocessing, spatial analysis or establishing results maps (Table 2). With unprecedented data resources, it is quite challenging to manage so much data in risk management and especially in disaster response, such as in the aforementioned 2015 Texas flooding event. Schumann et al. [33] suggest that the proactive assimilation of methodologies and tools into the mandated agencies are required in order to unlock the full potential of those various data. GIS, such as the most popular ArcGIS products and other commercial or open source software, are required to process the original remote sensing images and videos, and carry out spatial analysis, modeling and visualization. Meanwhile, statistics can draw solid conclusions from the satellite images data and GIS spatial analysis.

Hydrologic models can take full use of the remote sensing and GIS data and carry out lots of physical experiments and scenario analyses. They are able to provide a full spectrum of modeling what happened in the past and project what will happen in the future. The availability of model simulations over a long time period also allows for a robust estimate of low-probability events that were not recorded in ground observations [34]. This is especially so for the remote and mountainous areas where there are few or even none in situ observations for rainfall and stream flows; satellite-measured rainfall is normally used to drive a hydrological model to simulate historical flooding events, thus projecting current and future flooding risk [35].

In summary, remote sensing techniques have played increasingly important roles in the hydrologic community (Table 1). They can map the spatial and temporal distributions of water resources, quantitatively measure the hydrologic flux, and monitor the working conditions of hydraulic infrastructures, drought conditions and flooding inundation. GIS, statistics and numerical models together can unlock the potential of various remote sensing data resources, and make for better management of water resources, drought and flooding disasters. The following session 2 summarizes the 12 papers published in this special issue (Table 2), which present the best practices, cutting-edge technologies and applications of remote sensing, GIS and hydrological models for water resources mapping, satellite rainfall measurements, storm hyetography analysis, runoff and urban flooding simulation, water body and flooding inundation mapping, and risk management.

**Table 2.** Latest geographic information systems (GIS) and remote sensing technologies and hydrologic models applied in the 12 papers published in this special issue.

| Application Fields | Specific Contents | GIS, Algorithm, Model, Sensor or Satellites | Reference |
|---|---|---|---|
| Water resources mapping and management | Glaciers mapping | Landsat, ASTER GDEM, GIS, TIN 3D model. | [36] |
| | Soil moisture detection | GPR, CMP, FO, GIS spatial analysis | [16] |
| | Groundwater and subsidence analysis | GIS spatial analysis, GPS | [37] |
| | Irrigation planning | UAV, HTM for video image classification, GIS visualization | [32] |
| Rainfall measurements and design storm | Rainfall measurements | TRMM, GPM, GIS spatial analysis and visualization | [38] |
| | Design storm and urban flood modeling | Huff curve, SWMM, GIS data preprocess and visualization | [39] |
| Rainfall runoff prediction and flood forecasting | Flood modeling | GSSHA model, GPM IMERG, GIS visualization | [40] |
| | Rainfall Runoff simulation | RCM, LSM, CoLM, CoLM+LF, GIS data preprocess | [41] |
| | Flood inundation forecast | ARX regressor, MOGA algorithm, GIS visualization | [42] |
| Water body and flood mapping | Flash flood detection | TMPA real time 3B2RT, CT, CDFs, JFI, GIS spatial analysis | [43] |
| | Urban water body mapping | ZY-3 images, AUWEM, GIS spatial analysis | [44] |
| | Flood inundation mapping | ENVISAT, ASAR, GIS spatial analysis | [45] |

## 2. Summary of This Special Issue

### 2.1. Water Resources Mapping and Management

Mountain glaciers and snow in the Tianshan Mountains are critical water resources in arid and semi-arid Central Asia [46]. Glacier areas are defined as the extent in two horizontal dimensions (2D area) in the ice mass balance community [47], and often used to estimate the total ice volume by volume-area power law equations [48]. In the high Tianshan Mountains, most glaciers lie on steep slopes, and their actual surface extent (3D area) may be much larger than the 2D area. Wang et al. [36] in this special issue establish a 3D model to quantify glaciers' 3D and 2D area differences in the Muzart Glacier catchment and in Central Tianshan using ASTER GDEM data, CGI2 and Landsat images. They found that glaciers' 3D areas was 34.2% larger than their 2D areas in the Muzart Glacier catchment and by 27.9% in the entire Central Tianshan, where glaciers' 3D areas reduced by 115 km$^2$ between 2007 and 2013, being 27.6% larger than their 2D area reduction. This confirms that there is significantly large difference between glaciers' 3D and 2D areas in the steep Central Tianshan. As they remarked, "Those large areal differences remind us to re-consider a glacier's real topographic extent when discussing an alpine glacier's areal and volume changes, especially in calculating the glacier's surface energy balance and melting rates in the high Asian mountain glaciers with large surface slopes and strong solar radiation."

Surface soil is a critical boundary layer between atmosphere and land surface. Soil water content affects local agriculture, ecology, hydrology and climate. In the dry desert steppe, soil moisture is one of the main factors that control vegetation growth and ecosystem restoration. The common soil moisture measurement technologies, such as the gravimetric method, neutron method, Time Domain Reflector (TDR), Frequency Domain Reflectometry (FDR), and so on, provide point measurements with high accuracy, while being labor and time-consuming and may destroy the soil structures. Lu et al. [16] in this special issue present their study to measure the steppe soil moisture using Ground Penetrating Radar (GPR). The common-mid point (CMP) method and fixed offset (FO) method are used for sensitivity analysis, while the gravimetric soil moisture measurements are used to validate the accuracy of the GPR measurements. Their results show Topp's equation is more suitable than Roth's equation for processing GPR data in the desert steppe. Both CMP and FO methods show high accuracy in GPR soil moisture measurements. Vegetation affects the measurement precision, and precipitation reduces the effective sampling depth of the ground wave from 0.1 m to 0.05 m. Overall, the operation of GPR measurements is simple and does not damage the soil layer structure, while providing high accuracy and easy movement.

Groundwater is an important freshwater resource in mid-latitude and in arid and semiarid regions. Instead of directly measuring the soil moisture or aforementioned glaciers, Li et al. [37] in this special issue used GIS and statistical tools to study the geographic distribution of land subsidence, groundwater drawdown, and compressible layer thickness using in situ monitoring data in the metropolitan areas of Beijing, the capital of China. The Beijing Plain lies in the alluvial–pluvial plain fan built up by river deposits and belongs to the temperate continental monsoon climate with annual mean temperature of 10–15 °C and precipitation of 601 mm. Land subsidence is one of the critical threats to the sustainable development of Beijing. Multiple approaches including point (gravity center), line (major axes), and polygon (coverage) views are tested for analyzing spatial change patterns. Results show that the Chaoyang District of Beijing had the largest land subsidence and groundwater drawdown, both of which concentration trends were consistent and the principle orientation was southwest–northeast (SW–NE). The spatial distribution pattern of land subsidence was similar to that of the compressible layer. Those results are useful for assessing the distribution of land subsidence and managing groundwater resources.

Irrigation planning is an important component in water resource and precision agriculture management. Golf courses are one kind of precision agriculture, and their turfgrass has high water demand. Turfgrass irrigation is rapidly transitioning to reuse water because of the water price incentive and mandated water management policies. Therefore, knowing the turfgrass areas and growth conditions can help plan the water and treated sewage effluent needs exactly at a daily or weekly rate. Perea-Moreno et al. [32] in this special issue utilized UAV video images to extract automatically the turfgrass areas and growth conditions by a Hierarchical Temporal Memory (HTM) algorithm, and further assess the water needs for turfgrass irrigation. The extracted turfgrass area from video imagery classification could achieve an accuracy of 98%. They commented, "Technical progress in computing power and software has shown that video imagery is one of the most promising environmental data acquisition techniques available today. This rapid classification of turfgrass can play an important role for planning water management."

## 2.2. Rainfall Measurements and Design Storm

Rainfall is one of the most critical components of water cycle and water resources recharge. Heavy rainfall often causes devastating flood events. Typhoon-related heavy rainfall has unique structures in both time and space at mesoscale. Satellite rainfall estimate may better delineate the structures of heavy rainfall, which is helpful for early-warning systems and disaster management. Wang et al. [38] in this special issue compares the latest versions of two satellite rainfall products with ground rain gauge observations along the coastal region of China from 2014 to 2015. They are the GPM IMERG final run and TMPA 3B42V7. Overall, correlation coefficients (CCs) of both IMERG and TMPA with gauge observations for the eight typhoon events investigated are significant at the 0.01 level, but both TMPA and IMERG tend to underestimate the heavy rainfall against the gauge observation, especially around the storm center. The IMERG final run exhibits better performance than TMPA 3B42V7. In space, both products have the best applicability within the range of 50–100 km away from typhoon tracks, and the worst beyond the 300-km range. It is always a challenging task to measure accurately heavy rainfall by rain gauges, a ground radar network, or satellite sensors.

The temporal evolution of heavy rainfall over certain area is called the storm hyetograph. Given a total rain depth and duration over a certain return period, the storm hyetograph (also called design storm) determines the peak flooding volume and is critical for drainage design in storm water management [49]. The common design storms for drainage design include the Triangular curve [50], the Chicago curve [51] and the Soil Conservation Service (SCS) curve [52]. Pan et al. [39] in this special issue compared these curves and found that they tend to underestimate the peak rainfall in the metropolitan areas of Guangzhou, south China. The normalized time of peak rainfall is at 33% ± 5% for all storms in Guangzhou, and most storms (84%) are in the 1st and 2nd quartiles. Pan et al. [39] improved the Huff curve by separately describing rising and falling limbs and then combined them into a full storm hyetograph, instead of dividing the storms into four quartiles as in the original Huff curve analysis. The improved Huff curve can better represent the storm hyetographs in Guangzhou than the other three curves. It generates larger peak flooding volumes that match better with the street water inundation depth when they are input in the Storm Water Management Model (SWMM) for given heavy storm events. "The Improved Huff curve has great potential in storm water management such as flooding risk mapping and drainage facility design, after further validation." [39].

### 2.3. Rainfall Runoff Prediction and Flood Forecasting

Hydrological models are the backbones of climatic and hydrologic simulations, water resources management, and flood forecasting. Hydrological models originated from conceptual and clumped models, and are advancing to physically-based, distributed models, such as the Gridded Surface Subsurface Hydrologic Analysis (GSSHA). Sharif et al. [40] in this special issue utilized the GSSHA model to simulate a recent flood event to gain a better understanding of the runoff generation and spatial distribution of flooding in a very arid catchment of Hafr Al Batin City, north-eastern Saudi Arabia. The GPM IMERG rainfall products (the uncalibrated early run and calibrated final run) were used to drive the GSSHA model. This showed that 85% of the flooding was generated in the urbanized portion of the catchments for the simulated flood event. Urban storm drainage and catchment runoff were used in simulations by different models. The variable model grid sizes allowed the GSSHA model to be applied on large basins that include the entire catchment for a coarse grid size and urban centers that need to be modeled at very high resolutions. Thus, urban flooding can be simulated by a single physically-based and distributed model that could model the local heavy storm runoff in the urban areas and the regional rainfall runoff on a large river catchment, and the integrated urban flooding risk can be considered at the same time.

Compared to the fine hydrologic modeling of storm runoff at grid sizes of tens of meters and minutes or hourly intervals, runoff prediction in the regional climate models (RCM) such as the Land Surface Models (LSM) is much coarser, at tens of kilometers and daily or monthly scales. The original Common Land Model (CoLM) predicts runoff from net water at each computation grid even without the explicit Lateral Flow (LF) scheme. Lee and Choi [41] in this special issue proposed a CoLM+LF model to improve the runoff prediction by incorporating a set of lateral surface and subsurface runoff computations into the existing terrestrial hydrologic processes in CoLM. The CoLM+LF model was assessed in the Nakdong River Watershed of Korea using Earth observations at the 30-km resolution and daily time step. The simulated runoff by CoLM and the CoLM+LF was then compared with the daily stream flow observations at the Jindong stream gauge station in the study watershed during 2009. CoLM+LF can simulate the effect of runoff travel time over a watershed by an explicit lateral flow scheme, and can more effectively capture seasonal variations in daily streamflow than CoLM. It is expected to be a helpful and essential tool for water resource management and hydrological impact assessment.

Flood inundation forecast technology can generally be divided into either numerical simulation or black-box modeling. Numerical simulation is based on theoretical deduction and often has good accuracy, while demanding high computing resources and being difficult to use for the real-time forecasting of rapid disaster mitigation and rescue response in most conditions, such as during a typhoon and flash flooding. In contrast, the black-box model relies on different approaches by deeming the process from rainfall to inundation as a black box to simulate the relationship between input rainfall and output runoff and inundation [53]. It cannot explain the physical mechanism, but can correctly and effectively simulate the response after full calibration at much faster computing speed than physically-based models. Ouyang et al. [42] in this special issue proposed such a black-box model that combines non-sequential regressors for the ARX (Auto-Regressive model with eXogenous inputs)-based typhoon inundation forecast. The difficulty when using the model is finding an optimal combination of regressors to perform accurate prediction. They developed a novel approach to integrate a Multi-Objective Genetic Algorithm (MOGA) to transfer the search for the optimal combination of non-sequential regressors into an optimization problem. The results (tested in the northeastern Taiwan) showed that the optimal models acquired through this model had good inundation forecasting capabilities in terms of accuracy, time-shift error, and error distribution, thus providing practical benefits for decision making and rescue response during a typhoon landfall period.

## 2.4. Water Body and Flood Mapping

It is a challenge to forecast accurately flash flooding by hydrological models in arid regions of the Middle East like Saudi Arabia because of the sporadic storm events and scarce stream flow data. The vulnerability of arid and semi-arid regions to flash floods was thought to be similar to that of regions having heavy rain owing to the strong convective storms and the rapid formation of flash floods [54,55]. Tekeli [43] in this special issue examines the feasibility of flash flood detection over the city of Jeddah in western Saudi Arabia using TRMM TMPA Real Time (RT) 3B2RT data during 2000–2014. Three indices, constant threshold (CT), cumulative distribution functions (CDFs) and Jeddah flood index (JFI), were developed to detect flash flood events using the 3-h 3B42RT rainfall data. CDF worked best. It did not miss any flood event and had a hit rate of up to 94%. Compared to hydrological models using various variables, this approach seems promising in arid regions, although only rainfall data are used.

Water surface is easily detectable by remote sensing images in most conditions because of its low reflectance, while it is a challenging task to accurately extract urban water bodies from high-resolution images due to the shadowing effect of high-rise buildings and trees. To disentangle this problem, Yang et al. [44] in this special issue proposed an automatic urban water extraction method (AUWEM) to extract urban water bodies from high-resolution ZY-3 multi-spectral images. They first refined the Normalized Difference of Water Index (NDWI) algorithm by constructing two new indices, namely NNDWI1, which is sensitive to turbid water, and NNDWI2, which is sensitive to the water body interfered with by vegetation. Both indices were then used to map all water body and shaded areas by image threshold segmentation. An object-based technology was then developed to detect the shades, which were finally removed from the classified water bodies. This automated approach was tested by five images featuring different areas and environments including lakes and rivers in the cities of Beijing, Suzhou, Wuhan and Guangzhou, China. Compared to the Maximum Likelihood Method (MaxLike) and NDWI, AUWEM had a detection accuracy of 93%, against 86% for Maxlike and 89% for NDWI, and exhibited both smaller omission errors and commission errors. It even works better when detecting water edge and small rivers, and can effectively distinguish shadows of high buildings from water bodies to improve the overall accuracy.

Flood inundation mapping is similar to water body mapping, while facing more challenges, such as the short response time and cloud blockage. Flood mapping from various data sources can greatly improve disaster response, e.g., all optical images, UAV video images and radar images were applied in the May 2015 Texas flooding event [33]. Radar microwave images beat the optical images in flood mapping by their unique capability to penetrate through cloud. Frappart et al. [45] in this special issue use the ENVISAT ASAR images to recovery the flood extent between 2005 and 2008 in the Guayas watershed on the Pacific Coast of Ecuador, where floods are an annual phenomenon and become devastating during El Niño years. Flooded pixels present lower backscattering than bare soil or vegetation as the radar electromagnetic wave is specularly reflected by water surfaces. The core algorithm of the method is change detection using radar backscattering coefficients at the C-band between the wet and dry seasons. Mapping inundation water under tree canopy and other vegetation needs special consideration, since vegetation usually decreases the radar backscattering coefficients. In spite of the coarse spatial resolution (1 km) of these SAR images, the patio-temporal (monthly) dynamics of the flood in the Guayas watershed between 2005 and 2008 was mapped using ASAR images for the first time in this watershed. Moreover, other radar satellites launched in recent years, such as Sentinel-1A in April 2014, Sentinel-2A in June 2015 and Sentinel-1B in April 2016, satellite SAR (C-band) etc., can provide global coverage of flood inundation mapping every few days at unprecedented spatial resolution of tens of meters.

## 3. Conclusions

Remote sensing and GIS play critical roles in water resource and flood inundation mapping and risk management. Remote sensing provides critical data for mapping water resources (snow and glaciers, water bodies, soil moisture and groundwater), measuring hydrological fluxes (ET, precipitation and river discharge), and monitoring drought and flooding inundation; while GIS provides the best tools for water resource, drought and flood risk management and for hydrologic models' setup, input data processing, output analysis and visualization. This special issue presents the best practices, cutting-edge technologies and applications of remote sensing, GIS and hydrologic models for water resource mapping, satellite rainfall measurements, runoff and urban flood simulation, water body and flood inundation mapping, and risk management. The latest technologies applied include 3D model analysis and visualization of glaciers, UAV video image classification for turfgrass mapping and irrigation planning, ground penetration radar for soil moisture estimation, TRMM and GPM satellite rainfall measurements, storm hyetograph analysis, rainfall runoff and urban flooding simulation, and satellite radar and optical image detection for urban water bodies and flooding inundation. GIS is very versatile, but operating heroically behind the scenes at most times. GIS techniques are used in almost every paper published in this special issue for data preprocessing, spatial analysis or making results maps. The applications of those technologies are expected to greatly relieve the pressures on water resources and enable better mitigation of and adaptation to the disastrous impact of droughts and flooding.

**Author Contributions:** Xianwei Wang designed the article structure and wrote the manuscript. Hongjie Xie initiated the idea of this review article and revised the manuscript.

**Funding:** This study is funded by the Water Resource Science and Technology Innovation Program of Guangdong Province (#2016-19).

**Acknowledgments:** We are grateful for being invited to be the Guest Editor for this special issue by the Water Editorial Office who set up and took care of the editorial process. We thank all the authors who contribute to this special issue. We thank the reviewers who contributed their expertise and time on reviewing those articles. Without their support, it would be impossible to assess properly those manuscripts submitted.

**Conflicts of Interest:** The authors declare no conflict of interest.

# References

1.  Tang, Q.; Gao, H.; Lu, H.; Lettenmaier, D.P. Remote sensing: Hydrology. *Prog. Phys. Geogr.* **2009**, *33*, 490–509. [CrossRef]
2.  Hall, D.K.; Riggs, G.A.; Salomonson, V.V.; DiGirolamo, N.E.; Bayr, K.J. MODIS snow-cover products. *Remote Sens. Environ.* **2002**, *83*, 181–194. [CrossRef]
3.  Wang, X.; Xie, H. New methods for studying the spatiotemporal variation of snow cover based on combination products of MODIS Terra and Aqua. *J. Hydrol.* **2009**, *371*, 192–200. [CrossRef]
4.  Wang, X.; Zhu, Y.; Chen, Y.; Liu, H.; Huang, H.; Liu, K.; Liu, L. Influences of forest on MODIS snow cover mapping and snow variations in the Amur River basin in Northeast Asia during 2000–2014. *Hydrol. Process.* **2017**. [CrossRef]
5.  Guo, W.; Liu, S.; Xu, J.; Wu, L.; Shangguan, D.; Yao, X.; Wei, J.; Bao, W.; Yu, P.; Liu, Q. The second Chinese glacier inventory: Data, methods and results. *J. Glaciol.* **2015**, *61*, 357–372. [CrossRef]
6.  Nuimura, T.; Sakai, A.; Taniguchi, K.; Nagai, H.; Lamsal, D.; Tsutaki, S.; Kozawa, A.; Hoshina, Y.; Takenaka, S.; Omiya, S. The GAMDAM Glacier Inventory: A quality-controlled inventory of Asian glaciers. *Cryosphere Discuss.* **2015**, *8*, 849–864. [CrossRef]
7.  Pfeffer, W.T.; Arendt, A.A.; Bliss, A.; Bolch, T.; Cogley, J.G.; Gardner, A.S.; Hagen, J.O.; Hock, R.; Kaser, G.; Kienholz, C.; et al. The Randolph Glacier Inventory: A globally complete inventory of glaciers. *J. Glaciol.* **2014**, *60*, 522–537. [CrossRef]
8.  Aizen, V.B.; Kuzmichenok, V.A.; Surazakov, A.B.; Aizen, E.M. Glacier changes in the Tien Shan as determined from topographic and remotely sensed data. *Glob. Planet. Chang.* **2007**, *56*, 328–340. [CrossRef]
9.  Kaab, A.; Berthier, E.; Nuth, C.; Gardelle, J.; Arnaud, Y. Contrasting patterns of early twenty-first-century glacier mass change in the Himalayas. *Nature* **2012**, *488*, 495–498. [CrossRef] [PubMed]
10. Pieczonka, T.; Bolch, T.; Wei, J.; Liu, S. Heterogeneous mass loss of glaciers in the Aksu-Tarim Catchment (Central Tien Shan) revealed by 1976 KH-9 Hexagon and 2009 SPOT-5 stereo imagery. *Remote Sens. Environ.* **2013**, *130*, 233–244. [CrossRef]
11. Zhang, G.; Yao, T.; Xie, H.; Kang, S.; Lei, Y. Increased mass over the Tibetan Plateau: From lakes or glaciers? *Geophys. Res. Lett.* **2013**. [CrossRef]
12. Zhang, G.; Yao, T.; Piao, S.; Bolch, T.; Xie, H.; Chen, D.; Gao, Y.; O'Reilly, C.M.; Shum, C.K.; Yang, K.; et al. Extensive and drastically different alpine lake changes on Asia's high plateaus during the past four decades. *Geophys. Res. Lett.* **2017**, *44*. [CrossRef]
13. Entekhabi, D.; Jackson, T.J.; Njoku, E.; O'Neill, P.; Entin, J. Soil moisture active/passive (SMAP) mission concept. *Proc. SPIE* **2008**, *70850H*. [CrossRef]
14. Rodell, M.; Velicogna, I.; Famiglietti, J.S. Satellite-based estimates of groundwater depletion in India. *Nature* **2009**, *460*, 999–1002. [CrossRef] [PubMed]
15. Wang, X.; Linage, C.R.; Famiglietti, J.; Zender, C.S. Gravity Recovery and Climate Experiment detection of water storage changes in the Three Gorges Reservoir of China and comparison with in situ measurements. *Water Resour. Res.* **2011**, *47*, 1–13. [CrossRef]
16. Lu, Y.; Song, W.; Lu, J.; Wang, X.; Tan, Y. An Examination of Soil Moisture Estimation Using Ground Penetrating Radar in Desert Steppe. *Water* **2017**, *9*, 521. [CrossRef]
17. Steelman, C.M.; Endres, A.L.; Jones, J.P. High-resolution ground-penetrating radar monitoring of soil moisture dynamics: Field result, interpretation, and comparison with unsaturated flow model. *Water Resour. Res.* **2012**, *48*, 184–189. [CrossRef]
18. Wang, X.; Xie, H.; Sharif, H.; Zeitler, J. Validating NEXRAD MPE and Stage III precipitation products for uniform rainfall on the Upper Guadalupe River Basin of the Texas Hill Country. *J. Hydrol.* **2008**, *348*, 73–86. [CrossRef]
19. Wang, X.; Xie, H.; Mazari, N.; Sharif, H.; Zeitler, J. Evaluation of the near-real time NEXRAD DSP Product in the evolution of heavy rain events on the Upper Guadalupe River Basin, Texas. *J. Hydroinformat.* **2013**, *15*. [CrossRef]
20. Huffman, G.J.; Bolvin, D.T.; Nelkin, E.J.; Wolff, D.B.; Adler, R.F.; Gu, G.; Neikin, E.J.; Bowman, K.P.; Hong, Y.; Stocker, E.F.; et al. The TRMM multisatellite precipitation analysis (TMPA): Quasi-global, multiyear, combined-sensor precipitation estimates at fine scales. *J. Hydrometeorol.* **2007**, *8*, 38–55. [CrossRef]

21. Sorooshian, S.; AghaKouchak, A.; Arkin, P.; Eylander, J.; Foufoula-Georgiou, E.; Harmon, R.; Hendrickx, J.M.H.; Imam, B.; Kuligowski, R.; Skahill, B.; et al. Advanced concepts on remote sensing of precipitation at multiple scales. *Bull. Am. Meteorol. Soc.* **2011**, *92*. [CrossRef]

22. Wang, D.S.; Wang, X.; Liu, L.; Huang, H.; Pan, C.; Wang, D.G. Evaluation of CMPA precipitation estimate in the evolution of typhoon-related storm rainfall events in Guangdong province, China. *J. Hydroinform.* **2016**. [CrossRef]

23. Bastiaanssen, W.; Pelgrum, H.; Wang, J.; Ma, Y.; Moreno, J.; Roerink, G.; van der Wal, T. A remote sensing surface energy balance algorithm for land (SEBAL). *J. Hydrol.* **1998**, *212–213*, 198–212. [CrossRef]

24. Häuslera, M.; Conceiçãoc, N.; Tezzad, L.; Sáncheze, J.M.; Campagnolo, M.L.; Häuslerf, A.J.; Silvaa, J.M.N.; Warnekeg, T.; Heygsterg, G.; Ferreirab, M.I. Estimation and partitioning of actual daily evapotranspiration at an intensive olive grove using the STSEB model based on remote sensing. *Agric. Water Manag.* **2018**, *201*, 188–198. [CrossRef]

25. Feng, L.; Hu, C.; Chen, X.; Li, R.; Tian, L.; Murch, B. MODIS observations of the bottom topography and its inter-annual variability of Poyang Lake. *Remote Sens. Environ.* **2011**, *115*, 2729–2741. [CrossRef]

26. King, T.V.; Neilson, B.T.; Rasmussen, M.T. Estimating discharge in low-order rivers with high-resolution aerial imagery. *Water Resour. Res.* **2018**, *54*, 863–878. [CrossRef]

27. Smith, L.C.; Pavelsky, T.M. Estimation of river discharge, propagation speed, and hydraulic geometry from space: Lena River, Siberia. *Water Resour. Res.* **2008**, *44*, W03427. [CrossRef]

28. Tourian, M.J.; Sneeuw, N.; Bardossy, A. A quantile function approach to discharge estimation from satellite altimetry (ENVISAT). *Water Resour. Res.* **2013**, *49*. [CrossRef]

29. Caccamo, G.; Chisholm, L.A.; Radstock, R.A.; Puotinen, M.L. Assessing the sensitivity of MODIS to monitor drought in high biomass ecosystems. *Remote Sens. Environ.* **2011**, *115*, 2626–2639. [CrossRef]

30. Rhee, J.; Im, J.; Carbone, G.J. Monitoring agricultural drought for arid and humid regions using multi-sensor remote sensing data. *Remote Sens. Environ.* **2010**, *114*, 2875–2887. [CrossRef]

31. Wang, X.; Liu, M.; Liu, L. Responses of MODIS spectral indices to typical drought events from 2000 to 2012 in Southwest China. *J. Remote Sens.* **2014**, *18*, 433–442.

32. Perea-Moreno, A.-J.; Aguilera-Ureña, M.-J.; Meroño-De Larriva, J.-E.; Manzano-Agugliaro, F. Assessment of the Potential of UAV Video Image Analysis for Planning Irrigation Needs of Golf Courses. *Water* **2016**, *8*, 584. [CrossRef]

33. Schumann, G.J.-P.; Frye, S.; Wells, G.; Adler, R.; Brakenridge, R.; Bolten, J.; Murray, J.; Slayback, D.; Policelli, F.; Kirschbaum, D.B.; et al. Unlocking the Full Potential of Earth Observation during the 2015 Texas Flood Disaster. *Water Resour. Res.* **2016**. [CrossRef]

34. Giustarini, L.; Chini, M.; Hostache, R.; Pappenberger, F.; Matgen, P. Flood hazard mapping combining hydrodynamic modeling and multi annual remote sensing data. *Remote Sens.* **2015**, *7*, 14200–14226. [CrossRef]

35. Wing, O.E.J.; Bates, P.D.; Sampson, C.C.; Smith, A.M.; Johnson, K.A.; Erickson, T.A. Validation of a 30 m resolution flood hazard model of the conterminous United States. *Water Resour. Res.* **2017**, *53*. [CrossRef]

36. Wang, X.; Chen, H.; Chen, Y. Large Differences between Glaciers 3D Surface Extents and 2D Planar Areas in Central Tianshan. *Water* **2017**, *9*, 282. [CrossRef]

37. Li, Y.; Gong, H.; Zhu, L.; Li, X. Measuring Spatiotemporal Features of Land Subsidence, Groundwater Drawdown, and Compressible Layer Thickness in Beijing Plain, China. *Water* **2017**, *9*, 64. [CrossRef]

38. Wang, R.; Chen, J.; Wang, X. Comparison of IMERG Level-3 and TMPA 3B42V7 in Estimating Typhoon-Related Heavy Rain. *Water* **2017**, *9*, 276. [CrossRef]

39. Pan, C.; Wang, X.; Liu, L.; Huang, H.; Wang, D. Improvement to the Huff Curve for Design Storms and Urban Flooding Simulations in Guangzhou, China. *Water* **2017**, *9*, 411. [CrossRef]

40. Sharif, H.O.; Al-Zahrani, M.; Hassan, A.E. Physically, Fully-Distributed Hydrologic Simulations Driven by GPM Satellite Rainfall over an Urbanizing Arid Catchment in Saudi Arabia. *Water* **2017**, *9*, 163. [CrossRef]

41. Lee, J.S.; Choi, H.I. Improvements to Runoff Predictions from a Land Surface Model with a Lateral Flow Scheme Using Remote Sensing and In Situ Observations. *Water* **2017**, *9*, 148. [CrossRef]

42. Ouyang, H.T.; Shih, S.S.; Wu, C.S. Optimal Combinations of Non-Sequential Regressors for ARX-Based Typhoon Inundation Forecast Models Considering Multiple Objectives. *Water* **2017**, *9*, 519. [CrossRef]

43. Tekeli, A.E. Exploring Jeddah Floods by Tropical Rainfall Measuring Mission Analysis. *Water* **2017**, *9*, 612. [CrossRef]

44. Yang, F.; Guo, J.; Tan, H.; Wang, J. Automated Extraction of Urban Water Bodies from ZY-3 Multi-Spectral Imagery. *Water* **2017**, *9*, 144. [CrossRef]

45. Frappart, F.; Bourrel, L.; Brodu, N.; Riofrío Salazar, X.; Baup, F.; Darrozes, J.; Pombosa, R. Monitoring of the Spatio-Temporal Dynamics of the Floods in the Guayas Watershed (Ecuadorian Pacific Coast) Using Global Monitoring ENVISAT ASAR Images and Rainfall Data. *Water* **2017**, *9*, 12. [CrossRef]

46. Aizen, V.B.; Aizen, E.M.; Kuzmichonok, V.A. Glaciers and hydrological changes in the Tien Shan: Simulation and prediction. *Environ. Res. Lett.* **2007**, *2*, 45019. [CrossRef]

47. Cogley, J.G.; Hock, R.; Rasmussen, L.A.; Arendt, A.A.; Bauder, A.; Braithwaite, R.J.; Jansson, P.; Kaser, G.; Möller, M.; Nicholson, L. *Glossary of Glacier Mass Balance and Related Terms*; IHP-VII Technical Documents in Hydrology No. 86, IACS Contribution No. 2; UNESCO-IHP: Paris, France, 2011.

48. Bahr, D.B.; Meier, M.F.; Peckham, S.D. The physical basis of glacier volume-area scaling. *J. Geophys. Res. Solid Earth* **1997**, *102*, 20355–20362. [CrossRef]

49. Kang, M.S.; Goo, J.H.; Song, I.; Chun, J.A.; Her, Y.G.; Hwang, S.W.; Park, S.W. Estimating design floods based on the critical storm duration for small watersheds. *J. Hydro-Environ. Res.* **2013**, *7*, 209–218. [CrossRef]

50. Yen, B.C.; Chow, V.T. Design hyetographs for small drainage structures. *J. Hydraul. Div. ASCE* **1980**, *106*, 1055–1076.

51. Keifer, G.J.; Chu, H.H. Synthetic storm pattern for drainage design. *J. Hydraul. Div. ASCE* **1957**, *83*, 1–25.

52. SCS. *Urban Hydrology for Small Watersheds*; Technical Release 55; U.S. Department of Agriculture: Washington, DC, USA, 1986.

53. Karlsson, M.; Yakowitz, S. Rainfall-runoff forecasting methods, old and new. *Stoch. Hydrol. Hydraul.* **1987**, *1*, 303–318. [CrossRef]

54. Haggag, M.; El-Badry, H. Mesoscale numerical study of quasi-stationary convective system over Jeddah in November 2009. *Atmos. Clim. Sci.* **2013**, *3*, 73–86. [CrossRef]

55. Zipser, E.J.; Cecil, D.J.; Liu, C.; Nesbitt, S.W.; Yorty, D.P. Where are the most intense thunderstorms on earth? *Bull. Am. Meteorol. Soc.* **2006**, *87*, 1057–1071. [CrossRef]

*water*

MDPI

Article

# Large Differences between Glaciers 3D Surface Extents and 2D Planar Areas in Central Tianshan

**Xianwei Wang [1,\*], Huijiao Chen [1] and Yaning Chen [2]**

[1] School of Geography and Planning, and Guangdong Key Laboratory for Urbanization and Geo-Simulation, Sun Yat-sen University, Guangzhou 510275, China; chenhj37@mail2.sysu.edu.cn

[2] State Key Laboratory of Desert and Oasis Ecology, Xinjiang Institute of Ecology and Geography, Chinese Academy of Sciences, Urumqi 830011, China; chenyn@ms.xjb.ac.cn

\* Correspondence: Wangxw8@mail.sysu.edu.cn; Tel.: +86-20-84114623

Academic Editor: Y. Jun Xu

Received: 30 January 2017; Accepted: 14 April 2017; Published: 17 April 2017

**Abstract:** Most glaciers in China lie in high mountainous environments and have relatively large surface slopes. Common analyses consider glaciers' projected areas (2D Area) in a two-dimensional plane, which are much smaller than glacier's topographic surface extents (3D Area). The areal difference between 2D planar areas and 3D surface extents exceeds $-5\%$ when the glacier's surface slope is larger than $18°$. In this study, we establish a 3D model in the Muzart Glacier catchment using ASTER GDEM data. This model is used to quantify the areal difference between glaciers' 2D planar areas and their 3D surface extents in various slope zones and elevation bands by using the second Chinese Glacier Inventory (CGI2). Finally, we analyze the 2D and 3D area shrinking rate between 2007 and 2013 in Central Tianshan using glaciers derived from Landsat images by an object-based classification approach. This approach shows an accuracy of 89% when it validates by comparison of glaciers derived from Landsat and high spatial resolution GeoEye images. The extracted glaciers in 2007 also have an agreement of 89% with CGI2 data in the Muzart Glacier catchment. The glaciers' 3D area is 34.2% larger than their 2D area from CGI2 in the Muzart Glacier catchment and by 27.9% in the entire Central Tianshan. Most underestimation occurs in the elevation bands of 4000–5000 m above sea level (a.s.l.). The 3D glacier areas reduced by 30 and 115 $km^2$ between 2007 and 2013 in the Muzart Glacier catchment and Central Tianshan, being 37.0% and 27.6% larger than their 2D areas reduction, respectively. The shrinking rates decrease with elevation increase.

**Keywords:** glacier; 2D area; 3D area; Central Tianshan

## 1. Introduction

Mountain glaciers and snow are crucial water resources for the surrounding river, lake, oasis, cropland and urban life in arid Central Asia [1]. Glaciers' ice volumes are usually estimated by Volume-Area (V-A) power law equations since there are few in situ measurements of ice volume using modern techniques, such as sounding echo, ground radar or gravity methods [2–4]. The V-A scaling method is based on ice dynamics imposed by the climatic and topographic conditions in different glacierized regions, and has an inherent steady-state assumption [2]. This assumption is often violated, with many glaciers being out of equilibrium [5]. The volume estimation errors can exceed 50% for individual glaciers [6]. Moreover, glaciers' area change does not closely correspond to ice thickness changes (increase in the accumulation zone and decrease in the ablation zone), resulting in even larger errors, especially in estimating the ice volume changes by using glacier's areas in different years [2].

The glacier area is defined by the ice mass balance community as the extent in two horizontal dimensions (Figure 1), i.e., the extent/outline of the glacier is projected onto the surface of an ellipsoid Earth surface, rather than the real topographic surface/the slope normal [7]. The former is called 2D area (Figure 1, A1), and the latter is called 3D area (Figure 1, A2) in this study hereafter. Meanwhile, the ice/glacier thickness is defined as the vertical length (Figure 1, T1) measured parallel to the vertical axis of the ellipsoid Earth surface and not normal to the glacier surface [7]. Thus, the ice volume is the integral of the planar area and thickness. In contrast, the snow layer thickness (Figure 1, T2) is usually measured perpendicularly relative to the slope normal of the snow/land surface [8]. Both the glacier's 2D area and thickness values are close to the true values for flat ice sheets and glaciers with gentle slope (<18°), while greater difference exists for glaciers with larger slopes, although the two pairs of definitions for area and thickness relative to horizontal normal (2D area) or slope normal (3D area) make no difference for calculating ice volume together (Figure 1). The 3D area might be a better variable in the ice volume estimate using the V-A scaling method, since it considers the slope factor and reflects ice thickness changes. Moreover, glaciers' 3D surface extent could be a better variable in modeling their surface melting and sublimation [9].

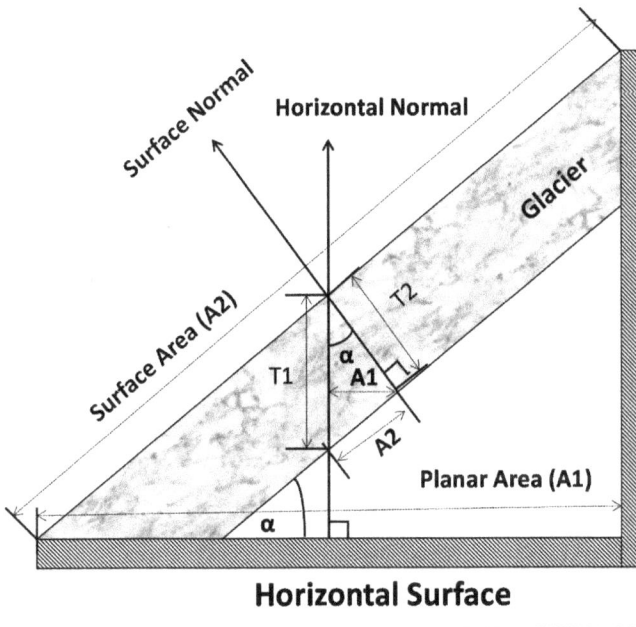

**Figure 1.** Schematic diagram of the definitions of glacier's area (A) and thickness (T) in a longitudinal glacier profile.

Most glaciers in Central Tianshan lie in high mountainous areas over 3000 m a.s.l. These alpine glaciers often have complex catchments, divisions and large slopes. For example, one of the large glaciers, the Muzart Glacier near the Tumor Peak, is highly labile with fluctuating length, area, volume, and shape [1,10,11], thus violating the steady state assumption of the V-A scaling method and leading to large uncertainties in the ice volume estimation.

Numerous studies have investigated glacier area changes in Central Tianshan based on satellite and airborne images and topographic data/DEM [1,12–16]. Most studies analyze 2D planar areas, while few studies discuss the difference between glacier's 2D areas and 3D areas [17], partially due to the unavailability of topographic data with relatively high spatial resolution. Therefore, the primary objective of this study is to compute glaciers' 2D and 3D areas and evaluate how the differences between them relate to changes in surface slope and elevation bands in the Muzart Glacier catchment and Central Tianshan.

## 2. Study Area and Data Analysis

### 2.1. Study Area

The Muzart Glacier catchment is located in the upper Muzart River Basin in Central Tianshan (also called Tien Shan in some literature) Mountains, and is the largest center of modern glaciation in the Tianshan Range (Figure 2). Locally, the Muzart Glacier also includes its northern division, or the northern Muzart Glacier catchment, which drains to the northern Muzart River, the upper tributary of the Tekes River and then Yili River. This study only focuses on the southern division of the Muzart Glacier, i.e., simplified as the Muzart Glacier catchment in this study. The (southern) Muzart River is more than 80% supported by snow/glacier melt water and is an important tributary of the Tarim River [11,18]. The snow/glacier melt water provides a critical water resource for the downstream piedmont oases. According to the Second Chinese Glacier Inventory, there are 318 glaciers with a total area of 1192 km$^2$ in the Muzart Glacier catchment, including hanging glaciers, cirque glaciers, single valley glaciers and compound valley glaciers [11]. The average slope in the Muzart glacial catchment is 31.4° with elevation ranging from 2500 to 7400 m a.s.l. Beyond the Muzart Glacial catchment, glaciers distribute above 3000 m a.s.l. in the entire Central Tianshan, and the contour of 2500 m a.s.l. is applied to constrain the analyzing ranges of glacier areas (Figure 2).

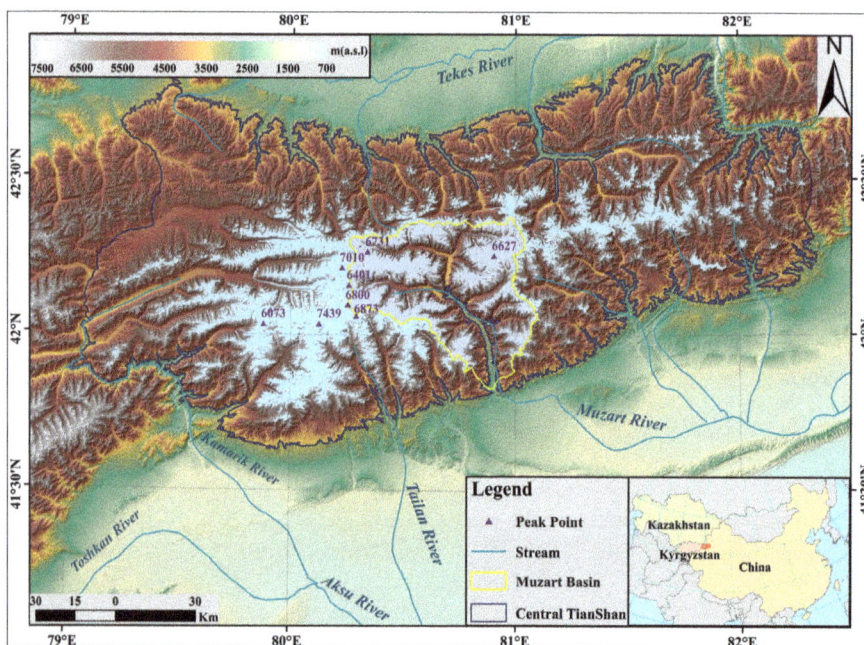

**Figure 2.** Study area in Central Tianshan Mountain and the Muzart Glacier catchment (yellow line) of the upper Muzart River Basin, China. The analyzing areas are constrained by the 2500 m elevation contour (dark blue line).

### 2.2. Data and Analysis

The Second Chinese Glacier Inventory (CGI2) data provided by the West Data Center for glaciology and geocryology, at Lanzhou, China [19] are used to analyze the spatial distribution and to compare the glacier maps classified from Landsat images in this study. The outline of CGI2 was derived using the band ratio segmentation method, and had extensive manual editions based on

218 Landsat TM/ETM+ scenes (30 m of spatial resolution) acquired mainly during 2006–2010. Glaciers positioning errors were about 10 m for clean-ice outlines and 30 m for debris-covered outlines, and area errors were 3.2% [10].

Four Landsat 5 TM images acquired on 24 August 2007, which are the same images used in CGI2, and four Landsat 8 OLI images acquired on 10 September 2013, are used to extract the glacier areas in the Muzart Glacier catchment and Central Tianshan. These scenes were cloud-free over glaciers and had minimum seasonal snow cover, which is best for glacier mapping [13]. The extracted glaciers are compared to CGI2 in the Muzart Glacier catchment. Also, a suitable high-resolution GeoEye (nominal 2-m spatial resolution) acquired on 20 April 2015 is downloaded to validate the glacier outline derived from Landsat images.

A semi-automatic methodology is utilized to delineate the glacier area using an object-based image classification approach on eCognition 9.0 (Trimble Inc., Sunnyvale, CA, USA) [12,20–22]. The specific procedures are illustrated in (Figure 3). The Landsat/GeoEye images are first segmented using multi-resolution segmentation which creates the image object based on spectral and shape characteristics [2]. Then, the class hierarchy is built with a focus on clean ice. Next, the classifier is trained and applied by using the Support Vector Machine (SVM) with a linear kernel [23]. The classified glaciers are manually corrected by visual comparison with images acquired in different years. Finally, the classified objects are merged and exported to vector polygons for further visually checking and manual edition on ArcMap, eliminating misclassified pro-glacial water, snow cover, and shadow areas by overlaying with DEM data and GoogleEarth images.

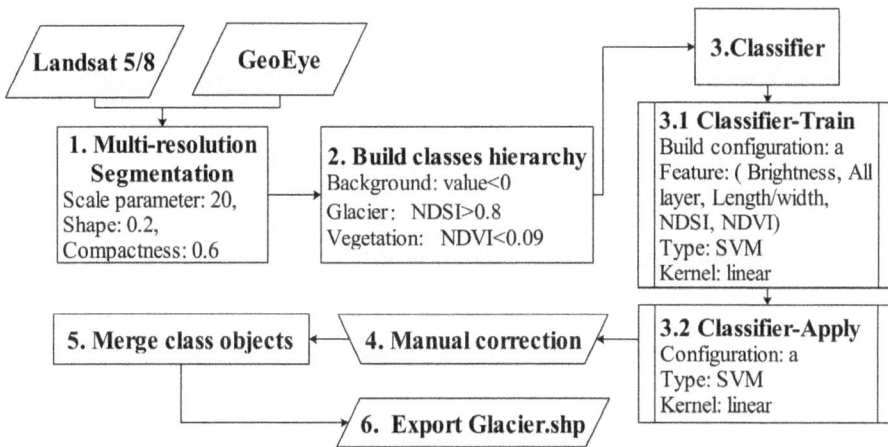

**Figure 3.** Flowcharts of glacier outline delineation using object-based image classification.

The ASTER GDEM V2 was downloaded from Japan Space Systems [24], and had a 30-m spatial resolution with reported vertical accuracies of less than 17 m and horizontal accuracies of 71 m. It is suitable for the compilation of topographic parameters in a glacier inventory [25,26], and is used to establish the 3D model, delineate the glacier catchment, compute the slope, and classify the slope zones and elevation bands in this study. All Landsat and GeoEye images, CGI2 and the ASTER GDEM V2 data sets are reprojected to the Universal Transverse Mercator (UTM) coordinate system, zone 44 before analysis.

The study areas are further divided into elevation bands with a 500-m interval and slope zones with a $10°$ interval based on the ASTER GDEM V2 data. The 3D surface areas in the entire area and different elevation bands and slope zones are estimated by raster-based methods based on the 3D model established from the ASTER GDEM V2 data. Similarly, the 2D project glacier areas are also analyzed in those elevation bands and slope zones. All areal calculations are carried out on ArcMap 10.3 (ESRI, Redlands, CA, USA).

## 3. Results

### 3.1. Glacier Outline Extractions

This study derives the glacier maps directly from Landsat images in 2007 and 2013 by utilizing the object-based classification approach, thus can detect the glacier's 2D and 3D area changes using consistent glacier maps between the two years. In glacier extraction, snow cover and glaciers are not separated, and it is not possible to distinguish them from optical images because of snow-covered glaciers. Meanwhile, both snow and glacier have a similar spectral signature in the optical wavelength range. Glacier/snow covered 65% of this sub-catchment with a total area of 50.5 $km^2$ on 20 April 2015 in the validation GeoEye image (Figure 4). Statistical results show that the glacier/snow classification accuracy is 89.3% (Table 1). Both glacier boundary lines match well with the underlying white glacier/snow of GeoEye images. Most differences are located in debris-covered glaciers, shaded glaciers and the glacier edges. Some of those small and scattering glaciers identified by GeoEye images were seasonal snow in the lower ranges.

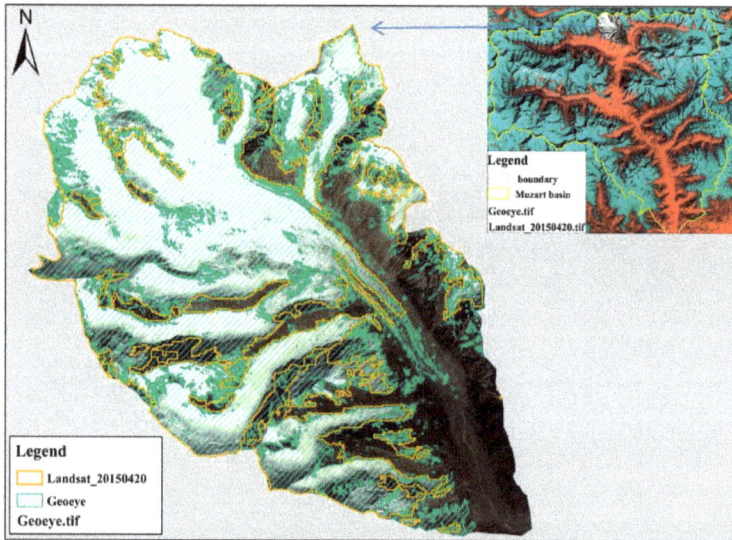

**Figure 4.** Glacier outlines derived from Landsat images (yellow polygons) and GeoEye images (blue polygon and background images) using object-based classification in the upper sub-catchment of the Muzart Glacier catchment on 20 April 2015.

**Table 1.** Error matrix of glacier mapping between Landsat and GeoEye01 images in the upper sub-catchment of the Muzart Glacier catchment on 20 April 2015.

| Image Classification | | Reference ($km^2$) | 2015 Landsat 8 OLI ($km^2$) | |
| --- | --- | --- | --- | --- |
| | | | Glacier | Non-Glacier |
| GeoEye01 | Glacier | 32.8 | 29.3 89.3% | 3.5 10.7% |
| | Non-glacier | 17.7 | 3.8 21.5% | 13.9 78.5% |
| | | Overall Accuracy | | 85.5% |

The extracted glacier areas (2D) are 89.3% of CGI2 in the Muzart Glacier catchment (Table 2), where the extracted glaciers' 3D areas are 91.9% of CGI2. Most of these lower estimates occur at the lower end of debris-covered glaciers and have gentle slope (Figures 5 and 6). Most of the debris-covered/mixed glacier tongues are not classified as glaciers in this comparison due to their low reflectance, while CGI2 manually edits them to be glaciers (Figure 5). The glaciers' area, as extracted from CGI2 in slope zones of less than 10°, is 22% of the total area for 2D areas and 17% of the total area for 3D areas (Figure 6a), while they are only 16% and 11%, respectively, for those extracted from Landsat images in this study (Figure 6b). The debris-covered glacier areas are around 5%–10% of the total areas according to statistical analysis of all CGI2 data in Central Tianshan [10].

**Figure 5.** Comparison of glacier outlines of CGI2 (blue polygon) and those derived from Landsat images (yellow polygons and background images) in this study using object-based classification in the Muzart Glacier catchment on 24 August 2007.

**Table 2.** Statistics of glacier areas (km$^2$) based on CGI2 and those extracted from Landsat images in the Muzart Glacier catchment and the entire Central Tianshan Mountain during 2007–2013. CGI2 does not cover the entire Central Tianshan; its statistics not given.

| Glaciers | Muzart Glacier Catchment | | | Central Tianshan Mountain | | |
|---|---|---|---|---|---|---|
| | 2D Area | 3D Area | Relative Difference (3D-2D)/2D | 2D Area | 3D Area | Relative Difference (3D-2D)/2D |
| CGI2 | 1160 | 1557 | 34.2% | | | |
| 24 August 2007 | 1036 | 1431 | 38.1% | 4518 | 5778 | 27.9% |
| Landsat/CGI2 | 89.3% | 91.9% | | | | |
| 10 September 2013 | 953 | 1316 | 38.1% | 4101 | 5244 | 27.9% |
| Dif. (2013–2007) | −81 | −111 | 37.0% | −418 | −533 | 27.6% |
| Dif. (2013–2007)/2007 | −7.76% | −7.82% | | −9.25% | −9.23% | |

## 3.2. Glacier Distributions

Glaciers distribute in a large range of slopes in the Muzart Glacier catchment (Figure 6a). According to CGI2, glaciers' 2D planar areas are 1160 km$^2$ in total, while their 3D surface extents are 1557 km$^2$. About half of the glaciers lie in slopes larger than 30°, causing great differences (397 km$^2$, 34.2%) between the 3D surface extents and 2D planar areas. When the slope is less than 10°, the absolute values of both 3D and 2D areas are similar, while their own frequency percentages reduce from 22% for the 2D area to 17% for the 3D area. When the slopes range from 10° to 20°, the 3D area is 5% larger than the 2D area. When the slopes range from 20° to 30°, the 3D area is 16% larger than the 2D area. When the slopes range from 40° to 50°, the 3D area is 47% larger than the 2D area. When the slope is larger than 50°, glaciers' 3D areas are nearly two times that of the 2D areas. In addition, the slope zones separate the glacier catchment into many fractional areas instead of continuous areas like elevation bands, leading to large distribution differences/fluctuations between two glaciers (CGI2 and glaciers derived from Landsat images in this study) in each slope zones (Figure 6b).

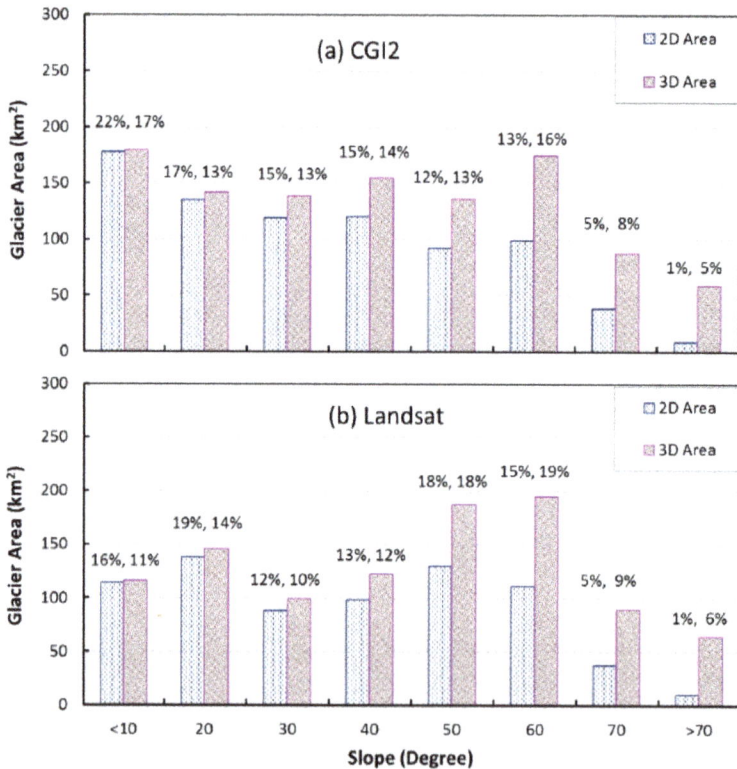

**Figure 6.** Histogram of glacier 2D and 3D areas within different slope zones based on the second Chinese Glacier Inventory (CGI2) data (**a**) and glaciers classified from Landsat images (**b**) on 24 August 2007 in the Muzart Glacier catchment. The numbers above the columns are the frequency percentages of glacier areas in each slope zones against total 2D and 3D areas, respectively.

The glacier maps extracted from Landsat images in 2007 are also analyzed in different elevation bands for their 2D and 3D areas in the Muzart Glacier catchment and the entire Central Tianshan (Figure 7). Their total glacier areas were 1036 km$^2$ (1431 km$^2$ for 3D) and 4518 km$^2$ (5778 km$^2$ for 3D) in 2007, and their 3D surface extents are 38.1% and 27.9% larger than the 2D planner areas, respectively

(Table 2). These ratios remained similar in 2013. Most glaciers (83%) distribute in elevation bands of 4000–4500 m (47%) and 4500–5000 m (36%) in the Muzart Glacier catchment (Figure 7a). By contrast, glaciers are more evenly distributed throughout the entire Central Tianshan, although the frequency components of glaciers in the elevation bands of 4000–4500 m (38%) and 4500–5000 m (29%) are also dominant (67%), but smaller than those in the Muzart Glacier catchment (Figure 7b).

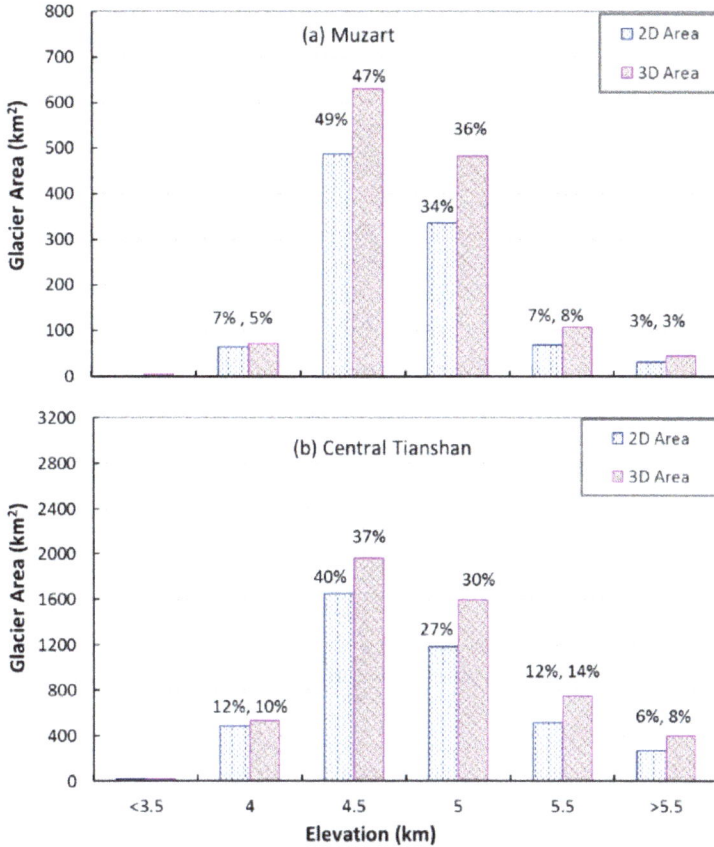

**Figure 7.** Histogram of Glacier 2D and 3D areas within different elevation bands in 2007 in the Muzart Glacier catchment (**a**) and Central Tianshan (**b**). The numbers above the columns are the frequency percentages of glacier areas in each elevation bands against the total 2D and 3D areas, respectively.

### 3.3. Glacier Area Changes

Aside from the glacier distributions in different slope zones and elevation bands, this study further compares the glaciers' 2D and 3D area changes between 2007 and 2013, and their spatial distributions in different elevation bands in the Muzart Glacier catchment and Central Tianshan. Overall, the glacier areas reduced between 2007 and 2013 (Table 2). In the Muzart Glacier catchment, the 2D planar area reduced by 81 km$^2$, while their actual 3D surface extents reduced by 111 km$^2$, which is 30 km$^2$ (37.0%) larger than the 2D planar area reduction, although their relative shrinking rates are quite similar. As expected, the shrinking rates decrease with elevation increase, and the dominant shrinking areas (42.8 km$^2$, 57%) occurred in the elevation band of 4000–4500 m, where the actual 3D shrinking areas were 16.2 km$^2$ (38%) larger than the 2D area (Figure 8a).

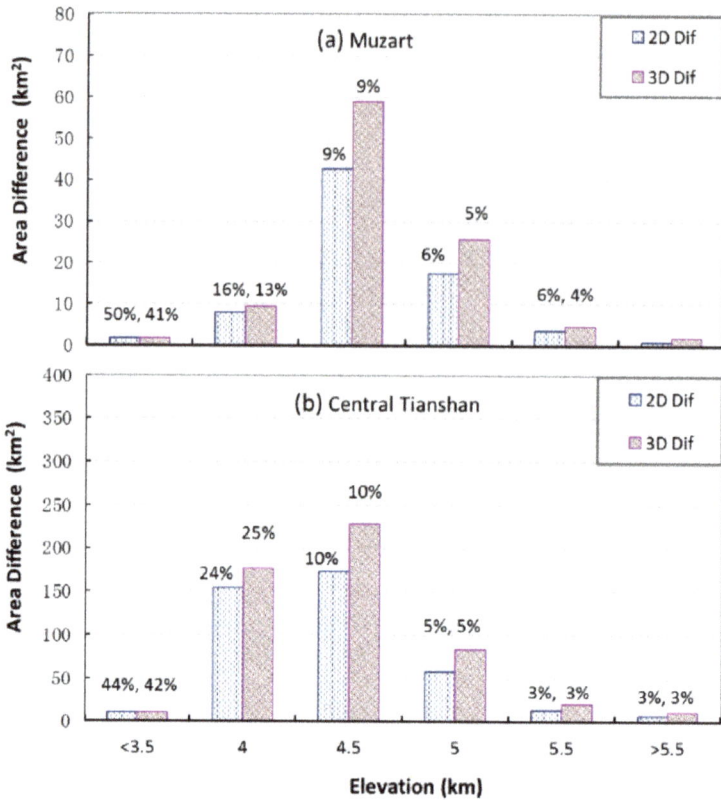

**Figure 8.** Histogram of Glacier 2D and 3D area difference between 2007 and 2013 (2013–2007) in different elevation bands in the Muzart Glacier catchment (**a**) and Central Tianshan (**b**). The numbers above columns are the area shrinking rates ((2013–2007)/2007) in each elevation bands.

In the entire Central Tianshan Mountains, the 2D planar area reduced by 418 km², while their actual 3D surface extents reduced by 533 km², which is 115 km² (27.6%) larger than the 2D planar area. Their relative shrinking rates are also similar, being slightly larger than that in the Muzart Glacier catchment (Table 2). The shrinking rates also decrease with elevation increase (Figure 8b), and the dominant shrinking areas occurred in the elevation bands of 3500–4000 m (37% for 2D area and 33% for 3D area) and 4000–4500 m (42% for 2D area and 36% for 3D area).

## 4. Discussion

### 4.1. Glacier Classifications

There are many factors that affect the accuracy of glacier classification using optical images, such as classification approach, seasonal snow, cloud, shadow, debris, and so on. The primary objective of this study is to investigate the differences between glacier's 2D planar areas and 3D surface extents using CGI2 data, instead of developing or evaluating different classification approaches. In order to quantify how the difference of 2D and 3D areas affects the change rates of glaciers between different years, consistent glacier area products are expected to cancel out their systematic errors between different glacier products.

There are several classification methods and band-combination options to extract the glacier outlines in literature. This study only utilizes the object-based classification method to extract glacier outline from Landsat images and GeoEye images using the software eCognition 9.0. This approach has been widely used for glacier mapping recently [12,20–22]. The common procedure provided in the software manual is adopted to delineate the glacier outlines like those in the literature (Figure 3). Meanwhile, manual corrections were intensive in some areas, such as those within debris-covered/mixed glaciers, shade, under cloud or seasonal snow cover, similar to those stated in making the second Chinese Glacier Inventory [10]. In such complex situations, we compare the images acquired for different years, seasons and time, and only retain the minimum outline. As shown in Figure 9, loading the Landsat image on the 3D surface could be better than the 2D planar image to determine the shaded area and debris-covered glacier tongue, leading to higher confidence and accuracy in manual editing.

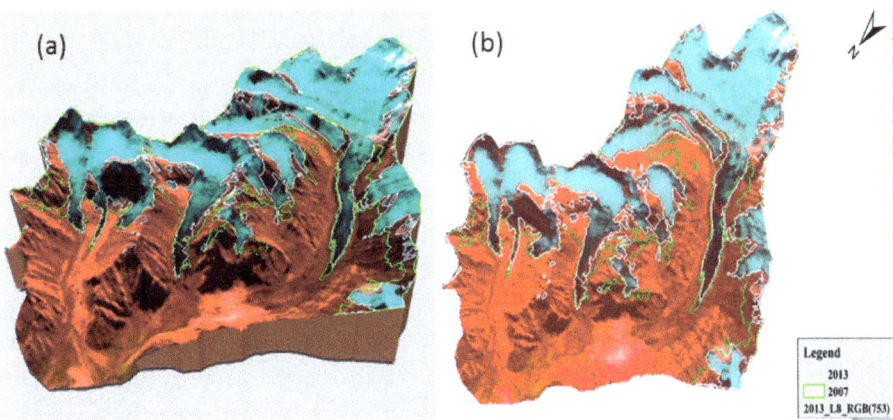

**Figure 9.** Comparison of 3D (**a**) and 2D (**b**) glacier outlines in 2007 (green lines) and 2013 (white lines) in a glacier sub-catchment of the Muzart Glacier catchment (mid-west). The background image is the Landsat 8 on 10 September 2013.

This classification approach is validated by comparing glacier outlines derived from high resolution GeoEye image on 20 April 2015, since there was no high spatial resolution image available on the date of the Landsat images, i.e., on 24 August 2007 and 10 September 2013. The selected validation image is constrained by the limited high spatial resolution images and the Landsat images on the same date in the study areas. There was much snow over glacier surface and nearby rocks on 20 April 2015, only when both the GeoEye and Landsat images were available and cloud-free in the study catchment. In practice, snow is not separated from glacier in classification, and a large part of the debris-covered glaciers are not included in our analysis as well (Figure 5). This explains why there is a larger area difference between 3D and 2D areas for our delineated glacier outlines (38.1%) than for those of CGI2 (34.2%) in the Muzart Glacier catchment (Table 2), since debris-covered glaciers have gentler slopes (Figure 5). The glacier 2D areas derived from the Landsat image on 20 April 2015 have an agreement of 89.3% with those from the GeoEye image (Table 1, Figure 4). The classified glacier 2D areas have 89.3% agreement with CGI2 in the Muzart Glacier catchment, both using the same Landsat images (Table 2, Figure 5). This accuracy is similar to those reported in the literature, e.g., 93% for clean ice, 83% for debris-covered glacier, and total accuracy of 91% [22]. This indicates that the classified glacier outlines are suitable for glaciers' 2D and 3D area analysis.

*4.2. Difference between Glacier 2D and 3D Areas*

The difference between glacier 2D and 3D areas increases with slope in the Muzart Glacier catchment for both CGI2 and the classified glaciers, revealing their geometric relationship (Figure 6). The slope zones divide the study areas into many small fractions, resulting in larger uncertainties than the elevation bands due to the edge/boundary issues overlaying with glacier areas (Figures 6 and 7). The 3D areas are 38.1% and 27.9% larger than the 2D areas in the Muzart Glacier catchment and Central Tianshan, respectively. This large difference is significant in calculating the total ice volume using the V-A scaling method [2–4], and computing surface energy balance and mass melting [9]. It is worthy of further investigation into whether or not 3D areas work better than 2D areas for estimating the total ice volume by the V-A scaling method.

The shrinking rates of glaciers' 2D areas are $-1.1\%.a^{-1}$ for Muzart Glacier catchment and $-1.3\%.a^{-1}$ for Central Tianshan in this study. They fall within the ranges of those reported in the literature (Table 3). The glaciers' area remained near constant in the Inylchek Glacier of Central Tianshan during 1999–2007 [27]. The largest shrinking rate was $-1.7\%.a^{-1}$ reported by Du and Li [28] in the Mt. Karlik of Eastern Tianshan during 2007–2013, then $-1.0\%.a^{-1}$ published by Kaldybayev [29] in the Karatal River Basin of Nothern Tianshan during 1989–2012. The mean glacier surface slope of CGI2 is 19.9°, while glaciers in the Central Tianshan, Pamir plateau, Qilian Mountains and Altun Mountains have the steepest glacier surfaces, over one-third of their surface slopes are greater than 30° [10]. Geometrically, the 3D area is 5% larger than the 2D area as the slope is larger than 18°, 15% larger for 30° and 41% larger for 45° (Figure 1). These large areal differences between 3D real surface extents and the projected virtual 2D area is significant not only in areal and volume calculation, but also in glaciers' precise surface energy budget and mass balance/melting modeling, especially in the high Asian mountain glaciers with large surface slope and strong solar radiation.

**Table 3.** The glaciers 2D area changes reported from different studies in Tianshan Mountains.

| Location | Region | Period | 2D Area Changes (%) | Change Rate (%.a$^{-1}$) | Document Source |
|---|---|---|---|---|---|
| Jinghe River Basin | Eastern Tianshan | 1964–2004 | −15.2 | −0.4 | [30] |
| Karatal River Basin | Northern Tianshan | 1989–2012 | −23.0 | −1.0 | [29] |
| Ak-Shyirak massif | Western Tianshan | 2003–2013 | −5.9 | −0.6 | [15] |
| Ili-Kungoy | Central Tianshan | 2007–2013 | −4.0 | −0.4 | [31] |
| Mt. Karlik | Eastern Tianshan | 2007–2013 | −9.9 | −1.7 | [28] |
| Inylchek Glacier | Central Tianshan | 1999–2007 | −0.3 | −0.1 | [27] |
| Muzart Glacier (south) | Central Tianshan | 2007–2013 | −7.8 | −1.1 | This study |
| Central Tianshan | Central Tianshan | 2007–2013 | −9.2 | −1.3 | This study |

## 5. Summary and Remark

This study utilizes the lastest relatively high-resolution global topographic data (ASTER GDEM V2) and CGI2 data to illustrate the large areal difference between glaciers' 3D real surface extents and their projected 2D planar area in the Muzart Glacier catchment and Central Tianshan. Besides the CGI2 data, this study also extracts the glacier outlines from Landsat images in 2007 and 2013 by an object-based classification approach, which is validated using GeoEye high-resolution images and shows an accuracy of 89.3%. The extracted glacier outlines in 2007 also had an agreement of 89.3% with CGI2 data in the Muzart Glacier catchment. Most of the differences are in the lower-end of glaciers covered by debris.

The difference between 3D surface extents and 2D planar areas from those extracted glacier outlines in 2007 and 2013 (38.1%) are slightly larger than those of CGI2 (34.2%) in the Muzart Glacier catchment and were 27.9% on average in the entire Central Tianshan. The difference between 3D areas and 2D areas for the shrunk glaciers were slightly smaller than those of existing glaciers in the Muzart Glacier catchment (37.0%), and the entire Central Tianshan (27.6%) since many of the shrunk ones were located on the lower end of glaciers and had a smaller slope. Consequently, their relative shrinking rates from 2007 to 2013 were similar in both Muzart Glacier catchment ($-7.8\%$, 30 km$^2$) and Central Tianshan ($-9.2\%$, 115 km$^2$), although there was a large difference between 3D areas and 2D area of those shrunk glaciers. Those large areal differences remind us to re-consider glacier's real topographic extent when discussing alpine glacier's areal and volume changes, especially in calculating the glaciers surface energy balance and melting rates in the high Asian mountain glaciers with large surface slope and strong solar radiation.

**Acknowledgments:** This study was funded by the Open Foundation of the State Key Laboratory of Desert and Oasis Ecology, Xinjiang Institute of Ecology and Geography, Chinese Academy of Sciences (#G2014-02-06), Natural Science Foundation of China (#41371404, #41630859), Fundamental Research Funds from Sun Yat-sen University (#15lgzd06), and the Water Resource Science and Technology Innovation Program of Guangdong Province (#2016-19). Those funds were used to cover the costs of publishing in open access. We thank the World Data Center for Glaciology and Geocryology in Lanzhou, China, US NASA, Japan Space Systems and the Google Earth and DigitalGlobe companies for providing the CGI2, Landsat and GeoEye images and ASTER GDEM data support of our research work. Comments from two anonymous reviewers greatly improve this manuscript and are highly appreciated.

**Author Contributions:** Xianwei Wang designed the experiments, analyzed the results and wrote the manuscript; Huijiao Chen processed the data, analyzed the results and wrote part of the manuscript; Yanning Chen improved the experiments and results analysis.

**Conflicts of Interest:** The authors declare no conflict of interest.

# References

1. Aizen, V.B.; Aizen, E.M.; Kuzmichonok, V.A. Glaciers and hydrological changes in the Tien Shan: simulation and prediction. *Environ. Res. Lett.* **2007**, *2*, 45019. [CrossRef]
2. Adhikari, S.; Marshall, S.J. Glacier volume-area relation for high-order mechanics and transient glacier states. *Geophys. Res. Lett.* **2012**, *39*, 6. [CrossRef]
3. Bahr, D.B.; Meier, M.F.; Peckham, S.D. The physical basis of glacier volume-area scaling. *J. Geophys. Res. Solid Earth* **1997**, *102*, 20355–20362. [CrossRef]
4. Chen, J.; Ohmura, A. Estimation of alpine glacier water resources and their change since the 1870s. In *Hydrology in Mountainous Regions. I: Hydrological Measurements, the Water Cycle, Proceedings of Two International Symposia, Lausanne, Switzerland, 27 August–1 September 1990*; Lang, H., Musy, A., Eds.; International Association of Hydrological Sciences (IAHS): Wallingford, UK, 1990; Volume 193, pp. 127–135.
5. Farinotti, D.; Huss, M.; Bauder, A.; Funk, M.; Truffer, M. A method to estimate the ice volume and ice-thickness distribution of alpine glaciers. *J. Glaciol.* **2009**, *55*, 422–430. [CrossRef]
6. Meier, M.F.; Dyurgerov, M.B.; Rick, U.K.; O'Neel, S.; Pfeffer, W.T.; Anderson, R.S.; Anderson, S.P.; Glazovsky, A.F. Glaciers Dominate Eustatic Sea-Level Rise in the 21st Century. *Science* **2007**, *317*, 1064–1067. [CrossRef] [PubMed]
7. Cogley, J.G.; Hock, R.; Rasmussen, L.A.; Arendt, A.A.; Bauder, A.; Braithwaite, R.J.; Jansson, P.; Kaser, G.; Möller, M.; Nicholson, L. *Glossary of Glacier Mass Balance and Related Terms*; IHP-VII Technical Documents in Hydrology No. 86, IACS Contribution No. 2; UNESCO-IHP: Paris, France, 2011.
8. Fierz, C.; Armstrong, R.L.; Durand, Y.; Etchevers, P.; Greene, E.; McClung, D.M.; Nishimura, K.; Satyawali, P.K.; Sokratov, S.A. *The International Classification for Seasonal Snow on the Ground*; IHP-VII Technical Documents in Hydrology No.83, IACS Contribution No.1; UNESCO-IHP: Paris, France, 2009.
9. Zhang, Y.; Luo, Y.; Sun, L.; Liu, S.; Chen, X.; Wang, X. Using glacier area ratio to quantify effects of melt water on runoff. *J. Hydrol.* **2016**, *538*, 269–277. [CrossRef]
10. Guo, W.; Liu, S.; Xu, J.; Wu, L.; Shangguan, D.; Yao, X.; Wei, J.; Bao, W.; Yu, P.; Liu, Q. The second Chinese glacier inventory: Data, methods and results. *J. Glaciol.* **2015**, *61*, 357–372. [CrossRef]
11. Zhao, J.; Song, Y.; King, J.W.; Liu, S.; Wang, J.; Wu, M. Glacial geomorphology and glacial history of the Muzart River valley, Tianshan Range, China. *Quat. Sci. Rev.* **2010**, *29*, 1453–1463. [CrossRef]
12. Bajracharya, S.R.; Maharjan, S.B.; Shrestha, F. The status and decadal change of glaciers in Bhutan from the 1980s to 2010 based on satellite data. *Ann. Glaciol.* **2014**, *55*, 159–166. [CrossRef]
13. Ke, L.; Ding, X.; Song, C. Heterogeneous changes of glaciers over the western Kunlun Mountains based on ICESat and Landsat-8 derived glacier inventory. *Remote Sens. Environ.* **2015**, *168*, 13–23. [CrossRef]
14. Paul, F.; Bolch, T.; Kääb, A.; Nagler, T.; Nuth, C.; Scharrer, K.; Shepherd, A.; Strozzi, T.; Ticconi, F.; Bhambri, R. The glaciers climate change initiative: Methods for creating glacier area, elevation change and velocity products. *Remote Sens. Environ.* **2015**, *162*, 408–426. [CrossRef]

15. Petrakov, D.; Shpuntova, A.; Aleinikov, A.; Kääb, A.; Kutuzov, S.; Lavrentiev, I.; Stoffel, M.; Tutubalina, O.; Usubaliev, R. Accelerated glacier shrinkage in the Ak-Shyirak massif, Inner Tien Shan, during 2003–2013. *Sci. Total Environ.* **2016**, *562*, 364–378. [CrossRef] [PubMed]

16. Wang, P.; Li, Z.; Wang, W.; Li, H.; Wu, L.; Huai, B.; Zhou, P.; Jin, S.; Wang, L. Comparison of changes in glacier area and thickness on the northern and southern slopes of Mt. Bogda, eastern Tianshan Mountains. *J. Appl. Geophys.* **2016**, *132*, 164–173. [CrossRef]

17. Zhang, Y.; Zhang, L.; Yang, C.; Bao, W.; Yuan, X. Surface area processing in GIS for different mountain regions. *Forest Sci. Pract.* **2011**, *13*, 311–314. [CrossRef]

18. Pieczonka, T.; Bolch, T.; Wei, J.; Liu, S. Heterogeneous mass loss of glaciers in the Aksu-Tarim Catchment (Central Tien Shan) revealed by 1976 KH-9 Hexagon and 2009 SPOT-5 stereo imagery. *Remote Sens. Environ.* **2013**, *130*, 233–244. [CrossRef]

19. World Data Center for Glaciology and Geocryology, Lanzhou. Available online: http://wdcdgg.westgis.ac.cn/ (accessed on 12 October 2015).

20. Kraaijenbrink, P.D.A.; Shea, J.M.; Pellicciotti, F.; Jong, S.M.D.; Immerzeel, W.W. Object-based analysis of unmanned aerial vehicle imagery to map and characterise surface features on a debris-covered glacier. *Remote Sens. Environ.* **2016**, *186*, 581–595. [CrossRef]

21. Nijhawan, R.; Garg, P.; Thakur, P. A comparison of classification techniques for glacier change detection using multispectral images. *Perspect. Sci.* **2016**, *8*, 377–380. [CrossRef]

22. Robson, B.A.; Nuth, C.; Dahl, S.O.; Hölbling, D.; Strozzi, T.; Nielsen, P.R. Automated classification of debris-covered glaciers combining optical, SAR and topographic data in an object-based environment. *Remote Sens. Environ.* **2015**, *170*, 372–387. [CrossRef]

23. Huang, C.; Davis, L.S.; Townshend, J.R.G. An assessment of support vector machines for land cover classification. *Int. J. Remote Sens.* **2002**, *23*, 725–749. [CrossRef]

24. ASTER Global Digital Elevation Model (GDEM). Available online: http://gdem.ersdac.jspacesystems.or.jp/ (accessed on 12 March 2016).

25. Frey, H.; Paul, F. On the suitability of the SRTM DEM and ASTER GDEM for the compilation of topographic parameters in glacier inventories. *Int. J. Appl. Earth Obs.* **2012**, *18*, 480–490. [CrossRef]

26. Mashimbye, Z.E.; Clercq, W.P.D.; Niekerk, A.V. An evaluation of digital elevation models (DEMs) for delineating land components. *Geoderma* **2014**, *213*, 312–319. [CrossRef]

27. Shangguan, D.H.; Bolch, T.; Ding, Y.J.; Kröhnert, M.; Pieczonka, T.; Wetzel, H.U.; Liu, S.Y. Mass changes of Southern and Northern Inylchek Glacier, Central Tian Shan, Kyrgyzstan, during ~1975 and 2007 derived from remote sensing data. *Cryosphere* **2015**, *9*, 703–717. [CrossRef]

28. Du, W.; Li, J. Mapping changes in the glaciers of the eastern Tienshan Mountains during 1977–2013 using multitemporal remote sensing. *J. Appl. Remote Sens.* **2014**, *8*, 689–697. [CrossRef]

29. Kaldybayev, A.; Chen, Y.; Vilesov, E. Glacier change in the Karatal river basin, Zhetysu (Dzhungar) Alatau, Kazakhstan. *Ann Glaciol* **2016**, *57*, 11–19. [CrossRef]

30. Wang, L.; Li, Z.; Wang, F.; Li, H.; Wang, P. Glacier changes from 1964 to 2004 in the Jinghe River basin, Tien Shan. *Cold Reg. Sci. Technol.* **2014**, *102*, 78–83. [CrossRef]

31. Narama, C.; Kääb, A.; Duishonakunov, M.; Abdrakhmatov, K. Spatial variability of recent glacier area changes in the Tien Shan Mountains, Central Asia, using Corona (~1970), Landsat (~2000), and ALOS (~2007) satellite data. *Glob. Planet. Chang.* **2010**, *71*, 42–54. [CrossRef]

*water*

**MDPI**

*Article*

# An Examination of Soil Moisture Estimation Using Ground Penetrating Radar in Desert Steppe

**Yizhu Lu [1,2], Wenlong Song [1,2,*], Jingxuan Lu [1,2], Xuefeng Wang [1] and Yanan Tan [1]**

[1] China Institute of Water Resources and Hydropower Research, Beijing 100038, China; Cynthialoo@sina.com (Y.L.); lujx@iwhr.com (J.L.); wangxf@iwhr.com (X.W.); tanyn@iwhr.com (Y.T.)

[2] Research Center on Flood & Drought Disaster Reduction of the Ministry of Water Resources, Beijing 100038, China

* Correspondence: songwl@iwhr.com; Tel.: +86-010-6878-1847

Received: 5 May 2017; Accepted: 10 July 2017; Published: 22 July 2017

**Abstract:** Ground penetrating radar (GPR) is a new technique of rapid soil moisture measurement, which is an important approach to measure soil moisture at the intermediate scale. To test the applicability of GPR method for soil moisture in desert steppe, we used the common-mid point (CMP) method and fixed offset (FO) method to evaluate the influence factors and the accuracy of GPR measurement with gravimetric soil moisture measurements. The experiments showed that Topp's equation is more suitable than Roth's equation for processing the GPR data in desert steppe and the soil moisture measurements by GPR had high accuracy by either CMP method or FO method. To a certain extent, the vegetation coverage affects the measurement precision and the soil moisture profile. The precipitation can reduce the effective sampling depth of the ground wave from 0.1 m to 0.05 m. The results revealed that GPR has the advantages of high measurement accuracy, easy movement, simple operation, and no damage to the soil layer structure.

**Keywords:** ground penetrating radar; GPR; soil moisture; desert steppe; gravimetric method; CMP

## 1. Introduction

Soil water is the basis of vegetation development, and soil moisture is an important indicator of climate, hydrology, ecology, and agriculture. The spatial and temporal distribution of soil moisture has a significant impact on precipitation infiltration, runoff, and other hydrological processes. In the desert steppe, soil moisture is the main factor that controls the vegetation growth and restoration, and it is the main factor to affect the ecosystem degradation and reversion. The monitoring of soil moisture in desert steppe is important to protect grassland vegetation, prevent desertification, improve the ecological environment, and provide the basis for grazing control and the prevention of grassland degradation.

Measurement technology of soil moisture can be roughly divided into three types according to the measuring scale. The first is the point scale method, including the gravimetric method, neutron method, TDR, FDR, and so on. The data determined by these methods can reflect the soil moisture of the observation point accurately, but it is time-consuming and laborious with destructive problems to the soil structure. Second, the intermediate scale method includes GPR and CRS. These are non-hazardous, non-contact, and non-destructive measurement methods that develop rapidly. The third is the coarser scale method mainly composed of satellite platforms. The point scale methods of measuring soil moisture are not capable of collecting large scale data rapidly. The remote sensing methods have the advantage of large coverage and repeated observations on a regular basis but at coarse scales [1]. Ground penetrating radar (GPR), as a nondestructive geophysical methods, has the ability to monitor soil moisture bridging the scale gap between point and remote sensing measurements, and to quantify the spatial variation of soil moisture [2]. The non-invasive character of GPR offers the mobility needed to map soil water content of large areas (up to 500 m × 500 m a day) [3]. GPR method for measuring soil moisture is widely used in irrigation experiments [4], monitoring soil moisture dynamic [5–7], agriculture [8–10], and observation of long time series [11,12]. FO method, CMP method, and Wide angle reflection and refraction (WARR) method are the three commonly used methods of GPR. By WARR and CMP methods, soil moisture can be estimated easily. However, measurements by WARR and CMP methods have low spatial resolution and require more time. FO method has the advantage of faster measurement and higher resolution by towing the antenna with a vehicle [4], but it is difficult to calibrate time zero and identify ground direct wave. Many studies on GPR method of soil moisture measurement by scholars have been carried out in the world. Grote [8] used CMP method to measure soil moisture by 450 MHz and 900 MHz GPR, and found an RMSE of 0.022 $m^3/m^3$ and 0.015 $m^3/m^3$, respectively. Huisman [13], Galagedara [4], and Weihermller [14] studied the optimum antenna spacing of FO method to separate ground direct wave from airwave by WARR method or CMP method. Huisman [13] and Weihermller [14] compared 225 MHz and 450 MHZ GPR with TDR to measure soil moisture, concluding that the soil moisture obtained by FO method had an RMSE of 0.018 $m^3/m^3$ and 0.011 $m^3/m^3$, respectively. Lunt [15] compared FO method by 100 MHz GPR with the neutron method and got an RMSE of 0.018 $m^3/m^3$. Stoffregen [16] extracted reflected wave to estimate soil moisture by 1 GHz GPR and got a standard deviation of 0.01 $m^3/m^3$ compared with lysimeter data. However, there is less research on GPR measurement of soil moisture applied in China area. Instead, there are more studies on GPR simulation experiments conducted by Chinese researchers. For example, Wang [17] used WARR method to determine the optimum antenna spacing of FO method and applied FO method to estimate soil moisture over large areas by 200 MHz GPR in arid area of China. Wang [17] found a deviation of only 0.015 $m^3/m^3$ and an effective depth of 0.20 m compared with TDR results. Qin [18] used 200 MHz GPR to monitor the spatial change of soil water before and after snow melt in desert, and the absolute error of GPR measurement was less than 0.03 $m^3/m^3$ compared with TDR results. Guo [19], Ma [20], and Li [21] studied on the relationship between soil moisture and GPR signal attributes by simulation experiments.

Research on GPR application for different soil types is still in the initial stage, and mainly used in laboratory test or in the desert area [22–25], and there is no research to focus on the applicability of soil water measurement by GPR in the desert steppe region. This research used the CMP method and FO method of GPR to measure soil moisture in desert steppe, verifying the measurement accuracy by gravimetric method synchronously. The application in different land cover types was also analyzed. The specific objective of this research was to study the suitability of monitoring soil moisture by GPR and its influencing factors for desert steppe: (1) To choose a more appropriate formula from Topp's equation and Roth's equation for calculating soil moisture in desert steppe; (2) To assess the accuracy of CMP and FO methods in desert steppe; (3) To analyze the influencing factors of GPR measurement in desert steppe. Research on the real-time monitoring of soil moisture in desert steppe contributes to the vegetation protection, and helps to prevent soil desertification and protect functional ecosystems. It also provides technical support for remote sensing calibration of soil moisture, agricultural production, and ecological restoration for desert steppe.

## 2. Materials and Methods

### 2.1. Study Site

The study was carried out at the experimental base of Institute of Water Resources for Pastoral in Xilamuren Town, Baotou City, Inner Mongolia region (41°22′ N, 111°12′ E). The base nearing the south of Tabu River covers an area of 150 hectares with the highest elevation of 1690.3 m and the lowest elevation of 1585.0 m. The study area belongs to the temperate semi-arid continental monsoon climate. The average annual precipitation is 284 mm, the average annual evaporation is 2305 mm, and the annual average temperature is 2.5 °C. The study area with the zonal soil of Kastanozems is located in the Wulanchabu desert steppe of central Inner Mongolia Region. The local vegetation in the base showing typical steppe characteristics is not disturbed from grazing and human activities (Figure 1). The vegetation edificator in the study area is *Stipa krylovii*, the dominant species is *Leymus chinensis*, and other important species are *Artemisia frigida*, *Cleistogenes*, *Agropyron cristatum*, etc.

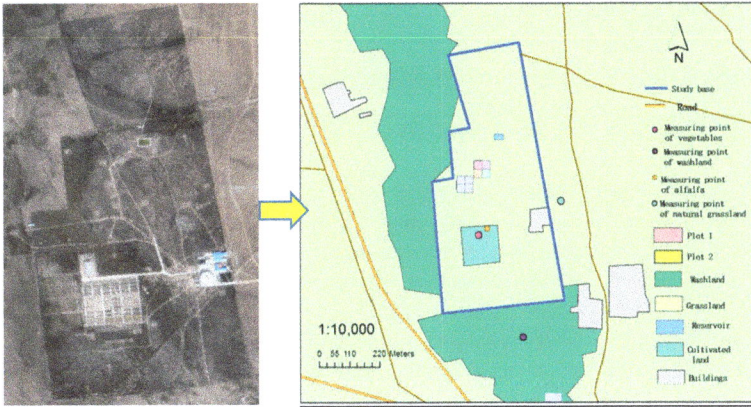

**Figure 1.** Distribution of land cover types in study area.

## 2.2. GPR Theory

### 2.2.1. Measuring Principle

Ground penetrating radar (GPR) is an electromagnetic detection technology which uses the high frequency electromagnetic wave to detect the inner structure and the characteristic of buildings in the center frequency ranging from 10 MHZ to 3 GHZ [15]. The electromagnetic wave received by the receiving transducer is divided into air wave, ground wave, reflected wave, and refraction wave. The propagation speed of radar waves in unsaturated soil depends on its relative dielectric constant. The velocity of ground wave and reflected wave can be extracted from GPR data to calculate relative dielectric constant of the soil. Then soil moisture can be obtained by the relationship between soil moisture and its relative dielectric constant. The depth of soil moisture is determined by the corresponding sampling depth of ground wave or reflected wave. In the low loss medium, the relationship between the electromagnetic wave velocity v and the relative dielectric constant $K$ is

$$K = (c/v)^2 \qquad (1)$$

where $c$ is the propagation velocity of electromagnetic wave in vacuum [26] and $v$ is the electromagnetic wave velocity in the low loss medium. The relationship between soil moisture constants ($\theta$) and relative dielectric constant of the soil ($\varepsilon$) can be described by the empirical formula, semi theoretical formula. This study used two common empirical formulas which are Topp's equation [27] and Roth's equation [28] to calculate soil water content

$$\text{Topp's equation: } \theta = -0.053 + 0.0293\varepsilon - 0.00055\varepsilon^2 + 0.0000043\varepsilon^3 \qquad (2)$$

$$\text{Roth's equation: } \theta = -0.078 + 0.0448\varepsilon - 0.00195\varepsilon^2 + 0.0000361\varepsilon^3 \qquad (3)$$

The effective depth of the soil water content calculated by the ground direct wave is related to the antenna frequency and soil type, which has not been clearly defined. However, some research attempted to establish the empirical formula of the effective depth of the ground direct wave, in which Sperl (1999) [29] proposed the function of the effective depth $Z$ and radar wave length $\lambda$.

$$Z \approx 0.145\lambda^{1/2} \qquad (4)$$

The experimental results of Huisman [3] showed that the effective depth of 225 MHZ and 450 MHZ GPR is 0.10 m, which is consistent with the conclusion of Sperl [29].

The measuring depth of the soil moisture by reflected wave is the sampling depth of reflection wave. When the soil moisture is calculated by the reflection wave velocity, the reflection layer is assumed continuous, horizontal, and with a clear interface. In this study, the soil profile structure was obtained by FO method, and the reflected wave velocity was measured by CMP method, and then the depth of the reflection layer was calculated.

### 2.2.2. GPR Methods

GPR can obtain relative dielectric constant of the soil by extracting the information of radar waves, and then retrieve the soil moisture content. According to the different measuring mode of GPR, the GPR method is mainly divided into the multi-offset reflection method (including CMP and WARR methods), FO method, surface reflection method, and transillumination method [30,31]. On the basis of obvious layered soil structure in study area, we applied the CMP method of GPR to soil moisture measurement, and FO method to obtain the soil profile structure. The CMP method is the measuring method in which the center point of the antennas is fixed and the receiving and transmitting antennas move in opposite directions at the same distance synchronously from the center point of the antennas (Figure 2) [32]. CMP method measures the propagation velocity of direct wave and reflected wave directly for soil moisture calculation with high accuracy and different depth. FO method is a method that the receiving and transmitting antenna spacing is fixed and moving at the same interval. (Figure 2) [32]. The soil profile structure in the study area can be obtained by FO method, which can be used to determine the soil layer, and combined with the results of the CMP method to determine the depth of the soil moisture content. When the appropriate antenna spacing of FO method is set, the ground direct wave and the airwave can be separated. Then the soil relative permittivity is calculated by antenna spacing ($x$), ground direct wave arrival time ($t_{GW}$), and airwave arrival time ($t_{AW}$) using the Formula (5) to estimate the soil moisture. The measured depth of FO method is the sampling depth of the ground wave.

$$\varepsilon = \left(\frac{c}{v}\right)^2 = \left(\frac{c(t_{GW} - t_{AW}) + x}{x}\right)^2 \tag{5}$$

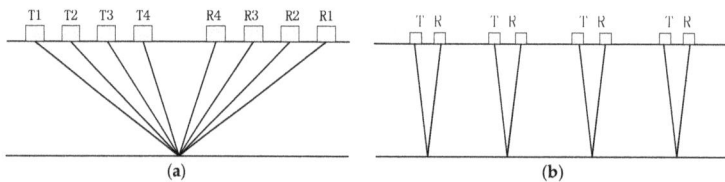

**Figure 2.** Schematic diagram of CMP method (**a**) and FO method (**b**).

### 2.3. Data Acquisition

Measurements were mainly carried out in two 30 m × 30 m plots chosen in the study area (named Plot 1 and Plot 2 in Figure 1) which had flat terrain, uniform underlying surface condition, and differences in vegetation growth. On 11 June 2015, 6 CMP measurements were collected in the middle of each row after raining in each plot. On 20 September 2015, we collected 18 measurements spaced 7.5 m apart by CMP method along six transects in each plot (Figure 3). We also selected two measuring points with exuberant vegetation in the Plot 2 measuring the soil moisture before and after weeding, and four measuring points of different land cover types (vegetables, alfalfa, natural grassland, washland; Figure 1) in the experimental area to obtain soil moisture by CMP method. The six transects of 30 m spaced 5 m apart along six rows were set for FO method. We used FO method to obtain soil structure on 20 September 2015 and measure soil moisture on 24 August 2016.

Higher frequencies have higher spatial resolutions and a higher attenuation which lead to a lower depth of penetrating. This study intended to obtain soil moisture at different depths, so the selected frequency of GPR should not be too high. Considering the portability of GPR and the feature of FO method, a high-frequency antenna with wheels should be used to improve the measuring efficiency of FO method. For the above reasons, this study used the pulse EKKO PRO GPR at a center frequency of 250 MHZ produced by Sensors & Software, a Canadian company, mainly consisting of the transmitting antenna, the receiving antenna, the Digital Video logger (DVL) and control module. The CMP measurements were made with antenna separations increasing from 0.38 m to 5.38 m with increments of 0.10 m, a time window of 100 ns, a sampling interval of 0.4 ns, and 32 stacks per trace. The collection parameters of FO method for soil structure included an antenna separation of 0.38 m, a sampling interval of 0.4 ns, trace spacing of 0.05 m, and 32 stacks per trace at each location to improve the signal to noise ratio. FO method for soil moisture measurement changed the antenna spacing to 1.5 m and other parameters remained the same. According to the Formula (4), the effective depth of 250 MHZ GPR ground direct wave is about 0.10 m. Combined with the research results of Huisman, we selected the soil moisture measurements of 0.10 m depth from gravimetric method for accuracy verification of GPR ground direct wave. Therefore, soil samples were collected adjacent to the locations of measuring points for CMP method at the depth of 0.10 m.

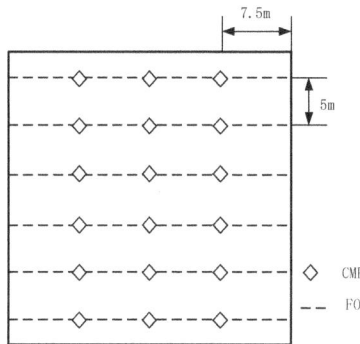

**Figure 3.** Disposition pattern of measuring points and lines in plots.

## 3. Results and Discussion

*3.1. Inspection of the GPR Measurement Accuracy*

Soil moisture measured by gravimetric method was used as the standard value to compare the GPR-derived soil moisture results estimated by Topp's equation to those by Roth's equation. The soil moisture estimated by the Topp's equation was much closer to the gravimetric result, and the soil moisture derived by Roth's equation was generally higher than the gravimetric data (Figure 4). The relative error and variation of Topp's equation was significantly lower than those of the Roth's equation (Figure 5). For desert steppe region, the Topp's equation is more stable and accurate when calculating the soil moisture measured by GPR. This paper chooses the Topp's equation to calculate the GPR-derived soil moisture for the following analysis.

For CMP method, the results showed that the average soil moisture at the depth of 0.10 m from ground wave in Plot 1 was 0.075 $m^3/m^3$, and that in Plot 2 was 0.094 $m^3/m^3$. The soil moisture extracted by ground wave had an RMSE of 0.0101 $m^3/m^3$ compared to the gravimetric measurements at the depth of 0.10 m.

Soil moisture calculated by CMP method and FO method was compared with gravimetric measurements at the depth of 0.10 m. The accuracy of FO method and CMP method were similarly high with an RMSE of 0.0068 m$^3$/m$^3$ and 0.0101 m$^3$/m$^3$ (Figure 6), which means the GPR measurement results are reliable in desert steppe.

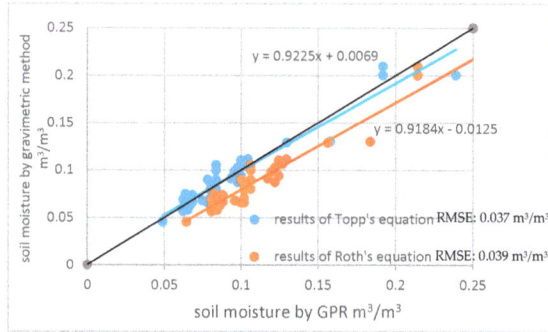

**Figure 4.** Soil moisture calculated by Topp's equation and Roth's equation in comparison with gravimetric method.

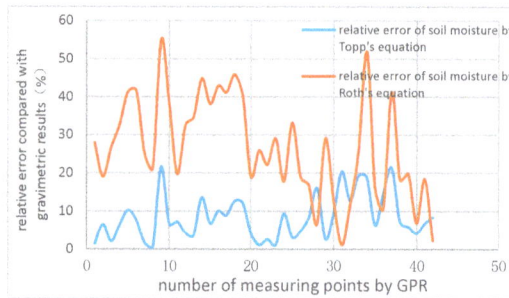

**Figure 5.** Comparison between relative error of soil moisture calculated by Topp's equation and that by Roth's equation.

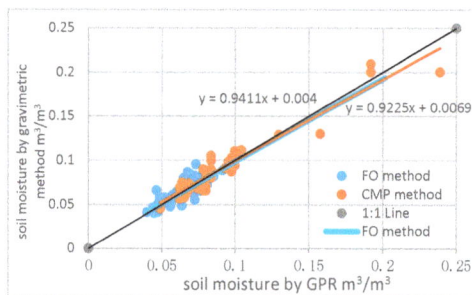

**Figure 6.** Soil moisture calculated by CMP method and FO method in comparison with gravimetric method.

### 3.2. Soil Moisture of Different Land Cover Types

In this experiment, the soil moisture of four different land cover types (Figure 7) at different depth obtained by ground wave and reflected wave are shown in Table 1. In the farmland area, the soil natural layered structure was destroyed, and the measured results were consistent with the actual conditions. Because alfalfa and vegetable regions are located in the farmland area with drip irrigation facilities, the soil moisture of alfalfa and vegetable regions were significantly higher than other land cover regions, suggesting that the measurement results were in line with the actual conditions. The change of soil water content in the soil profile of vegetable and alfalfa was larger than that of grassland in the soil profile, and the soil water was accumulated in the soil surface. In addition to the area of farmland, the soil moisture of grassland and its fluctuation at depths from 0 m to 0.40 m between different regions were relatively close, while the soil moisture at depths below 0.40 m in grassland between regions had larger fluctuation than that in surface soil.

(a)    (b)

(c)    (d)

**Figure 7.** Vegetables (**a**), alfalfa (**b**), natural grassland (**c**), washland (**d**).

**Table 1.** Soil moisture of four different land cover types and two plots obtained by ground wave and reflected wave.

| Land Cover Type | Soil Moisture by Ground Wave ($m^3/m^3$) | Effective Depth (m) | Soil Moisture above the First Reflected Layer ($m^3/m^3$) | Effective Depth (m) | Soil Moisture above the Second Reflected Layer ($m^3/m^3$) | Effective Depth (m) |
|---|---|---|---|---|---|---|
| Alfalfa | 0.1919 | 0.10 | 0.0978 | 0.63 | | |
| Vegetables | 0.1578 | 0.10 | 0.1961 | 0.25 | 0.1612 | 0.66 |
| Washland | 0.0801 | 0.10 | 0.0622 | 0.40 | 0.1243 | 0.71 |
| Natural Grassland | 0.0819 | 0.10 | 0.0652 | 0.36 | 0.1021 | 0.75 |
| Plot 1 | 0.0750 | 0.10 | 0.0975 | 0.41 | | |
| Plot 2 | 0.0939 | 0.10 | 0.0716 | 0.43 | | |

### 3.3. Effects of Vegetation Coverage on Soil Moisture and Its Measuring Accuracy

Vegetation coverage has a great impact on soil moisture and surface evapotranspiration. Due to the difference in root distribution, the response to this impact is not the same at different depths of the soil profile. The vegetation coverage of two plots had obvious differences, where the average normalized difference vegetation index (NDVI) of Plot 1 and Plot 2 were 0.30 and 0.56. The average surface soil moisture of Plot 1 and Plot 2 at the depth of 0.10 m were 0.0750 $m^3/m^3$ and 0.0939 $m^3/m^3$ respectively, of which the soil moisture of Plot 2 was significantly higher. However, for deeper soil moisture (about 0.42 m), the average soil moisture in Plot 1 was 0.0975 $m^3/m^3$, and higher than that in Plot 2 with 0.0716 $m^3/m^3$.

The reason for this phenomenon is that vegetation coverage was obviously different in two plots (Figure 8). The vegetation coverage of Plot 2 was higher than that of Plot 1. The higher vegetation coverage lead to less evaporation of water on the soil surface, so the surface soil moisture of Plot 2 at the depth of 0.10 m is higher. Whereas the vegetation evapotranspiration is derived from the water absorbed by roots from the soil moisture. The vegetation root system of Plot 2 was more developed and dense than that in Plot 1 because of the higher vegetation coverage. The vegetation root in Plot 2 absorbed more water from deep soil because of higher evapotranspiration than that in Plot 1 under the same condition. As a result, the deep soil moisture of Plot 1 was higher.

The comparison of soil moisture measured by ground wave in Plot 1 and Plot 2 (Figure 9) showed that the measurements of Plot 1 were more similar to gravimetric results at the depth of 0.10 m than that of Plot 2, with an RMSE of 0.0059 m$^3$/m$^3$ and 0.0130 m$^3$/m$^3$, respectively. The measuring accuracy of Plot 2 was significantly higher than Plot 1 perhaps because the high vegetation coverage affects the air refraction of the GPR measurements. When the vegetation coverage was high, a large amount of air refraction wave was generated during the process of radar wave propagation, which interfered with the waveform of the ground wave and was reflected wave in the radar profile, affecting the extraction accuracy of wave velocity. At the same time, too much air refraction wave can cause rapid attenuation of radar wave, and hinder the GPR measurement of soil moisture at deep depth. Therefore, the accuracy of GPR measurements can be improved if weeding. For two selected measuring points with dense vegetation in Plot 2 (Figure 10), the relative error of soil moisture compared with gravimetric method decreased from 19.24% and 12.80%, respectively, to 4.22% and 6.74% by weeding. The experimental results further showed that the vegetation coverage to a certain extent affected the accuracy of GPR measurements.

(a)　　　　　　　　　　　　　　　(b)

**Figure 8.** Vegetation condition in Plot 1 (**a**) and Plot 2 (**b**).

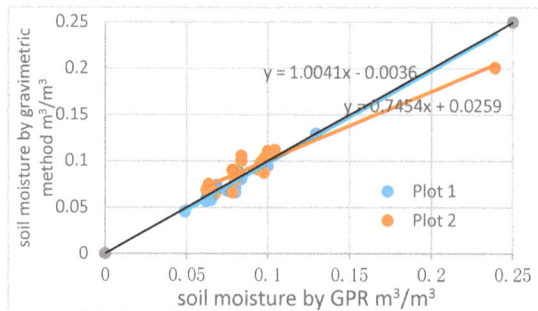

**Figure 9.** Comparison of the soil moisture measured by GPR and gravimetric method in Plot 1 and Plot 2.

(a) Before weeding         (b) After weeding

**Figure 10.** Contrast diagram before (**a**) and after (**b**) weeding (marking area by dotted line is weed control part).

### 3.4. Effect of Precipitation on GPR Measurements

After raining during 11 June 2016, we applied CMP method to measure soil moisture of two plots. Ground wave velocity was extracted to estimate surface soil moisture from radar profile, obtaining the average soil moisture of 0.1873 m$^3$/m$^3$ and 0.1563 m$^3$/m$^3$. Compared with gravimetric soil moisture at depths of 0.05, 0.10, and 0.15 m, the effective depth of the ground wave by CMP method was 0.05 m. The average relative error of GPR measurement compared with gravimetric soil moisture at the depth of 0.05 m was 9.45%. The depth of soil moisture extracted by ground wave became smaller after raining, that is, the effective depth of ground wave became smaller due to the influence of the precipitation. Because of less rain in the experimentation area for a long time, soil moisture was generally low. Precipitation makes surface soil moisture much higher than that of the lower soil layer, forming high speed propagation layer of radar wave in the soil layer about 0.05 m. The radar wave was spread on the interface formed by the difference of soil moisture, which makes the effective depth of the ground wave smaller.

### 4. Conclusions

As a nondestructive measuring method, ground penetrating radar (GPR) was used in the fields of soil water monitoring, and soil moisture dynamic. In this paper, GPR was used to measure soil moisture in Inner Mongolia desert steppe. The accuracy of GPR measurement was verified by gravimetric method. The influence of vegetation coverage and precipitation on GPR measurement was analyzed. The research showed that GPR can accurately measure the soil moisture of desert steppe and meet the actual demand of field monitoring.

1.  For desert steppe region, the Topp's equation is more accurate than the Roth's equation in calculating the soil moisture of GPR data.
2.  The soil moisture measurements by GPR were consistent with gravimetric results, with the RMSE of only 0.0101 m$^3$/m$^3$. Compared with the traditional gravimetric method and TDR, GPR can quickly measure the soil moisture at different depths and obtain soil stratification condition without destroying soil layer structure by virtue of the portable and operational characteristics.
3.  The vegetation coverage affects the accuracy of GPR measurement and also affects the profile distribution of soil water content. When vegetation coverage is high, the air refraction wave interferes with the ground wave and reflected wave in the radar profile, which can reduce the accuracy of GPR measurement.
4.  Under certain conditions, precipitation reduces the effective depth of the ground wave, and further affects the depth of the soil moisture measured by the GPR ground wave.

5. The accuracy comparison of GPR measurement in different soil types, the application of different GPR methods in desert steppe, and the combination of GPR and soil water model are to be further studied.

**Acknowledgments:** This work was supported by the National Key Research and Development Plan of China (2016YFC0400106-2), by the Natural Science Foundation of China (51609259), and by the Research Program of China Institute of Water Resources and Hydropower Research (JZ0145B472016, JZ0145B862017). We would like to thank the Institute of Water Resources for Pastoral for providing the research field and the field working condition.

**Author Contributions:** Yizhu Lu and Jingxuan Lu designed the experiments; Yizhu Lu, Wenlong Song, and Jingxuan Lu performed the experiments; Yizhu Lu and Xuefeng Wang processed the GPR data; Yizhu Lu and Yanan Tan participated in the analysis of the data; Yizhu Lu and Wenlong Song wrote the paper.

**Conflicts of Interest:** The authors declare no conflict of interest.

## Abbreviations

| | |
|---|---|
| TDR | Time domain reflector |
| FDR | Frequency domain reflectometer |
| GPR | Ground penetrating radar |
| CRS | Cosmic-ray sensing probe |
| FO method | Fixed offset method |
| CMP method | Common-midpoint method |
| WARR method | Wide angle reflection and refraction method |

## References

1. Wang, L.; Qu, J.J. Satellite remote sensing applications for surface soil moisture monitoring: A review. *Front. Earth Sci.* **2009**, *3*, 237–247. [CrossRef]
2. Ardekani, M.R.M. Off- and on-ground GPR techniques for field-scale soil moisture mapping. *Geoderma* **2013**, *200–201*, 55–66. [CrossRef]
3. Huisman, J.A.; Sperl, C.; Bouten, W.; Verstraten, J.M. Soil water content measurements at different scales: Accuracy of time domain reflectometry and ground-penetrating radar. *J. Hydrol.* **2001**, *254*, 48–58. [CrossRef]
4. Galagedara, L.W.; Parkin, G.W.; Redman, J.D.; von Bertoldi, P.; Endres, A.L. Field studies of the GPR ground wave method for estimating soil water content during irrigation and drainage. *J. Hydrol.* **2005**, *245*, 182–197. [CrossRef]
5. Hubbard, S.; Grote, K.; Rubin, Y. Mapping the volumetric soil water content of a California vineyard using high-frequency GPR ground wave data. *Lead. Edge* **2002**, *21*, 552–559. [CrossRef]
6. Steelman, C.M.; Endres, A.L. Assessing vertical soil moisture dynamics using multi-frequency GPR common-midpoint soundings. *J. Hydrol.* **2012**, *436–437*, 51–66. [CrossRef]
7. Ma, Y.; Zhang, Y.; Zubrzycki, S.; Guo, Y.; Farhan, S.B. Hillslope-scale variability in seasonal frost depth and soil water content investigated by GPR on the southern margin of the sporadic permafrost zone on the Tibetan Plateau. *Permafr. Periglac. Process.* **2015**, *26*, 321–334. [CrossRef]
8. Grote, K.; Hubbard, S.; Rubin, Y. Field-scale estimation of volumetric water content using ground-penetrating radar ground wave techniques. *Water Resour. Res.* **2003**, *39*, 1321–1335. [CrossRef]
9. Wijewardana, Y.G.N.S.; Galagedara, L.W. Estimation of spatio-temporal variability of soil water content in agricultural fields with ground penetrating radar. *J. Hydrol.* **2010**, *391*, 24–33. [CrossRef]
10. Pan, X.; Zhang, J.; Huang, P.; Roth, K. Estimating field-scale soil water dynamics at a heterogeneous site using multi-channel GPR. *Hydrol. Earth Syst. Sci.* **2012**, *16*, 4361–4372. [CrossRef]
11. Overmeeren, R.A.V.; Sariowan, S.V.; Gehrels, J.C. Ground penetrating radar for determining volumetric soil water content: Results of comparative measurements at two test sites. *J. Hydrol.* **1997**, *197*, 316–338. [CrossRef]
12. Steelman, C.M.; Endres, A.L.; Jones, J.P. High-resolution ground-penetrating radar monitoring of soil moisture dynamics: Field result, interpretation, and comparison with unsaturated flow model. *Water Resour. Res.* **2012**, *48*, 184–189. [CrossRef]

13. Huisman, J.A.; Snepvangers, J.J.J.C.; Bouten, W.; Heuvelink, G.B.M. Mapping spatial variation in surface soil water content: Comparison of ground-penetrating radar and time domain reflectometry. *J. Hydrol.* **2002**, *269*, 194–207. [CrossRef]

14. Weihermüller, L.; Huisman, J.A.; Lambot, S.; Herbst, M.; Vereecken, H. Mapping the spatial variation of soil water content at the field scale with different ground penetrating radar techniques. *J. Hydrol.* **2007**, *340*, 205–216. [CrossRef]

15. Lunt, I.A.; Hubbard, S.S.; Rubin, Y. Soil moisture content estimation using ground-penetrating radar reflection data. *J. Hydrol.* **2005**, *307*, 254–269. [CrossRef]

16. Stoffregen, H.; Yaramanci, U.; Zenker, T.; Wessolek, G. Accuracy of soil water content measurements using ground penetrating radar: Comparison of ground penetrating radar and lysimeter data. *J. Hydrol.* **2002**, *267*, 201–206. [CrossRef]

17. Wang, Q.; Zhou, K.; Sun, L.; Qin, Y.; Li, G. A study of fast estimating soil water content by ground penetrating radar. *J. Nat. Resour.* **2013**, *28*, 881–888. (In Chinese)

18. Qin, Y.; Chen, X.; Zhou, K.; Sun, L.; Zhang, J. Using GPR to sound the spatial and temporal distributions of dune surface soil water contents before and after snowmelt in the early spring. *J. Glaciol. Geocryol.* **2012**, *34*, 690–697. (In Chinese)

19. Guo, X.; Wang, M.; Zhang, G.; Hou, L.; Sun, P.; Meng, Q. Nondestructive and quick in-situ testing of unsaturated sandy soil water content using ground penetrating radar refection method. *Period. Ocean Univ. China* **2010**, *40*, 141–145. (In Chinese)

20. Ma, F.; Lei, S.; Yang, S.; Zhen, F.; Wang, Y. Study on the relationship between soil water content and ground penetrating radar signal attributes. *Chin. J. Soil Sci.* **2014**, *45*, 809–815. (In Chinese)

21. Li, H.; Zhong, R. Numerical study on the relationship between amplitudes of ground penetrating radar wave and water content in soil. *J. Appl. Sci.* **2015**, *33*, 41–49. (In Chinese)

22. Mangel, A.R.; Moysey, S.M.J.; Ryan, J.C.; Tarbutton, J.A. Multi-offset ground-penetrating radar imaging of a lab-scale infiltration test. *Hydrol. Earth Syst. Sci.* **2011**, *16*, 4009–4022. [CrossRef]

23. Moysey, S. Hydrologic trajectories in transient ground-penetrating radar reflection data. *Geophysics* **2010**, *75*, 211–219. [CrossRef]

24. Cui, F.; Wu, Z.; Wang, L.; Wu, Y. Application of the ground penetrating radar ARMA power spectrum estimation method to detect moisture content and compactness values in sandy loam. *J. Appl. Geophys.* **2015**, *120*, 26–35. [CrossRef]

25. Tran, A.P.; André, F.; Lambot, S. Validation of near-field ground-penetrating radar modeling using full-wave inversion for soil moisture estimation. *IEEE Trans. Geosci. Remote Sens.* **2014**, *52*, 5483–5497. [CrossRef]

26. Davis, J.L.; Annan, A.P. Ground penetrating radar for high resolution mapping of soil and rock stratigraphy. *Geophys. Prospect.* **1989**, *37*, 531–551. [CrossRef]

27. Topp, G.C.; Davis, J.L.; Annan, A.P. Electromagnetic determination of soil water content: Measurements in coaxial transmission lines. *Water Resour. Res.* **1980**, *16*, 574–582. [CrossRef]

28. Roth, C.H.; Malicki, M.A.; Plagge, R. Empirical evaluation of the relationship between soil dielectric constant and volumetric water content as the basis for calibrating soil moisture measurements by TDR. *Soil Sci.* **1992**, *43*, 1–13. [CrossRef]

29. Sperl, C. Erfassung der Raum-Zeitlichen Variation des Boden-Wassergehaltes in Einem Agrarokosystem mit dem Ground-Penetrating Radar. Ph.D. Thesis, Technische Universitat, München, Germany, 1999; p. 182.

30. Huisman, J.A.; Hubbard, S.S.; Redman, J.D.; Annan, A.P. Measuring soil water content with ground penetrating radar: A review. *Vadose Zone J.* **2003**, *2*, 476–491. [CrossRef]

31. Tosti, F.; Slob, E. Determination by using GPR of the volumetric water content in strucures substructures foundations and soil. In *Civil. Engineering Applications of Ground Penetrating Radar*; Springer International Publishing: Cham, Switzerland, 2015.

32. Lu, Y.; Song, W.; Lu, J.; Su, Z.; Liu, H.; Tan, Y.; Han, J. Soil water measurement by ground penetrating radar and its scale features. *South.-to-North. Water Transf. Water Sci. Technol.* **2017**, *15*, 37–44. (In Chinese)

*water*

MDPI

*Article*

# Measuring Spatiotemporal Features of Land Subsidence, Groundwater Drawdown, and Compressible Layer Thickness in Beijing Plain, China

**Yongyong Li [1,2,3], Huili Gong [1,2,3,*], Lin Zhu [1,2,3,*] and Xiaojuan Li [1,2,3]**

1   College of Resource Environment and Tourism, Capital Normal University, Beijing 100048, China;
    cnulyy0921@cnu.edu.cn (Y.L.); xiaojuanli@vip.sina.com (X.L.)
2   Base of the State Key Laboratory of Urban Environmental Process and Digital Modeling,
    Capital Normal University, Beijing 100048, China
3   Beijing Laboratory of Water Resources Security, Capital Normal University, Beijing 100048, China
*   Correspondence: gonghl_1956@sina.com (H.G.); hi-zhulin@163.com (L.Z.); Tel.: +86-10-6890-3139 (L.Z.)

Academic Editor: Hongjie Xie
Received: 29 September 2016; Accepted: 12 January 2017; Published: 22 January 2017

**Abstract:** Beijing is located on multiple alluvial-pluvial fans with thick Quaternary unconsolidated sediments. It has suffered serious groundwater drawdown and land subsidence due to groundwater exploitation. This study aimed to introduce geographical distribution measure methods into land subsidence research characterizing, geographically, land subsidence, groundwater drawdown, and compressible layer thickness. Therefore, we used gravity center analysis and standard deviational ellipse (SDE) methods in GIS to statistically analyze their concentration tendency, principle orientation, dispersion trend, and distribution differences in 1995 (1999), 2007, 2009, 2011, and 2013. Results show that they were all concentrated in Chaoyang District of Urban Beijing. The concentration trend of land subsidence was consistent with that of groundwater drawdown. The principle orientation of land subsidence was SW–NE, which was more similar with that of the static spatial distribution of the compressible layer. The dispersion tendency of land subsidence got closer to that of the compressible layer with its increasing intensity. The spatial distribution difference between land subsidence and groundwater drawdown was about 0.2, and that between land subsidence and compressible layer thickness it decreased from 0.22 to 0.07, reflecting that the spatial distribution pattern of land subsidence was increasingly close to that of the compressible layer. Results of this study are useful for assessing the distribution of land subsidence development and managing groundwater resources.

**Keywords:** land subsidence; groundwater drawdown; compressible layer; gravity center; standard deviational ellipse

## 1. Introduction

Regional land subsidence is a geological process occurring in a long-run equilibrium and inter-coordination between anthropogenic activity and the hydrogeological environment [1,2]. In most areas worldwide, compressible sediments are the material basis and it unbalances the starting point of land subsidence; groundwater drawdown is an inherent drive and its spatial diversity induces an uneven development process of land subsidence [3–11]. Land subsidence has increased the risk of other disasters and threatened the properties of the society [12–15]. Mapping and quantifying how, and to what extent, the groundwater drawdown and compressible layer influence non-uniform land subsidence based on multiple time-series displacement and hydrogeological data are of concern to many scholars [16–20].

Land subsidence in Beijing Plain is mainly triggered by over-exploitation of groundwater, and its magnitude and extent is affected by heterogeneity of compressible layers [21]. Integrated subsidence-monitoring programs with multiple surveying methods were designed to clarify both hydrological and mechanical processes of land subsidence [22]. The persistent scatterer interferometry (PSI) technique was adopted to quantify the dynamic evolution of land subsidence in the Beijing north plain, and to determine the spatial relationship with its triggering factors [23]. The small baseline interferometric synthetic aperture radar (InSAR) [24] technique was employed to investigate the relationship between land subsidence and groundwater level, active faults, cumulated soft soil thickness, different aquifer types, and the distance to pumping wells [10]. These studies proved that the spatial extent and magnitude of land subsidence in Beijing Plain has both spatial variability and inheritance. They paid much attention to adopting a GIS spatial overlay or visualization, focusing on its spatial extent, magnitude, and spatial correlation with groundwater drawdown, and the geological structural control at the macro scale, but it has been rarely reported that Geographic Distribution Measuring methods [25] were used to measure the distribution of subsidence-related temporal-spatial datasets that allows one to quantify their concentration tendency, development orientation, dispersion trend, and distribution differences and track their changes over time.

Gravity center analysis is an aggregated statistical method in geographical space. The gravity center dynamic could reveal the spatial concentration of geographical phenomena. It was widely used to assess spatial distribution evolvement in many fields, like population, economics, and employment [26,27]. The deviation direction indicates the adjustment of high intensity, and the deviation distance indicates the degree of equilibrium or adjustment magnitude [26]. The standard deviational ellipse (SDE) was first proposed by Lefever [28] to reveal characteristics of geospatial distribution [29–32]. It has been widely used in urban science [33,34], ecology [35,36], geology [37,38], and infectious disease distribution [39]. Mapping distributional trends for a set of violent events might identify a relationship to particular features, like ethnicity, terrain, land cover, targets, and separatist tradition [40]; comparing the size, shape, and overlap of ellipses for population, gross domestic product, and topography might provide insights regarding economic spatial variation [41]. Plotting ellipses across time series for $PM_{2.5}$ (aerosol particles smaller than 2.5 μm in diameter that are suspended in the air) concentrations might characterize the overall spatial dynamic process [42].

Therefore, this study adopted the two Geographic Distribution Measuring methods to calculate the gravity center and SDE of land subsidence, groundwater drawdown, and compressible layer thickness. Then, by comparing their parameters across time series, the understanding of their spatial distribution evolution and spatial correlation was improved. The main objective of this current study is three-fold: to introduce geographic distribution measuring methods into land subsidence research based on GIS spatial technology; to quantify the spatiotemporal distribution of land subsidence and its influencing factors in the development center, principle orientation, and dispersion, tracking their spatial distribution characteristics; and to distinguish their spatial distribution differences.

## 2. Materials and Methods

### 2.1. Study Area

The Beijing Plain (excluding Yanqing region) is located in the southeast part of Beijing, covering a total area of 6390 km², about 38% of Beijing (Figure 1). It lies in the alluvial-pluvial plain fan which was built up by the river deposits primarily from five rivers, including Yongding, Chaobai, Wenyu, Dashi and Jiyun Rivers. Urban districts include downtown, Chaoyang, Haidian, Shijingshan, and Fengtai. The average annual temperature is about 10–15 °C, and precipitation is 601.7 mm, which belongs to the temperate continental monsoon climate. Land subsidence is one of the critical threats to the sustainable development of Beijing city.

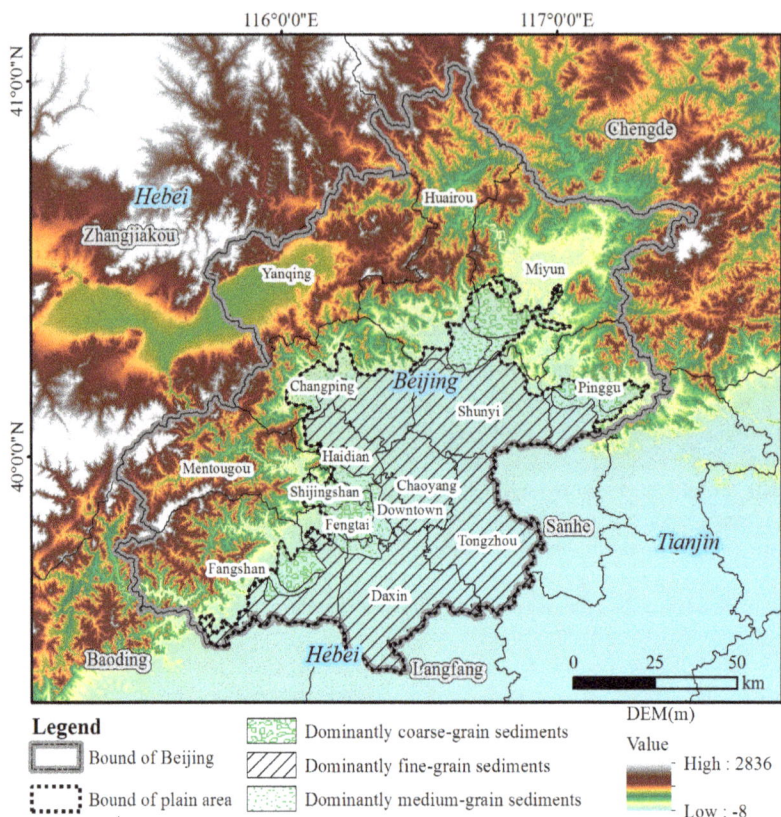

**Figure 1.** Location, type of main sediments, and digital elevation model of the study area.

## 2.2. Available Datasets

Cumulative subsidence contours (the starting time was 1955) in 1999, 2007, 2009, 2011, and 2013 were derived from the Beijing land subsidence monitoring network (Figure 2). The total mean square error of leveling per kilometer and the mean square error of point locations conform to the national norms of primary leveling and second-class leveling (GB12897-91) [43]. This level monitoring network includes 278 monitoring points that provide elevation observation, and monitoring frequency is once a year. By comparing current observations and previous ones, land subsidence of the observation points were identified and then a cumulative land subsidence contour map was plotted. By 2013, the area where cumulative subsidence was greater than 100 mm reached 4942 km$^2$. A, B, C, D, E, and F referred to typical funnel zones and their maximum values are shown in Table 1. The sedimentary time, genetic type, lithology, structure, thickness, and physical and mechanical properties of Quaternary strata are complicated and affect the occurrence and development of land subsidence in Beijing Plain. The thickness map (Figure 3) was derived after grouping silty clay, clay, silt, and other compressible sediments into a compressible layer [44]. Groundwater level observations in 1965, 1995, 2007, 2009, 2011, and 2013 were collected to delineate the changes of the groundwater seepage field (Figure 4). They are retrieved from official reports published by the Beijing Water Authority and the data are accurate and reliable [45]. Compared with that in 1965 in the natural state, fluctuating with terrain changes, the groundwater level in 2013 dropped significantly. Typical funnels of greater than −15 m have formed especially in the regions where Changping, Shunyi, and Chaoyang join together.

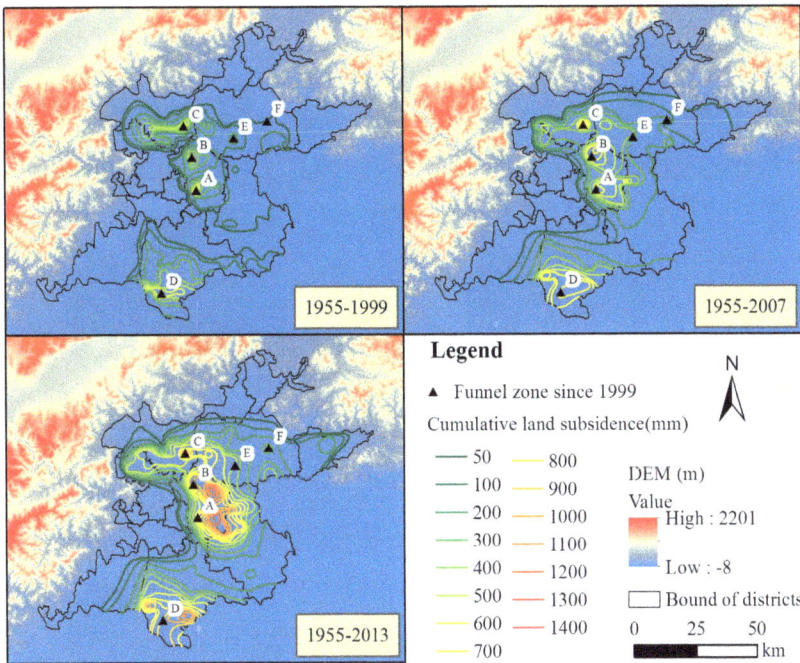

**Figure 2.** Cumulative land subsidence in Beijing Plain. For simplicity, only three of six collected datasets are shown, and the others are shown in Figure S1.

**Table 1.** Maximum values of the subsidence funnel zone (mm).

|   | Subsidence Funnel Zone | 1999 | 2005 | 2007 | 2009 | 2011 | 2013 |
|---|---|---|---|---|---|---|---|
| A | Balizhuang-Dajiaoting in Chaoyang | 700 | 750 | 750 | 800 | 1050 | 1300 |
| B | Laiguangying in Chaoyang | 500 | 650 | 800 | 950 | 1000 | 1400 |
| C | Shahe-Baxianzhaung in Changpiing | 650 | 1050 | 1100 | 1150 | 1200 | 1400 |
| D | Lixian-Yufa in Daxing | 650 | 800 | 850 | 950 | 1050 | 1200 |
| E | Pinggezhuang in Shunyi | 250 | 400 | | | | |
| F | Yang town in Shunyi | | 150 | 200 | 250 | 300 | 400 |

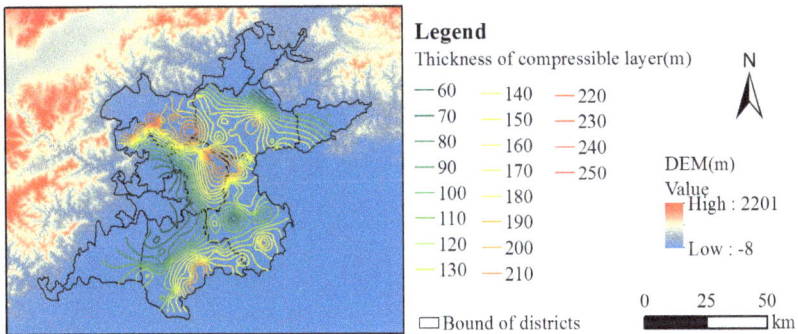

**Figure 3.** Compressible layer thickness.

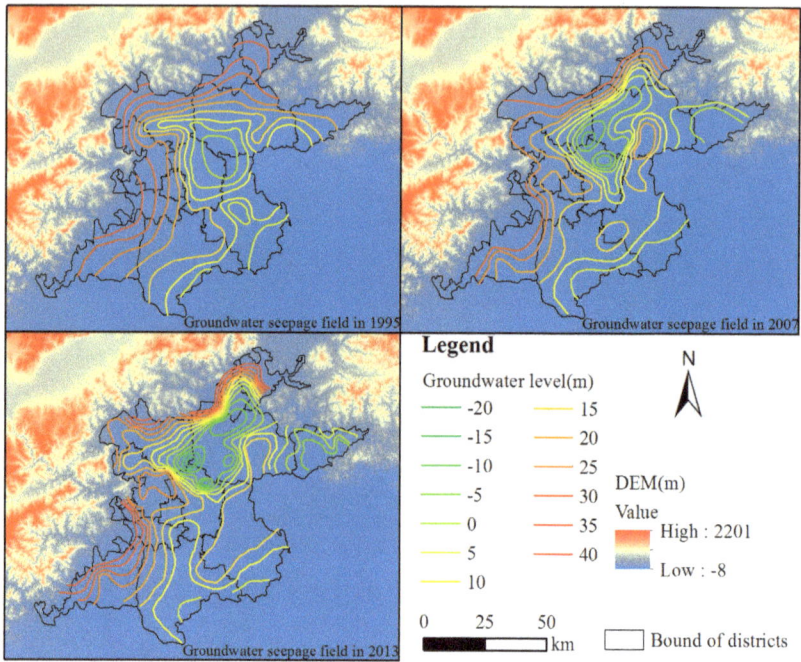

**Figure 4.** Groundwater seepage field. For simplicity, only three of six collected datasets are shown, and the others are shown in Figure S2.

Considering the smallest distance between two original isolines is larger than about 100 m in the area with the most serious land subsidence, the contour map was interpolated as a raster map with a 100 m resolution to keep the overall trend reflected by the map. Spatial sampling was performed to derive 3872 samples at an interval of 1000 m according to the completeness and coverage of data. The discrete point set could be used for spatial-temporal statistics calculation and assessment (Figure 5). Each sampling point can be regarded as a case with geographic coordinates $(x_i, y_i)$ $(i = 1, 2, \dots, n)$. The weight of geographic phenomena corresponding to each case was $w_i$. The groundwater drawdown was the groundwater level in 1965 minus that of the target year, because the groundwater seepage field in 1965 can be approximately regarded as that of 1955.

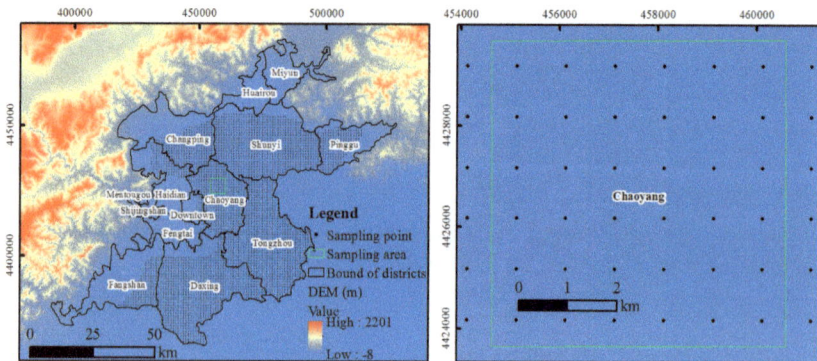

**Figure 5.** Discrete point set of land subsidence.

*2.3. Gravity Center Analysis*

The position of the gravity center was calculated using a combination of geographical coordinates and geographic space phenomena. It was extended from the concept of the spatial mean [46] and expressed as follows [26,27]:

$$\begin{cases} \bar{x} = \sum\limits_{i=1}^{n} w_i x_i / \sum\limits_{i=1}^{n} w_i \\ \bar{y} = \sum\limits_{i=1}^{n} w_i y_i / \sum\limits_{i=1}^{n} w_i \end{cases}, \tag{1}$$

where $\bar{x}$ and $\bar{y}$ represent the longitude and latitude coordinates (respectively) of gravity center.

According to Equation (1), if sampling points are evenly distributed and geographic space phenomena are homogeneous, the gravity center is equivalent to the regional geometric center. When the gravity center of the geographic space phenomena shows a significant offset from its regional geometric center, it indicates its disproportionate spatial distribution or gravity center deviation. In our study, gravity center motion is caused by uneven development of land subsidence. Based on tracking the gravity center derived from long-term monitoring data, its direction and distance of deviation can reflect the adjustment direction and magnitude of land subsidence. Similarly, the method can also be used to measure the influential factors of land subsidence in spatial variation features and, by comparing their movement, the influencing mode will be clarified from the macroscopic prospective.

*2.4. Standard Deviational Ellipse Analysis*

SDE [28] is based on the average center of a set of discrete points, and the calculation of the standard distance of other points away from the average center separately in the $x$ and $y$ directions. These two measures define the axes of an ellipse encompassing the distribution of features. This calculated ellipse covers the spatial center, extent, orientation, shape, and other aspects, with the specific indicators are represented by the average center, the SDE, major and minor axes, and azimuth (Figure 6a). When the SDE was weighted by an attribute value associated with the features, it was termed as a weighted SDE, and the weighted average center can be expressed as shown in Equation (1). The other main parameters and corresponding equations of the weighted SDE are shown in Table 2. $\theta$ refers to the azimuth of the ellipse; $\sigma$ refers to the standard deviation. $\widehat{x}_i$ and $\widehat{y}_i$ refers to the distance of the point $i$ away from the average center separately in the $x$ and $y$ directions. The SDE represents elements in the main distribution area; the major axis and minor axis correspond to the dispersion degree of geographical features in the principle and secondary direction; the azimuth reflects the main trend directions, allowing one to see if the distribution of features is elongated and, hence, has a particular orientation. The orientation is the rotation of the long axis measured clockwise from north. While one can get a sense of the orientation by drawing the features on a map, calculating the SDE makes the trend clear [42].

By comparing SDEs across time series, it is possible to characterize the overall spatial dynamic process. The dynamic of the center reveals the overall evolutionary track of elements; changes in the dimensions of the major and minor axes of an ellipse indicates an expansion or contraction of a specific spatial direction; and changes in the azimuth reflect the changes of overall elements in a particular spatial direction. In addition, the spatial differentiation coefficient was defined [41] to characterize the differentiation degree between different geospatial phenomena (Figure 6b). For instance, $I_{B/A}$ referring to the spatial differentiation coefficient of geospatial phenomena of B relative to $A$ can be expressed as:

$$I_{B/A} = \frac{C_B(A \cap B)}{B}, \tag{2}$$

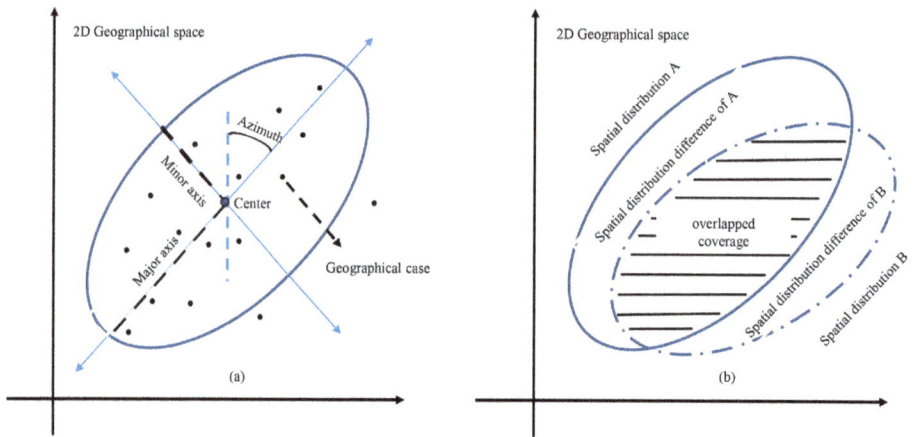

**Figure 6.** Space expression of the standard deviational ellipse (SDE): (**a**) basic parameters and (**b**) distribution difference [41].

**Table 2.** The main parameters and corresponding equations of weighted standard deviational ellipse (SDE).

| Parameter | Equation | |
|---|---|---|
| Azimuth angle | $\tan\theta = \dfrac{\left(\sum\limits_{i=1}^{n} w_i^2 \widetilde{x}_i^2 - \sum\limits_{i=1}^{n} w_i^2 \widetilde{y}_i^2\right) + \sqrt{\left(\sum\limits_{i=1}^{n} w_i^2 \widetilde{x}_i^2 - \sum\limits_{i=1}^{n} w_i^2 \widetilde{y}_i^2\right)^2 + 4\left(\sum\limits_{i=1}^{n} w_i^2 \widetilde{x}_i \widetilde{y}_i\right)^2}}{2\sum\limits_{i=1}^{n} w_i^2 \widetilde{x}_i \widetilde{y}_i}$ | |
| Standard deviation | $\sigma_x = \sqrt{\dfrac{\sum\limits_{i=1}^{n}\left(w_i \widetilde{x}_i \cos\theta - w_i \widetilde{y}_i \sin\theta\right)^2}{\sum\limits_{i=1}^{n} w_i^2}}$ | $\sigma_y = \sqrt{\dfrac{\sum\limits_{i=1}^{n}\left(w_i \widetilde{x}_i \sin\theta - w_i \widetilde{y}_i \cos\theta\right)^2}{\sum\limits_{i=1}^{n} w_i^2}}$ |

## 3. Results

### 3.1. Gravity Center Evolution Analysis

#### 3.1.1. Gravity Center of Land Subsidence

The gravity centers of land subsidence in Beijing Plain were located to the southeast of downtown in Chaoyang District from 1999 to 2013, and they were distributed between Fourth Ring Road and Fifth Ring Road (Figure 7). From Figure 2 and Table 1, we can see that (1) three main subsidence funnels were located within or nearby Chaoyang district; and (2) the Lixian-Yufa subsidence area was located at the southernmost tip and far away from the urban district. This indicates that the spatial pattern of land subsidence determines the gravity center location.

Table 3 shows that the gravity center of land subsidence experienced an accelerating-stabilizing-reducing move north by east yearly, approaching the geometric center. From 2007 to 2011, the gravity center moved to the northeast at a higher speed. After 2011, its speed slowed and the direction was more biased to the geometric center. Figure 2 shows that the first five typical subsidence funnels have taken shape and the whole extent has kept stable since 1999; the two funnels in Chaoyang mainly moved eastwards; the Lixian-Yufa subsidence area expanded north. From 2003 to 2010, the new Yang town funnel formed, owing to continuous overexploitation and it developed to the northeast compared with the old funnel [23]; the maximum of funnels in Chaoyang district reached 110 mm/year, while that in Changping district was moderate [10]. Generally speaking, the movement of the gravity center depended on the development trend of land subsidence. It produced a small shift from southwest to northeast, presenting a total stability of spatial distribution.

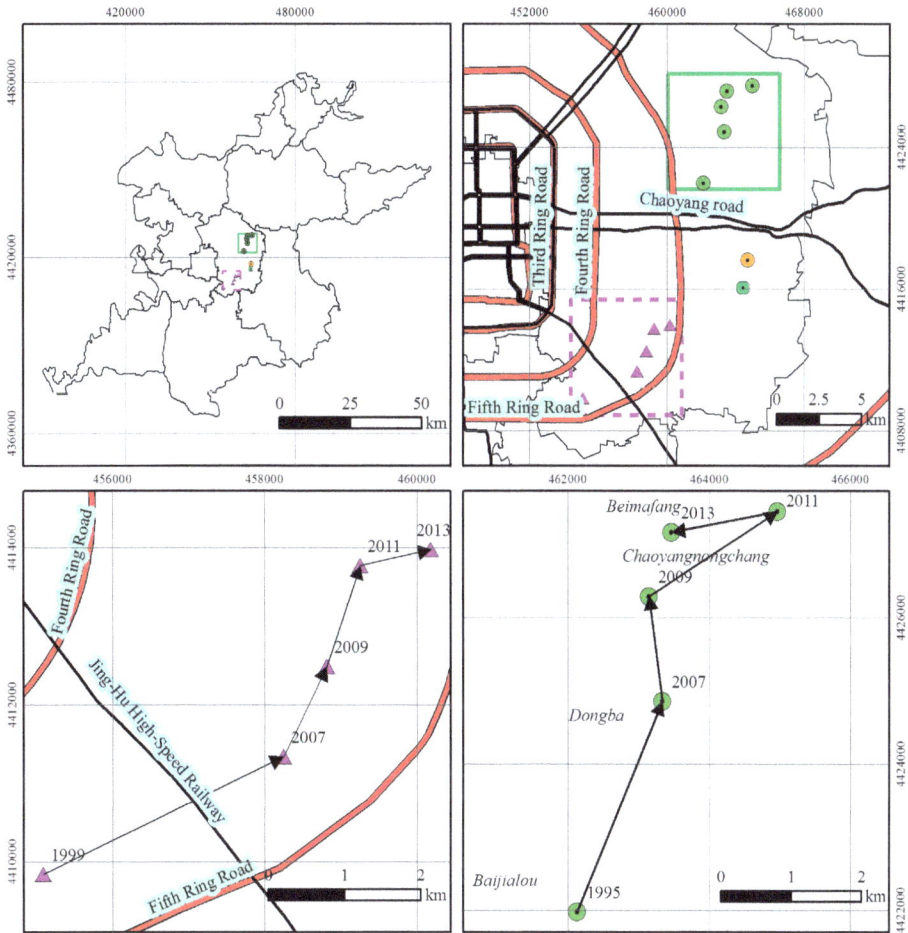

**Figure 7.** Gravity center of cumulative land subsidence, groundwater drawdown, and the compressible layer.

**Table 3.** Gravity center (the relative distance and direction are relative to the geometric center).

|  | Year | *x* | *y* | Distance (m) | Direction | Rate (m/Year) | Relative Distance (m) | Relative Direction |
|---|---|---|---|---|---|---|---|---|
| | 1999 | 455,082.73 | 4,409,843.56 | | | | 11,221.13 | South by West 33.70° |
| Cumulative land subsidence | 2007 | 458,249.82 | 4,411,339.05 | 3502.42 | North by East 64.72° | 437.80 | 7773.48 | South by West 37.49° |
| | 2009 | 458,814.36 | 4,412,488.92 | 1280.98 | North by East 26.15° | 640.49 | 6650.08 | South by West 32.58° |
| | 2011 | 459,259.27 | 4,413,771.85 | 1357.88 | North by East 19.13° | 678.94 | 5647.37 | South by West 24.01° |
| | 2013 | 460,186.49 | 4,413,978.36 | 949.93 | North by East 77.44° | 474.97 | 4720.14 | South by West 26.30° |
| | 1995 | 462,135.93 | 4,421,975.89 | | | | 6331.38 | North by West 21.13° |
| Groundwater drawdown | 2007 | 463,338.85 | 4,424,859.60 | 3124.55 | North by East 22.64° | 260.38 | 8855.55 | North by West 7° |
| | 2009 | 463,147.88 | 4,426,294.86 | 1447.90 | North by West 7.58° | 723.95 | 10,303.39 | North by West 7.08° |
| | 2011 | 464,963.60 | 4,427,455.10 | 2154.77 | North by East 57.42° | 1077.38 | 11,398.14 | North by East 2.74° |
| | 2013 | 463,472.92 | 4,427,165.65 | 1518.53 | South by West 10.99° | 759.27 | 11,135.78 | North by West 4.87° |
| Compressible layer thickness | | 464,698.89 | 4,417,620.02 | | | | 1575.25 | North by East 10.28° |
| Geometry center | | 464,417.87 | 4,416,070.03 | | | | | |

### 3.1.2. Gravity Center of Groundwater Drawdown and Compressible Layer Thickness

The gravity centers of groundwater drawdown in Beijing Plain were also located in Chaoyang District, to the northeast of downtown from 1999 to 2013, and they were to the northeast of the intersection of Chaoyang Road and Fifth Ring Road (Figure 7). This is because its main funnels were always distributed in the northeast of Chaoyang District and its extent fluctuated, but basically covered this area (Figure 4). The gravity center movement also experienced an accelerating-steady-reducing process (Figure 7 and Table 3). It moved north by east from Baijialou in 1995 to Dongba in 2007. From 2007 to 2009, its movement speed increased more than three times and the direction was biased to the west. From 2009 to 2011, its direction returned to north by west. After that, it moved south by west to Beimafang and the movement rate was slowed down.

The groundwater variation determined the gravity center movement dynamic. With the continuous overexploitation of groundwater through many years, the groundwater depression expanded in the northeast part of Beijing Plain, resulting in the gravity center moving northward. According to the statistics of the Beijing Water Authority, groundwater storage increased only in 2012 and it decreased more or less in other years (Figure 8). In 2012, the average annual precipitation of the whole city was 708 mm, 28% more than that in 2011, and 21% more than the average for previous years. Considering that precipitation is the main recharge for groundwater in Beijing Plain, a plentiful supply of groundwater in 2012 led to a rebound of 0.67 m of the average groundwater level compared with 2011. The estimated groundwater storage increased by $3.4 \times 10^9$ m³ and the dramatic groundwater decline trend was alleviated.

The gravity center of the compressible layer was located in east-central Chaoyang District, close to the geometric center (Figure 7). It was located on the northeast part of the Yongding alluvial-pluvial fan-fringe area, and also close to the of Chaobai fan-fringe area with mass compressible deposits, reflecting the spatial concentration of the compressible layer (Figure 3).

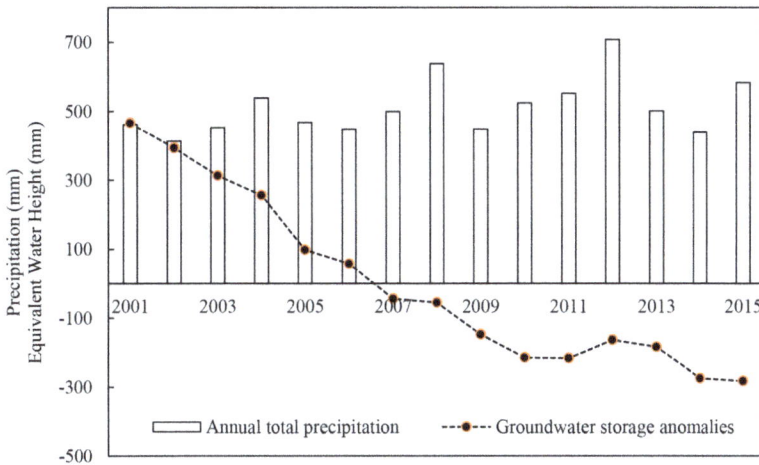

**Figure 8.** The 2001–2015 yearly time series precipitation, and groundwater storage anomalies in Beijing Plain. Note that these data were retrieved from the Beijing Water Resources Bulletin from 2001 to 2015 [45] issued by the Beijing Water Authority.

### 3.1.3. Coupling Analysis of the Gravity Center

The gravity center dynamic demonstrated the adjusting direction and intensity of land subsidence, groundwater drawdown, and compressible layer thickness. The length of the connecting line (Table 4) between them can reflect their relative adjusting trend. For land subsidence and groundwater drawdown, it increased in 2011, and then decreased. For land subsidence and compressible layer thickness, it decreased in 2013. This suggested that the concentration trend of land subsidence showed a larger difference with that of groundwater drawdown yearly until 2011, but it decreased sharply from 2011 to 2013. In light of the largely increasing groundwater recharge from precipitation in 2012, the decrease of groundwater drawdown slowed subsidence development. The concentration trend of land subsidence illustrated a smaller difference with that of the compressible layer thickness yearly, tending to be similar with that of the latter. This reflected their increasing spatial agreement.

**Table 4.** Coupling parameters of the gravity center. Note the cumulative land subsidence from 1955 to 1999 and the groundwater drawdown from 1965 to 1995 were compared because of the missing of same period. The same process was also seen in below.

| Year | 1995 (1999) | 2007 | 2009 | 2011 | 2013 |
|---|---|---|---|---|---|
| Land subsidence vs. groundwater drawdown (m) | 14,033.57 | 14,446.58 | 14,470.08 | 14,824.67 | 13,590.63 |
| Land subsidence vs. compressible layer thickness (m) | 12,367.04 | 9002.28 | 7807.42 | 6663.16 | 5798.57 |

Groundwater drawdown is a dynamic factor, which triggers land subsidence. The compressible layer is a static factor, which provides a potential medium for land subsidence. Under the context of urbanization, the high intensity of groundwater exploitation gradually moved to the upper part of the alluvial-pluvial plain fan, with more groundwater sources. Although the compressible sediments reduced relatively, it still exists and breeds severe land subsidence [23]. Thus, the synchronized behavior of land subsidence and groundwater drawdown made the length of the connecting line change slightly. With the development of land subsidence and the groundwater drawdown, partially- or fully-developed subsidence existed in the static compressible layer. Owing to the positive correlation between the magnitude of land subsidence and the thickness of the compressible layer, the more developed the land subsidence, the thicker the compressible layer. The length of the corresponding connecting line is, therefore, getting shorter. We can make an assumption that when the potential provided by the compressible layer is exhausted, the development of land subsidence will hit a plateau. This still needs continuous observation and further study.

*3.2. Development Orientation Comparison Analysis*

The change tendency of the principle orientation can be analyzed by the azimuth of the ellipse's major axis. From 1999 to 2013, the spatial direction of land subsidence development was SW–NE, showing a generally increasing trend north by east (Figure 9). From 1999 to 2007, the azimuth was increased with an average annual offset angle of 0.94°; from 2007 to 2009, the azimuth was enabled to maintain at about 18.70°; from 2011 to 2013, the azimuth was enabled to maintain at about 20.50° (Table 5). For groundwater drawdown weighted SDE, the azimuth was 24.96° north by east in 1995; it increased to 31.52° in 2007, at the ratio of 0.55°/year to the east. From 2007 to 2011, the azimuth increased and the rate was 0.58°/year, which was the same with that in the previous period; from 2011 to 2013, there was a sharp counter-clockwise shift (Table 5).

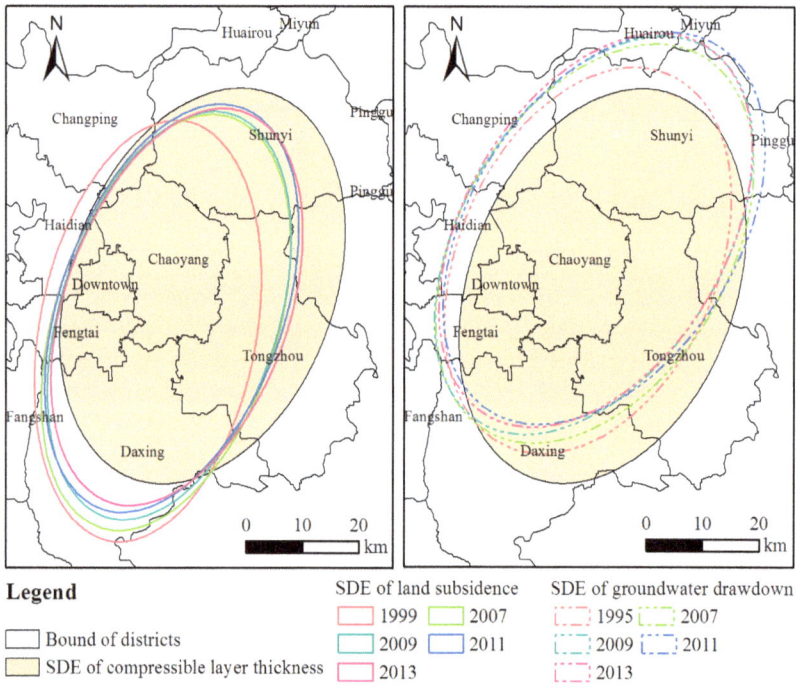

**Figure 9.** SDE of cumulative land subsidence, groundwater drawdown, and compressible layer thickness. The map on the left shows SDE of land subsidence; the map on the right shows SDE of water drawdown.

By comparison, the azimuth of the cumulative subsidence weighted SDE changed with that of groundwater drawdown, but was confined by that of the compressible layer thickness. The azimuth of the cumulative subsidence weighted SDE was at least 11° smaller than that of groundwater drawdown in the same year. They all had a growing tendency to north by east with some fluctuations and a peak in 20.58°. The principle direction of the cumulative subsidence weighted SDE rotated north by east, and the largest in 2009 was close to, but less than, that of the compressible layer thickness weighted SDE. This suggested that land subsidence was driven by the groundwater drawdown, but confined by the compressible layer in extent and magnitude because the thicker compressible layer is the material for land subsidence.

**Table 5.** SDE parameters of land subsidence, groundwater drawdown, and compressible layer thickness.

| SDE Parameters | Year | Short Axis (m) | Long Axis (m) | Rotation Angle (°) | Long Axis/Short Axis | Area (km$^2$) |
|---|---|---|---|---|---|---|
| Cumulative land subsidence | 1999 | 19,346.19 | 36,550.69 | 11.19 | 1.89 | 2221.28 |
| | 2007 | 19,789.50 | 36,998.78 | 18.72 | 1.87 | 2300.04 |
| | 2009 | 19,348.78 | 36,293.81 | 18.67 | 1.88 | 2205.97 |
| | 2011 | 19,875.39 | 36,567.86 | 20.58 | 1.84 | 2283.12 |
| | 2013 | 19,904.22 | 35,492.24 | 20.51 | 1.78 | 2219.19 |
| Groundwater drawdown | 1995 | 22,751.85 | 34,974.82 | 24.96 | 1.54 | 2499.73 |
| | 2007 | 23,714.93 | 37,426.64 | 31.52 | 1.58 | 2788.20 |
| | 2009 | 22,851.41 | 37,940.96 | 32.96 | 1.66 | 2723.58 |
| | 2011 | 23,416.54 | 37,325.74 | 33.87 | 1.59 | 2745.69 |
| | 2013 | 23,009.21 | 37,143.34 | 31.94 | 1.61 | 2684.74 |
| Compressible layer thickness | | 23,473.52 | 35,233.51 | 21.62 | 1.50 | 2598.10 |

### 3.3. Dispersion Tendency Comparison Analysis

The major axis and its ratio with the minor axis characterized the spatial dispersion tendency. From 1999 to 2007, the major axis length of the land subsidence weighted SDE increased at a rate of 56 m/year; from 2007 to 2009, it decreased by 704.97 m; and from 2011 to 2013, there was another stronger spatial contraction. The corresponding ratio was reduced from 1.89 in 1999 to 1.78 in 2013 (Table 5). The spatial contraction of the ellipse suggested the central region experienced more serious land subsidence than that in the external ellipse most of the time.

From 1995 to 2007, the major axis of the groundwater drawdown weighted SDE increased at the annual rate 204.32 m/year; since 2007, its length was generally kept above 37,000 m, and there was a peak in 2009 (Table 5). Since precipitation is the main recharge of groundwater, the groundwater drawdown varied with its fluctuation and there was an apparent trough in the total precipitation and corresponding total groundwater storage especially in 2009 (Figure 8). This suggested that the length of the major axis can be regarded as an important index to measure the groundwater drawdown.

By comparison, the major axis length of the land subsidence weighted SDE decreased and approached that of the compressible layer thickness. In addition, from 1999 to 2013, the ratio between the length of the major axis and the minor axis of the land subsidence weighted SDE was reduced more closely to that of the compressible layer thickness. This suggested that the land subsidence distribution pattern got closer to that of the compressible layer with the increase of its development intensity.

### 3.4. Spatial Differentiation Coefficient Comparison Analysis

The spatial differentiation coefficient between coverage areas can determine their spatial differential degree. The coverage area of the land subsidence weighted SDE fluctuated from 1999 to 2013 (Table 5). It almost covered four of the five funnels, except for Shahe-Baxianzhaung, whose development intensity was high, but was located along the direction of minor axis and close to the gravity center. Therefore, it had little influence in the spatial pattern, comparatively. The coverage area of the groundwater drawdown weighted SDE fluctuated from 1995 to 2013. Since 2007, it was maintained at about 2700 km$^2$, and the smallest appeared in 2013, associated with the precipitation peak of the previous year. The coverage area of the compressible layer thickness weighted SDE was roughly less than that of groundwater drawdown, and larger than that of land subsidence.

The spatial differentiation coefficient between the cumulative subsidence and groundwater drawdown moved up and down between 0.20 and 0.22; and that between the cumulative subsidence and the compressible layer thickness gradually decreased from 0.22 in 1999 to 0.07 in 2013 (Figure 10). This meant that (1) the spatial pattern of land subsidence is different with that of groundwater drawdown, but their difference degree is stable; and (2) the spatial pattern of land subsidence was increasingly close to that of the compressible layer. It can, therefore, be inferred that (1) land subsidence was triggered by groundwater drawdown, but confined and diversified by the compressible sediments in spatial magnitude and extent; and (2) the extent of land subsidence approached the extent of the compressible layer distribution area. Whether the land subsidence in the whole area might hit a plateau after its strong growth, or keep expanding, it needs further observations and confirmation. Continuous attention to this point is very necessary.

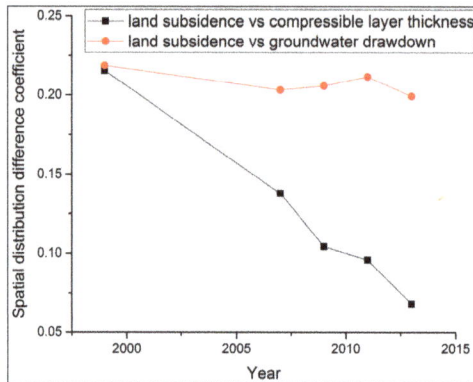

**Figure 10.** Spatial distribution difference coefficient.

## 4. Discussion

Our study made an attempt to introduce geographic distribution measuring methods into land subsidence synthetic research. It is a geographical approach to evaluate land subsidence, groundwater drawdown, and compressible layer thickness from multiple perspectives, including point (gravity center), line (major axes), and polygon (coverage) views. The compressible sediment is a typical type of natural endowment, invariable in a long time scale and providing the material basis of land subsidence. Its non-uniform distribution can accelerate or decelerate the development of land subsidence, unbalancing the starting point of spatial evolvement of land subsidence. Groundwater drawdown is mainly caused by artificial extraction, which is a typical type of anthropogenic activity. It is the inherent drive and its spatial diversity induces unbalanced development of land subsidence. According to Krugman's theory [47], the compressible sediment and groundwater drawdown can be regarded as the "first nature" and the "second nature". Therefore, it is of significance to understand regional land subsidence from a geographic view. In addition, it is a benefit that the method can be implemented by using the reliable commercial GIS software ArcGIS (ESRI, Redlands, CA, USA), which provides a wide variety of spatial analysis interfaces, and allows us to focus on the geographic distribution measured in the GIS environment.

Our result distinguished the spatial distribution differences between land subsidence, groundwater drawdown, and compressible layer thickness. They suggest that (1) land subsidence developed from the southwest to the northeast and has the same concentration trend to groundwater drawdown; (2) land subsidence was stable in extent, but increasing in magnitude; and (3) with the increasing intensity, the spatial pattern of land subsidence was more similar to that of the compressible layer. This can reflect that land subsidence in Beijing is triggered by groundwater overexploitation and influenced by spatial heterogeneity of compressible layers. This is in good agreement with the conclusions from Jia et al. [21] and Zhu et al. [23].

We quantified the development center, principle orientation, dispersion, and tracking of the spatial distribution characteristics of relative study subjects. This can be achieved thanks to the ongoing multiple monitoring data based on groundwater level surveying, and ground-based geological and geodetic measurements, which enable us to acquire sufficient data and draw a scientific conclusion. However, conventional technologies cannot meet the needs of both spatial and temporal resolutions such that we can only focus on general trends, neglecting seasonal variations and funnel features. For instance, only yearly spatial distribution characteristics of land subsidence can be detected owing to the high cost of leveling; groundwater drawdown does not fully explain land subsidence, considering the heterogeneity of specific storage. In this context, remote sensors bring useful information. The InSAR technique can be used to detect land subsidence with higher spatiotemporal resolutions [48–52]. The Gravity Recovery and Climate Experiment (GRACE) technique can reveal the large-scale time series water storage change and the groundwater storage loss [53,54]; most notably, it has a potential to detect heterogeneous groundwater storage variations at the subregional scale, smaller than the typical GRACE footprint (200,000 km$^2$) [55], and to detect anthropogenic signals over regions with high levels of groundwater consumption [56]. Beijing will receive about $1 \times 10^9$ m$^3$ of water through the south-to-north water diversion project and the northern part of Beijing Plain is designed for groundwater recharge. Meanwhile, ongoing urbanization will change the underlying surface condition which can decrease the quantity of precipitation infiltration. Under the above context, groundwater drawdown and land subsidence will take on a different spatial development and pattern. Our study will provide a good reference.

## 5. Conclusions

This paper proposed a comprehensive geographic measurement to improve the understanding of spatiotemporal distribution features of land subsidence, groundwater drawdown, and compressible layer thickness in Beijing Plain.

Land subsidence, groundwater drawdown, and compressible layer thickness were all concentrated in Chaoyang District. The concentration of land subsidence moved from southwest to northeast, which was basically consistent with groundwater drawdown. The compressible layer thickness was concentrated in the east-central Chaoyang District on, or close to, the alluvial-pluvial fan-fringe areas with mass compressible deposits.

The principle direction of land subsidence was SE–NE. It changed with that of the groundwater drawdown, but was getting closer to that of the compressible layer. The length ratio between the major and minor axes suggested that the dispersion tendency of the land subsidence became closer to that of the compressible layer with its increasing development intensity. The spatial contraction of the ellipse suggested the Chaoyang District, in the central region of the study area, experienced more serious subsidence than that in the surrounding areas most of the time.

The spatial distribution difference between land subsidence and groundwater drawdown was about 0.2, and that between land subsidence and compressible layer thickness decreased from 0.22 to 0.07, reflecting that the spatial pattern of land subsidence was increasingly close to the spatial distribution pattern of the compressible layer. Depending on development trends, the spatial pattern of land subsidence in the whole area might reach a plateau after its strong growth, which needs further study.

Generally speaking, land subsidence continued to develop with the groundwater drawdown, and its development was characterized by the inheritance owing to the compressible layer. They enabled the concentration trend of land subsidence in Beijing Plain to change little, and the spatial distribution difference remained stable. This paper focuses on the general distribution features of land subsidence, groundwater drawdown, and compressible layer thickness using GIS methods. If more detailed information on land subsidence and groundwater drawdown can be derived by cutting-edge remote sensing technologies, their distribution features can be finely depicted by the adopted methods, and the results will be more accurate and practical. Meanwhile, the error and reliability of the geographic measurement results will be broadly acceptable with the improvement of the temporal and spatial resolution of the collected monitoring datasets. Geographic distribution measurement approaches enriched the methodologies for studying land subsidence. This study will help in risk assessment of land subsidence under the "new normal" of south-to-north water diversion in Beijing Plain.

**Supplementary Materials:** Cumulative land subsidence (the starting time was 1955) in 2005, 2009, 2011, in Beijing Plain; groundwater seepage field in 1965, 2009, 2011, in Beijing Plain. Available online at www.mdpi.com/2073-4441/9/1/64/s1.

**Acknowledgments:** This work was supported by the National Natural Science Foundation of China (Grant number 41201420, 41171335 and 41130744), and the National Program on Key Basic Research Project (973 Program) (Grant number 2012CB723403). The authors would like to acknowledge all sources of data used in the present work, including cumulative subsidence contours and compressible layer thickness maps provided by the Beijing Institute of Hydrogeology and Engineering Geology, as well as groundwater level contours, precipitation and groundwater storage data published by the Beijing Water Authority. The authors wish to express their appreciation to all the members of the Water Editorial Office, and the two anonymous reviewers for their invaluable comments and constructive suggestions that helped considerably improve the quality of the paper.

**Author Contributions:** Yongyong Li, Huili Gong and Lin Zhu derived the methods, performed data analysis, and wrote the manuscript. Xiaojuan Li made important suggestions on data processing and analysis, and discussed the results.

**Conflicts of Interest:** The authors declare no conflict of interest.

## References

1. Galloway, D.L.; Burbey, T.J. Review: Regional land subsidence accompanying groundwater extraction. *Hydrogeol. J.* **2011**, *19*, 1459–1486. [CrossRef]
2. Gambolati, G.; Teatini, P. Geomechanics of subsurface water withdrawal and injection. *Water Resour. Res.* **2015**, *51*, 3922–3955. [CrossRef]
3. Calderhead, A.I.; Therrien, R.; Rivera, A.; Martel, R.; Garfias, J. Simulating pumping-induced regional land subsidence with the use of InSAR and field data in the Toluca valley, Mexico. *Adv. Water Resour.* **2011**, *34*, 83–97. [CrossRef]
4. Shi, X.Q.; Fang, R.; Wu, J.C.; Xu, H.X.; Sun, Y.Y.; Yu, J. Sustainable development and utilization of groundwater resources considering land subsidence in Suzhou, China. *Eng. Geol.* **2012**, *124*, 77–89. [CrossRef]
5. Modoni, G.; Darini, G.; Spacagna, R.L.; Saroli, M.; Russo, G.; Croce, P. Spatial analysis of land subsidence induced by groundwater withdrawal. *Eng. Geol.* **2013**, *167*, 59–71. [CrossRef]
6. Raspini, F.; Loupasakis, C.; Rozos, D.; Adam, N.; Moretti, S. Ground subsidence phenomena in the delta municipality region (northern Greece): Geotechnical modeling and validation with persistent scatterer interferometry. *Int. J. Appl. Earth Obs.* **2014**, *28*, 78–89. [CrossRef]
7. Brown, S.; Nicholls, R.J. Subsidence and human influences in mega deltas: The case of the Ganges-Grahmaputra-Meghna. *Sci. Total Environ.* **2015**, *527–528*, 362–374. [CrossRef] [PubMed]
8. Mahmoudpour, M.; Khamhechiyan, M.; Nikudel, M.R.; Ghassemi, M.R. Numerical simulation and prediction of regional land subsidence caused by groundwater exploitation in the southwest plain of Tehran, Iran. *Eng. Geol.* **2015**, *201*, 6–28. [CrossRef]
9. Castellazzi, P.; Arroyo-Domínguez, N.; Martel, R.; Calderhead, A.I.; Normand, J.C.L.; Gárfias, J. Land subsidence in major cities of central Mexico: Interpreting InSAR-derived land subsidence mapping with hydrogeological data. *Int. J. Appl. Earth Obs.* **2016**, *47*, 102–111. [CrossRef]
10. Chen, M.; Tomás, R.; Li, Z.H.; Motagh, M.; Li, T.; Hu, L.Y.; Gong, H.L.; Li, X.J.; Yu, J.; Gong, X.L. Imaging land subsidence induced by groundwater extraction in Beijing (China) using satellite radar interferometry. *Remote Sens.-Basel* **2016**, *8*, 468. [CrossRef]
11. Zou, L.; Kent, J.; Lam, N.; Cai, H.; Qiang, Y.; Li, K.N. Evaluating Land Subsidence Rates and Their Implications for Land Loss in the Lower Mississippi River Basin. *Water* **2016**, *8*, 10. [CrossRef]
12. Syvitski, J.; Kettner, A.; Overeem, I.; Hutton, E.; Hannon, M.; Brakenridge, G.; Day, J.; Vörösmarty, C.; Saito, Y.; Giosan, L.; et al. Sinking deltas due to human activities. *Nat. Geosci.* **2009**, *2*, 681–686. [CrossRef]
13. Erban, L.E.; Gorelick, S.M.; Zebker, H.A. Groundwater extraction, land subsidence, and sea-level rise in the Mekong delta, Vietnam. *Environ. Res. Lett.* **2014**, *9*, 1–6. [CrossRef]

14. Sušnik, J.; Vamvakeridou-Lyroudia, L.S.; Baumert, N.; Kloos, J.; Renaud, F.G.; La Jeunesse, I. Interdisciplinary assessment of sea-level rise and climate change impacts on the lower Nile delta, Egypt. *Sci. Total Environ.* **2015**, *503–504*, 279–288. [CrossRef] [PubMed]

15. Yin, J.; Yu, D.P.; Wilby, R. Modelling the impact of land subsidence on urban pluvial flooding: A case study of downtown Shanghai, China. *Sci. Total Environ.* **2016**, *544*, 744–753. [CrossRef] [PubMed]

16. Hoffmann, J.; Zebker, H.A.; Galloway, D.L.; Amelung, F. Seasonal subsidence and rebound in Las Vegas valley, Nevada, observed by synthetic aperture radar interferometry. *Water Resour. Res.* **2001**, *37*, 1551–1566. [CrossRef]

17. Bell, J.W.; Amelung, F.; Ramelli, A.R.; Blewitt, G. Land subsidence in Las Vegas, Nevada, 1935–2000: New geodetic data show evolution, revised spatial patterns, and reduced rates. *Environ. Eng. Geosci.* **2002**, *8*, 155–174. [CrossRef]

18. Teatini, P.; Tosi, L.; Strozzi, T.; Carbognin, L.; Wegmüller, U.; Rizzetto, F. Mapping regional land displacements in the Venice coastland by an integrated monitoring system. *Remote Sens. Environ.* **2005**, *98*, 403–413. [CrossRef]

19. Teatini, P.; Ferronato, M.; Gambolati, G.; Gonella, G. Groundwater pumping and land subsidence in the Emilia-Romagna coastland, Italy: Modeling the past occurrence and the future trend. *Water Resour. Res.* **2006**, *42*, W01406. [CrossRef]

20. Anderssohn, J.; Wetzel, H.U.; Walter, T.R.; Motagh, M.; Djamour, Y.; Kaufmann, H. Land subsidence pattern controlled by old alpine basement faults in the Kashmar valley, northeast Iran: Results from InSAR and levelling. *Geophys. J. Int.* **2008**, *174*, 287–294. [CrossRef]

21. Jia, S.M.; Wang, H.G.; Zhao, S.S.; Luo, Y. A tentative study of the mechanism of land subsidence in Beijing. *City Geol.* **2007**, *2*, 20–26. (In Chinese)

22. Zhang, Y.Q.; Gong, H.L.; Gu, Z.Q.; Wang, R.; Li, X.J.; Zhao, W.J. Characterization of land subsidence induced by groundwater withdrawals in the plain of Beijing city, China. *Hydrogeol. J.* **2014**, *22*, 397–409. [CrossRef]

23. Zhu, L.; Gong, H.L.; Li, X.J.; Wang, R.; Chen, B.B.; Dai, Z.X.; Teatini, P. Land subsidence due to groundwater withdrawal in the northern Beijing plain, China. *Eng. Geol.* **2015**, *193*, 243–255. [CrossRef]

24. Galloway, D.L.; Hudnut, K.W.; Ingebritsen, S.E.; Phillips, S.P.; Peltzer, G.; Rogez, F.; Rosen, P.A. Detection of aquifer system compaction and land subsidence using interferometric synthetic aperture radar, Antelope Valley, Mojave Desert, California Water. *Resour. Res.* **1998**, *34*, 2573–2585. [CrossRef]

25. Mitchell, A. *The ESRI Guide to GIS Analysis Volume 2: Spatial Measurements & Statistics*; ESRI Press: Redlands, CA, USA, 2005; pp. 21–61.

26. Li, X.B. Visulizing spatial equality of development. *Sci. Geogr. Sin.* **1999**, *19*, 254–257. (In Chinese)

27. Xu, J.H.; Yue, W.Z. Evolvement and comparative analysis of the population center gravity and the economy gravity center in recent twenty years in china. *Sci. Geogr. Sin.* **2001**, *21*, 385–389. (In Chinese)

28. Lefever, D.W. Measuring geographic concentration by means of the standard deviational ellipse. *Am. J. Soc.* **1926**, *32*, 88–94. [CrossRef]

29. Furfey, P.H. A note on Lefever's "standard deviational ellipse". *Am. J. Soc.* **1927**, *33*, 94–98. [CrossRef]

30. Warntz, W.; Neft, D. Contributions to a statistical methodology for areal distributions. *Reg. Sci.* **2006**, *2*, 47–66. [CrossRef]

31. Robert, S. The standard deviational ellipse; an updated tool for spatial description. *Geogr. Ann.* **1971**, *53*, 28–39.

32. Wang, B.; Shi, W.Z.; Miao, Z.L. Confidence analysis of standard deviational ellipse and its extension into higher dimensional Euclidean space. *PLoS ONE* **2015**, *10*, e0118537. [CrossRef] [PubMed]

33. Bashshur, R.L.; Metzner, C.A. The application of three-dimensional analogue models to the distribution of medical care facilities. *Med. Care* **1970**, *8*, 395–407. [CrossRef] [PubMed]

34. Vanhulsel, M.; Beckx, C.; Janssens, D.; Vanhoof, K.; Wets, G. Measuring dissimilarity of geographically dispersed space-time paths. *Transportation* **2011**, *38*, 65–79. [CrossRef]

35. Yue, T.X.; Fan, Z.M.; Liu, J.Y. Changes of major terrestrial ecosystems in China since 1960. *Glob. Planet. Chang.* **2005**, *48*, 287–302. [CrossRef]

36. Li, C.; Liu, J.P.; Liu, Q.F.; Yu, Y. Dynamic change of wetland landscape patterns in Songnen plain. *Wetl. Sci.* **2008**, *6*, 167–172. (In Chinese)

37. Wang, B.J.; Shi, B.; Inyang, H.I. GIS-based quantitative analysis of orientation anisotropy of contaminant barrier particles using standard deviational ellipse. *Soil Sediment Contam.* **2008**, *17*, 437–447.

38. Mamuse, A.; Porwal, A.; Kreuzer, O.; Beresford, S. A new method for spatial centrographic analysis of mineral deposit clusters. *Ore Geo. Rev.* **2009**, *36*, 293–305. [CrossRef]

39. Eryando, T.; Susanna, D.; Pratiwi, D.; Nugraha, F. Standard Deviational Ellipse (SDE) models for malaria surveillance, case study: Sukabumi district-Indonesia, in 2012. *Malar. J.* **2012**, *11*. [CrossRef]

40. O'Loughlin, J.; Witmer, F.D.W. The localized geographies of violence in the north Caucasus of Russia, 1999–2007. *Ann. Assoc. Am. Geogr.* **2011**, *101*, 178–201. [CrossRef]

41. Zhao, L.; Zhao, Z.Q. Projecting the spatial variation of economic based on the specific ellipses in China. *Sci. Geogr. Sin.* **2014**, *34*, 979–986. (In Chinese)

42. Peng, J.; Chen, S.; Lü, H.L.; Liu, Y.X.; Wu, J.S. Spatiotemporal patterns of remotely sensed PM 2.5, concentration in China from 1999 to 2011. *Remote Sens. Environ.* **2016**, *174*, 109–121. [CrossRef]

43. Jia, S.M.; Ye, C.; Liu, J.R.; Zhao, S.S.; Wang, R.; Yang, Y.; Wang, H.G.; Dong, D.W. *Investigation of Land Subsidence in Beijing*; Beijing Institute of Hydrogeology and Engineering Geology: Beijing, China, 2006; p. 5. (In Chinese)

44. Liu, Y.; Ye, C.; Jia, S.M. Division of water-bearing zones and compressible layers in Beijing's land subsidence areas. *City Geol.* **2007**, *2*, 10–15. (In Chinese) [CrossRef]

45. Beijing Water Authority, Government Affairs, Statistical Information. Available online: http://www.bjwater.gov.cn/pub/bjwater/zfgk/tjxx/ (accessed on 5 May 2016).

46. Griffith, D. Theory of Spatial Statistics. In *Spatial Statistics and Models*; Gaile, G.L., Willmott, C.J., Eds.; D. Reidel Publishing Company: Dordrecht, The Netherlands, 1984; pp. 3–15.

47. Krugman, P. First nature, second nature, and metropolitan location. *Reg. Sci.* **1993**, *33*, 129–144. [CrossRef]

48. Chen, F.; Lin, H.; Yeung, K.; Cheng, S.L. Detection of Slope Instability in Hong Kong Based on Multi-baseline Differential SAR Interferometry Using ALOS PALSAR Data. *GISci. Remote Sens.* **2010**, *47*, 208–220. [CrossRef]

49. Ng, H.M.; Ge, L.L.; Li, X.J.; Abidin, H.Z.; Andreas, H.; Zhang, K. Mapping land subsidence in Jakarta, Indonesia using persistent scatterer interferometry (PSI) technique with Alos Palsar. *Int. J. Appl. Earth Obs.* **2012**, *18*, 232–242. [CrossRef]

50. Strozzi, T.; Teatini, P.; Tosi, L.; Wegmüller, U.; Werner, C. Land subsidence of natural transitional environments by satellite radar interferometry on artificial reflectors. *Geophys. Res. Earth Surf.* **2013**, *118*, 1177–1191. [CrossRef]

51. Samsonov, S.V.; D'Oreye, N.; González, P.J.; Tiampo, K.F.; Ertolahti, L.; Clague, J.J. Rapidly accelerating subsidence in the greater Vancouver region from two decades of ERS-ENVISAT-RADARSAT-2 DInSAR measurements. *Remote Sens. Environ.* **2014**, *143*, 180–191. [CrossRef]

52. Dehghan-Soraki, Y.; Sharifikia, M.; Sahebi, M.R. A comprehensive interferometric process for monitoring land deformation using ASAR and PALSAR satellite interferometric data. *GISci. Remote Sens.* **2015**, *52*, 58–77. [CrossRef]

53. Yeh, P.J.-F.; Swenson, S.C.; Famiglietti, J.S.; Rodell, M. Remote sensing of groundwater storage changes in Illinois using the Gravity Recovery and Climate Experiment (GRACE). *Water Resour. Res.* **2006**, *42*, W12203. [CrossRef]

54. Wang, X.W.; de Linage, C.; Famiglietti, J.; Zender, C.S. Gravity Recovery and Climate Experiment (GRACE) detection of water storage changes in the Three Gorges Reservoir of China and comparison with in situ measurements. *Water Resour. Res.* **2011**, *47*, W12502. [CrossRef]

55. Huang, Z.Y.; Pan, Y.; Gong, H.L.; Yeh, P.J.-F.; Li, X.J.; Zhou, D.M.; Zhao, W.J. Subregional-scale groundwater depletion detected by GRACE for both shallow and deep aquifers in North China Plain. *Geophys. Res. Lett.* **2015**, *42*, 1791–1799. [CrossRef]

56. Pan, Y.; Zhang, C.; Gong, H.L.; Yeh, P.J.-F.; Shen, Y.J.; Guo, Y.; Huang, Z.Y.; Li, X.J. Detection of human-induced evapotranspiration using GRACE satellite observations in the Haihe River basin of China. *Geophys. Res. Lett.* **2016**, *43*. [CrossRef]

*water*

MDPI

*Article*

# Assessment of the Potential of UAV Video Image Analysis for Planning Irrigation Needs of Golf Courses

Alberto-Jesús Perea-Moreno [1,*], María-Jesús Aguilera-Ureña [1], José-Emilio Meroño-De Larriva [2] and Francisco Manzano-Agugliaro [3,4]

[1] Department of Applied Physics, University of Cordoba, CEIA3, Campus de Rabanales, 14071 Córdoba, Spain; fa1agurm@uco.es

[2] Department of Graphic Engineering and Geomatics, University of Cordoba, CEIA3, Campus de Rabanales, 14071 Córdoba, Spain; jemerono@uco.es

[3] CIAIMBITAL (Research Center on Agricultural and Food Biotechnology), University of Almeria, 04120 Almeria, Spain; fmanzano@ual.es

[4] Department of Engineering, University of Almeria, CEIA3, 04120 Almeria, Spain

* Correspondence: aperea@uco.es; Tel.: +34-957-212-633; Fax: +34-957-212-068

Academic Editors: Hongjie Xie and Xianwei Wang
Received: 9 October 2016; Accepted: 2 December 2016; Published: 8 December 2016

**Abstract:** Golf courses can be considered as precision agriculture, as being a playing surface, their appearance is of vital importance. Areas with good weather tend to have low rainfall. Therefore, the water management of golf courses in these climates is a crucial issue due to the high water demand of turfgrass. Golf courses are rapidly transitioning to reuse water, e.g., the municipalities in the USA are providing price incentives or mandate the use of reuse water for irrigation purposes; in Europe this is mandatory. So, knowing the turfgrass surfaces of a large area can help plan the treated sewage effluent needs. Recycled water is usually of poor quality, thus it is crucial to check the real turfgrass surface in order to be able to plan the global irrigation needs using this type of water. In this way, the irrigation of golf courses does not detract from the natural water resources of the area. The aim of this paper is to propose a new methodology for analysing geometric patterns of video data acquired from UAVs (Unmanned Aerial Vehicle) using a new Hierarchical Temporal Memory (HTM) algorithm. A case study concerning maintained turfgrass, especially for golf courses, has been developed. It shows very good results, better than 98% in the confusion matrix. The results obtained in this study represent a first step toward video imagery classification. In summary, technical progress in computing power and software has shown that video imagery is one of the most promising environmental data acquisition techniques available today. This rapid classification of turfgrass can play an important role for planning water management.

**Keywords:** water management; golf course; memory-prediction theory; object-based classification; unmanned aerial vehicle

## 1. Introduction

As a case of precision agriculture, golf courses can be considered; this is called precision turfgrass in the literature [1]. The huge dimensions of maintained turfgrass can be highlighted by the fact that it is estimated to cover 20 million ha in the USA [2]. Spatio-temporal variation of soil, climate, plants and irrigation requirements are new challenges for precision agriculture and, above all, complex turfgrass sites [3]. The irrigation of golf courses is a major concern in this crop maintenance, especially in a Mediterranean climate, both in the USA and in Europe [4]. Golf courses in the southwestern United States are rapidly transitioning to reuse water (treated sewage effluent), as municipalities provide price incentives or mandate the use of reuse water for irrigation purposes [5]. So, when reuse water of poor quality is used, as on golf courses in the arid southwestern United States, proper irrigation management is critical [6], so greenkeepers should pay attention to irrigation strategies employed on reuse water irrigated golf courses to properly manage for higher nitrogen and salt loads. In Spain, it is estimated that water consumption for a golf course is 6.727 $m^3$/ha per year (this is due to the use of poor water, 2.5 dS/m) [7].

Recently, unmanned aerial vehicles (UAVs) have provided a technological breakthrough with potential application in PA [8,9]. UAVs enable the quick production of cartographic material because they rely on different technologies, including cameras, video and GPS (Global Positioning System) [10]. Even though an UAV has very restricted, heavy limitations, the minimization of the sensors during the last year is allowing the use of lighter vehicles, or the use of more features and sensors to a given platform [11]. The opportunity offered by UAVs to observe the world from the sky provides the opportunity to study crops or turfgrass from an unusual viewpoint, allowing the visualization of details that cannot be easily seen from the ground [12,13].

Regarding agricultural purposes, aerial photography and colour video from UAVs presents an alternative to imagery from satellite and aerial platforms [14], which are often difficult to obtain or expensive [15]. Hassan et al. [15] highlight the problem of conventional methods in the classification process, using high resolution images to overcome or minimize the difficulty in classification of the mixed pixel areas. A huge number of applications are achieved using UAVs to monitoring the health of crops through spectral information, e.g., stressed or damaged crops change their internal leaf structure which could be rapidly detected by a thermal infrared sensor [16], therefore, this information is very important to detect stress such as water and nutrient deficiency in growing crops [17].

On the other hand, texture measurements from images obtained by UAVs have been integrated in object-oriented classification, specifically in the classification and management of agricultural land cover [18]. Likewise, there are studies that demonstrate the feasibility of using UAVs with thermal multispectral cameras for estimating crop water requirements, determining the ideal time for watering and saving water consumption without affecting productivity [19–21]. Therefore, the technology is versatile and capable of producing very useful cartographic material for working with PA; the technique also facilitates working with aerial photography in addition to LIDAR (Laser Imaging Detection and Ranging) or video cameras [22,23]. UAVs on golf courses have been used for some time to monitor certain agronomic variables, such as nitrogen [24], and should be considered as a valuable tool to monitor plant nutrition. In this case study, a rapid classification of turfgrass, among others, can play an important role to determine the water requirements of the different areas in order to plan water use.

For this purpose, information of important use can be analyzed and extracted from the images through the employment of powerful and automatic software. The object-based image classification techniques are applied not only for a high level of adaptability but automation as well. These techniques overcome some limitations of pixel-based classification by creating objects on the image through segmentation, using adjacent pixels with a spectral similarity [25]. Subsequently, object-based classification combines spectral contextual information for these objects to perform more complex classifications. These techniques have been successfully applied to images obtained by UAVs in agricultural [26,27], aquatic ecosystems [28] and urban [29] areas.

Therefore, for golf courses, irrigation need planning, especially if it is employed in large areas, and has to be monitored more frequently than other crops. UAVs, due to low cost and fast response time, are the technology that allows this monitoring. A monitoring system based on the video image analysis and classification, will allow a real-time control of crops. Thus, this research is the first step to show the technical viability of real-time control of crops.

Thus, given the positive results previously obtained in the classification of images and given that the applications developed using the HTM algorithm are capable of analyzing video images, the objective of the current study is to develop a recognition methodology for golf courses in real-time using video images taken by an UAV based in a HTM for possible application in planning irrigation needs in order to maximize the water use efficiency and help to plan water requirements of reuse water.

## 2. Material and Methods

### 2.1. UAV and Sensor Description

The material used in this work included images taken by an UAV DJI Phantom 2 Vision+ (Figure 1a) with a flight control system Naza-M V2, that has a range of 700 m and an altitude of 300 m; a HD integrated camera; and a 3-axis gimbal correcting movements in any axis and direction [30]. The Phantom 2 Vision+ is a rigid quadcopter with a maximum ascent rate of 6 m/s, a maximum descent rate of 2 m/s and a maximum flight speed of 15 m/s.

The camera of the Vision+ (Figure 1b) has a 140° FOV (Field of View), F2.8 connected to a 2.3'' sensor with 14 megapixels that can capture images in Adobe DNG and JPEG format as well as recording 1080-pixel and 30-fps videos or recording in a slow camera mode at 60-fps 720-pixel resolution [30]. The equipment also features a streaming video and telemetry data with a range of up to 700 m to a phone, tablet or computer and has a 5200-mAh lithium battery that can hold the quadcopter in the air for up to 25 min. The operator can control the camera using Wi-Fi to manage pan, tilt and camera light sensitivity, video or image modes. The Wi-Fi computer camera system is a very important element that allows for real-time viewing of everything being seen by the camera and the obtainment of video images in real-time.

The equipment also includes an inertial sensor and a barometric altimeter to measure altitude and latitude.

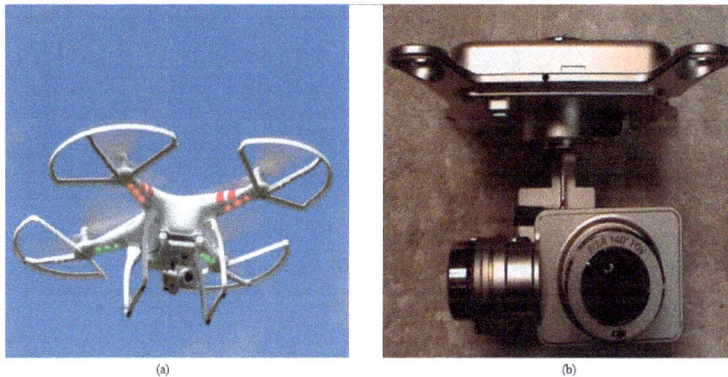

(a)　　　　　　　　　　(b)

**Figure 1.** Details of the DJI Phantom 2 Vision+. (**a**) General image of the quadcopter; and (**b**) the details of the camera.

### 2.2. Study Site

The first stage of this study is to propose a new methodology for analysing geometric patterns of video data acquired from UAVs using a new Hierarchical Temporal Memory (HTM) algorithm.

The information used in this research during the training phase includes simple and short videos, as these videos represent a first step in the integration of UAV video cameras into this technology and the search for the most suitable videos for the proposed purposes. The analyzed patterns to check the accuracy of the method were grapes (*Vitis vinifera*; see Figure 2) and other non-agricultural uses, namely urban and wood areas.

For each of these categories, 300 training videos and 150 testing videos with a total duration of 60 min were used. The videos were obtained in different areas of Redwood City, San Mateo County, California, United States (37°30.128' N 122°12.758' W; Figure 3).

**Figure 2.** Sample image of the video sequences studied.

**Figure 3.** Location map.

## 2.3. HTM Methodology

In recent years, the technology involved in remote sensing and object recognition has considerably advanced [31,32], with diverse applications ranging from recognition and vehicle classification [33] to the facial recognition of individuals [34]. Studies on detection and object recognition can be classified into two categories: keypoint-based object detection [35] and hierarchical and cascaded classifications [36]. Parallel to this development, a new technology applicable to the classification of digital pictures emerged: the Hierarchical Temporal Memory (HTM) learning algorithm. This classification technology is based on both neural networks and Bayesian networks but involves a particular algorithm based on a revolutionary model of human intelligence—the memory-prediction theory developed by Jeff Hawkins [37]. This theory is based on the workings of the human cerebral cortex, which has a structure in the form of "layers" in which information flows bidirectionally from the senses to the brain. From this operating hierarchy, a hypothesis of how the human mind works is created. The key point of this algorithm is found in the duality of the information received. All information we perceive has a spatial component and a temporal one; information is received by the human brain not as an isolated pattern but as a succession of patterns. The cerebral cortex stores the patterns that we perceive and how they are ordered in time. In light of that concept, the memory-prediction theory states that the cerebral cortex stores the new patterns and their evolution over time so that once these sequences stabilize, the brain can make predictions (or inferences) enabling it, without observing a full sequence, to know what pattern it is observing because it knows the sequence in which the patterns occur over time [37].

Thus, this new technology developed by Jeff Hawkins not only presents a new model of how human intelligence functions but also models a neural network system capable of emulating this theory. This classification technology is not specific to image analysis but is versatile for any type of information (from medical information to economic data), with a dual role: learning and pattern recognition in data flows and classifying unknown data according to the training received. Currently, we can find this technology integrated into the free software application NuPIC developed by NUMENTA® (Redwood City, CA, USA), which is used to classify data streams [38]. These data can be of many types, ranging from sign language [39] to eye retinal images for biomedical purposes [40]. There are open areas of research using HTM as a classifier for land planning, which is where our work focuses. In a previous study, Perea et al. [41] conducted an analysis of high-resolution images for classification and land planning in agricultural environments; starting from images from a UltracamD® (Graz, Austria) photo sensor of a region of southern Spain, classification results were obtained that recognized the ground cover up to 90.4%. In a similar fashion, using HTM in the recognition and object-oriented classification, the technology was successfully applied in the recognition of urban areas and green areas; the classification results obtained were approximately 93.8% [42].

Objects with a hierarchical structure, in both space and time, compose the world; this same concept is used by HTM to generate a series of interconnected nodes organized in a tree hierarchy [43]. Thus, the HTM presents a hierarchical structure either in space or time and represents the structure of the world [44].

The HTM learning algorithm implemented in the HTM Camera Toolkit free Application Programming Interface (API) was used in this experiment. This API allows easy implementations of HTM learning algorithms using real world images. Although this API can be used in a variety of contexts, in this paper we focus only on visual recognition applications (i.e., inputs are UAV videos). This API is built and configured by writing Python scripts, allowing researchers to design and configure the hierarchy of nodes based on their input data. To improve the accuracy based on node parameters configuration it is necessary to work with iterations.

As commented before, the principal objective of this first stage in this investigation is to propose a new methodology for analysing geometric patterns of video data acquired from UAVs using a new Hierarchical Temporal Memory (HTM) algorithm. For this purpose, the parameterization and structure of the HTM algorithm for learning and inference have been analyzed and constructed.

Figure 4 shows the overall methodology for HTM design and implementation. There are five phases in this methodology, from the definition and configuration of the data and HTM network to the training and its evaluation.

Once the data to be used have been defined, two steps were necessary to create this network: the creation of the architecture using the Python programming language and the formation of a set of training patterns.

Based on the experience of the research group in previous work [41,42,45], the HTM network was defined in three levels: the first two levels are composed by two sub-levels (a sub-level which analyses the spatial component and another sub-level that analyses the temporal component), and finally there is a classifier which sorts the images into common categories. The level 1 or input level is composed of $8 \times 8$ pixel input nodes, each associated to a single pixel. Nodes from the first level go through the raw image and receive a characteristic of the training pattern image, creating an entry vector formed by digital levels of $8 \times 8$ pixels. Level 2 is composed of 16 nodes that receive the information from the previous level; therefore, each level 2 node is formed by four primary child nodes (arranged in a 2 × 2 region). Finally, level 3 or higher comprises a single node, and it has 16 child nodes (arranged in a 4 × 4 region) and a receptive field of 64 pixels. In Figure 5, the downward connection of one node per level is shown. This system operates in two phases: the training phase and the inference phase. During the training phase, the network is exposed to training patterns and builds a model that categorizes patterns. During the inference phase, new patterns will be distributed in these categories. All nodes (except the initial node) process information in the same manner and consist of two modules: temporal and spatial [44]. Understanding an HTM node involves understanding the operation of these modules during the learning and training phases.

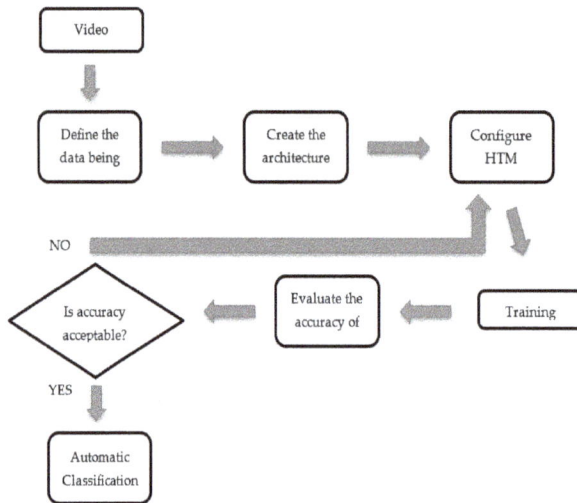

**Figure 4.** Overall methodology diagram.

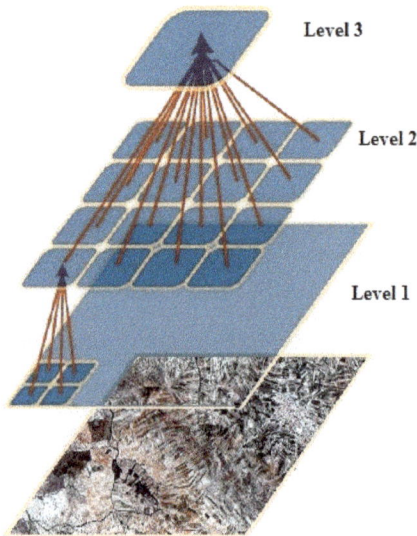

**Figure 5.** Details of the HTM structure. Level 1 is composed of 64 nodes; Level 2 is composed of 16 nodes and Level 3 comprises a single node.

### 2.3.1. Training Phase

During the training phase, the spatial module learns to classify input data based on the spatial coincidence of the elements that compose them. The input vector is compared with other vectors already stored. The exit of the spatial module (temporal module entrance) occurs in terms of their matches and can be seen as a pre-processing stage for the temporal module, simplifying entry. The temporal module learns temporary groups, which are groups of coincidences that frequently occur [44].

### Spatial Module

The spatial modules of the input nodes receive raw data from the sensor; spatial modules of the upper nodes receive the output data from their lower nodes. The input of the spatial module in the upper layers is the concatenation of the order set by the output of the nodes below. Its input is represented by a series of vectors, and the function of the spatial module is to build a matrix (match matrix) of input vectors that have recently occurred. There are several algorithms for the spatial modules, such as the Gaussian and Product algorithms. The Gaussian algorithm is used for nodes in the input level, and the top nodes of the hierarchy use the spatial module Product.

The Gaussian algorithm compares the input vector without dealing with the existing matches in the match matrix. If the Euclidean distance between the input vector and the existing match is sufficiently small, then the entry is considered as the same match, and the match count is incremented and stored in the memory. The distance between an input vector and previously stored vectors is:

$$d^2\left(x,\ w_j\right) = \sum_{i=1}^{D} \left(x_i - w_j\right)^2 \tag{1}$$

where $D$ is the dimension of the vector (64 in the first level), $x_i$ is the $i$th element of the input vector and $w_j$ is the position $i$ of the vector $j$ in the match matrix $W$. The match threshold of an input vector to an existing match is the Maximum distance parameter.

The product algorithm calculates the probability of similarity ($belief_i$), Equation (2), between an input in the inference and a vector that had been previously memorized by the spatial module:

$$belief_i = \prod_{j=1}^{nchildren} y_i\left(child_j\right) * x\left(child_j\right) \tag{2}$$

where *nchildren* is the number of secondary nodes (previous level) that the parent node has, $x$ is the input vector, $y_i$ are the vectors previously stored by the spatial module and $(child_j)$ is the part of a vector obtained from *nchildren* secondary nodes.

### Temporal Module

The temporal module forms groups of matches in time, called temporal groups. Subsequently, a temporal match matrix is built. After the training phase, the temporal module uses this matrix to create the temporal groups. This module uses the sum algorithm, which takes the best representations of all groups to classify new input patterns during inference. When a new input vector is presented during the training phase, the spatial module represents the input vector as one of the learned matches. This process increases the elements $(j, i)$ of the temporal match matrix and is controlled by the *transitionMemory* parameter. This increment ($It$) is calculated as follows, Equation (3):

$$It = transitionMemory - t + 1 \tag{3}$$

where $t$ is the training; the HTM time is in seconds between the current match and the past match.

2.3.2. Inference Phase

After training a node, the network transitions to the inference mode. When the complete network is trained, all of the nodes are in the inference state, and the network is capable of performing inference with new input patterns. Initially, a probability distribution is generated for the categories that were used during training.

Spatial Module

When an input pattern arrives to the spatial module, the network will generate a distribution of beliefs about the categories that have been created in the training phase. Both the Gaussian spatial module and the Product spatial module work differently during the inference stage, but both turn an input vector into a belief vector around the matches.

In the Gaussian spatial module, the distance between an input vector $x$ and each of the trained matches $w_j$ is calculated using Equation (1).

This distance becomes a probability vector considering $x$ as a random sample drawn from a set of multi-dimensional Gaussian probability distributions, all of them based in one of the trained matches. All of these distribution probabilities have the same constant variance in all dimensions, controlled by the *Standard Deviation* (SD) parameter, which is the square root of the variance. Each element $i$ of the probability vector $b$, which represents the probability of the input vector $x$ having the same cause as the match $i$, is calculated using the following equation:

$$y_i = exp\left\{-\frac{d^2\left(x, w_j\right)}{2SD^2}\right\} \tag{4}$$

where $d^2$ is defined in Equation (1) and $w_j$ is the match of the position $j$ in the match matrix $W$.

The algorithm of the Product spatial module divides the input vector at the outputs of each one of its subgroups. The algorithm uses the dot product with the same parts of the match and then calculates the products of these numbers, resulting in a probability vector element in matches in the match matrix.

Temporal Module

During the inference phase, the temporal module receives a probability vector concerning the matches in the spatial module. Subsequently, the module calculates the probability distribution of the groups. A choice is made between two different algorithms in the temporal module during the inference: *maxProp* and *sumProp*, controlled by the *PoolerAlgorithm* time parameter. These algorithms are defined in detail in [46].

*2.4. HTM Design and Implementation*

As commented before, we used the HTM Camera Toolkyt API, developed by Numenta® (Redwood City, CA, USA), in order to design the HTM network used in this investigation.

Once the network has been built, the second step is to configure the information handling and training process. Here, the key parameter is the number of iterations performed with the training images. In this case, 2000 iterations were performed at three levels. Experiments have demonstrated that increasing up to double the number of iterations (4000) does not result in a significant increase in the accuracy of the analysis [41,42,45].

In Table 1, the most relevant parameters of the network-training phase are presented, as are the starting values of the core network as recommended by [43].

Figure 6 presents images of each level of the network structure. Each image that is contained in a video is analyzed by the network. As a pre-treatment, all frames are rescaled to a specific resolution as many times as the parameter *ScaleCount* indicates (Figure 6a—original image, Figure 6b—rescaled image), after which the information goes through the first level of nodes (level 1). This first level is called the S1 layer, and it uses a filter (Gabor filter) to help in recognizing input patterns and making a selection among a series of categories based on geometric and temporal similarities. To extract features and analyze texture, Gabor filters are used [47].

**Table 1.** Parameters used during training.

| Parameter | Description | Values |
|---|---|---|
| maxDistance (*maxDist*) | Minimum Euclidean distance for storing a pattern as a new category, in the lower level of the training phase. | 1 |
| Scale factor (*ScaleCount*) | Number of scales of the same image that the sensor introduces into the network. | 1 |
| Spatial reference (*spatialRF*) | Size of the information reception field with respect to the total. | 0.2 |
| Temporal groups (*requestGroupCount*) | Sets the maximum number of temporal groups that will be created. | 24 |
| Spatial overlap (*spatialOverlap*) | Overlap between nodes of the same level according to the information received from child nodes. | 0.5 |
| Scale reference (*ScaleRF*) | Number of scales of which the node receives information. | 2 |
| Categories (*outputElementCount*) | Number of categories. | 3 |

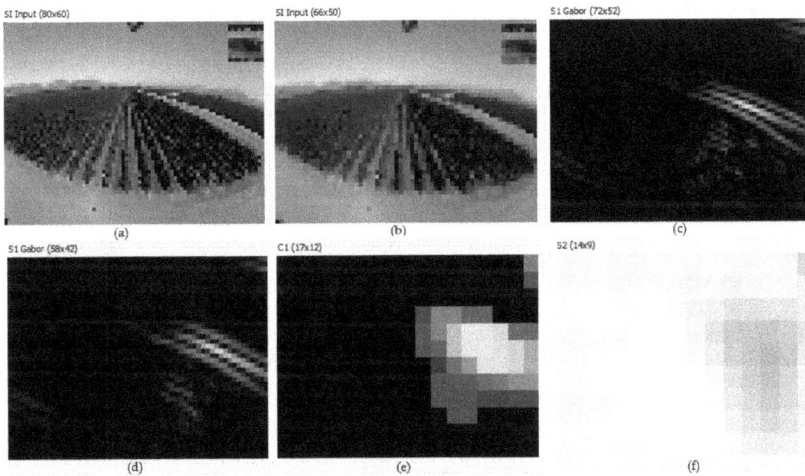

**Figure 6.** Operation of network levels during the training phase. (**a**) Input with 80 × 60 pixels resolution; (**b**) input with 66 × 50 pixels resolution; (**c**) S1 layer with 72 × 52 pixels resolution (Gabor filter); (**d**) S1 layer with 58 × 42 pixels resolution (Gabor filter); (**e**) C1 layer with 17 × 12 pixels resolution; and (**f**) S2 layer with 14 × 9 pixels resolution.

Due to this initial screening, we generate a database of the most common patterns and reduce the infinite number of patterns that we could receive in each image to a limited number. This level produces a set of patterns that are common or that are strongly present in the image as an output.

The input for Level 2 (Figure 6c—obtained from the image, Figure 6d—obtained from the rescaled image), designated C1 Layer (Figure 6e), is the output of the previous level (S1). Level 2 is where the clustering of time sequences occurs. In this level, grouping is performed based on the information of the previous layer, with the base patterns (equivalent to the invariant representations of the HTM theory) creating pattern sequences or pattern clusters using geometric criteria. These sequences are stored, generating more complex patterns.

The information travels up the network to level 3, called the S2 layer (Figure 6f), where information from the preceding level 2 (C1) arrives. Level 3 is where an initial classification is performed. During the training phase, a set of prototypical patterns are memorized through the sequences received from the classification made by the lower layer (C1). When the network is trained, the new data stream in this sub-layer will be compared to the memorized sequences performing an initial classification.

*2.5. Inference Phase*

Once the network has been trained with the data set that was provided, indicating the categories that we want it to recognize, we move to the inference phase. In the inference phase, we supply the network with a set of unknown images for it to classify according to what the network learned and memorized in the previous phase.

Figure 7 presents the system working in the inference phase. The status of any of the nodes of the different levels (Figure 7a–e) can be visualized while the network is processing the information. Finally, we have the C2 Layer, in which the process of grouping already classified patterns is repeated. This process is performed to convert the information into a probability vector, which collects the sequences with the maximum response to the classification process. Behind this last layer, we have a support vector machine (SVM) classifier. SVM, as a kernel learning method, is used for classification problems, performing a non-linear classification [48]. This classifier memorizes the categories with which we are working; these categories were defined during the training phase. This classifier is responsible for assigning the class to which each classified image belongs (Figure 7f). Once the inference stage is completed, a confusion matrix is obtained.

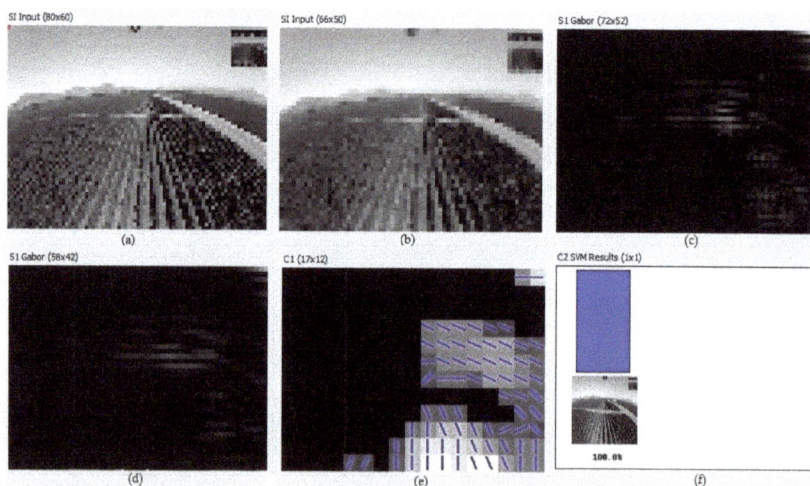

**Figure 7.** Operation of network level in the training phase. (**a**) Input with 80 × 60 pixels resolution; (**b**) input with 66 × 50 pixels resolution; (**c**) S1 Layer with 72 × 52 pixels resolution (Gabor filter); (**d**) S1 layer with 58 × 42 pixels resolution (Gabor filter); (**e**) C1 Layer with 17 × 12 pixels resolution; (**f**) SVM classifier.

## 3. Results and Discussion

During the experiments, internal network parameters that affect the learning process were modified, with the main goal of obtaining an optimal methodology for the recognition of video image patterns.

As mentioned above, the *maxDist* parameter defined the Euclidean distance between a known pattern and a new one, which is critical in the recognition and classification of patterns. An optimal value is essential for the successful creation of temporal groups during the training phase. A high value of the *maxDist* parameter contributes to the formation of fewer temporal groups, which could seriously impact the total recognition accuracy. On the other hand, a low value of the *maxDist* parameter generates a high number of temporal groups, which on top of the large memory demand, also results in poor recognition performance. To avoid these undesirable effects, it is very important to evaluate the optimal value for *maxDist* to achieve the best accuracy in the classifications.

In the original configuration, the *maxDist* parameter has a starting value of 1, and the influence of this parameter on the overall accuracy values in the different classifications was studied. The *maxDist* values (Table 2) used in this experiment were defined based on the results of the initial studies performed [41,42,45].

Table 2 presents the *maxDist* parameter values with respect to the overall accuracy obtained for each of the test classifications. The maximum accuracy value was 96% and was obtained at an intermediate value for a *maxDist* of 3. After this value, there is nearly a linear drop in the overall accuracy of the classifications. This drop is due to the number of coincidences detected during the training phase and the temporal groups formed.

**Table 2.** Overall accuracy and average number of coincidences and temporal groups learned in the 64 bottom nodes for different values of *maxDist*.

| *maxDist* | Overall Accuracy (%) | Number of Coincidences | Number of Temporal Groups |
|---|---|---|---|
| 1 | 86.77 | 55.00 | 25.00 |
| 3 | 96.00 | 44.79 | 20.00 |
| 6 | 83.13 | 17.94 | 13.65 |
| 9 | 76.37 | 12.20 | 8.21 |
| 12 | 64.50 | 10.12 | 5.67 |

For the previously mentioned optimal value of *maxDist*, the Urban class was the class that obtained the largest number of misclassified frames, as seen in Table 3, whereas the Grape class reached the highest accuracy of all the classes during classification.

**Table 3.** Confusion matrix for the optimum value of *maxDist*.

| Classes | Grapes | Urban | Woods |
|---|---|---|---|
| Grapes | 985 | 5 | 10 |
| Urban | 6 | 913 | 81 |
| Woods | 8 | 10 | 982 |

Looking at the second and third columns of Table 2, a large number of matches was not related to a greater overall accuracy of classification, as the number of matches in input patterns might be unrealistic, classifying new similar patterns in different categories. For example, if we set a low value for the parameter *maxDist*, it is forcing the creation of many different, but similar, groups. So, several categories may correspond to the same pattern.

For the case with *maxDist* of 3, which can be considered optimal, the number of matches obtained was 44.79. On the other hand, the effect of the value of the *maxDist* parameter on the creation of temporal groups during the training phase of the network can be seen in Table 2; the smaller the *maxDist* parameter, the greater the number of temporal groups was obtained, leading similar patterns to be classified in different classes. Conversely, increasing the value of the *maxDist* parameter reduces the formation of temporal groups, an effect that is not conducive in any way to obtaining an optimal accuracy in the classification, as the images of wineries and images of forest areas are classified in the same category (Table 4). For the case with the optimal *maxDist* value of 3, the number of temporal groups obtained was 20.

The effect of the *SD* parameter on the accuracy of the classification was verified. This parameter is calculated as the square root of the *maxDist*. This value is a reasonable starting value for *SD* because the distances between the matches are calculated as the square of the Euclidean distance instead of the normalized Euclidean distance.

**Table 4.** Confusion matrix for a *maxDist* value of 12.

| Classes | Grapes | Urban | Woods |
|---------|--------|-------|-------|
| Grapes | 600 | 5 | 395 |
| Urban | 157 | 767 | 76 |
| Woods | 407 | 25 | 568 |

Figure 8 presents the overall accuracy values obtained for different *SD* values. Similar to the *maxDist* parameter, there is growth in the overall accuracy value until it reaches a maximum of 96% for an *SD* value of 1.73. Smaller *SD* parameter values cause high beliefs to be assigned only to matches that are very close to the inferred pattern. Conversely, when using lower *SD* values, between 1 and 1.73, all of the matches receive high belief values independent of their distance to the inferred pattern.

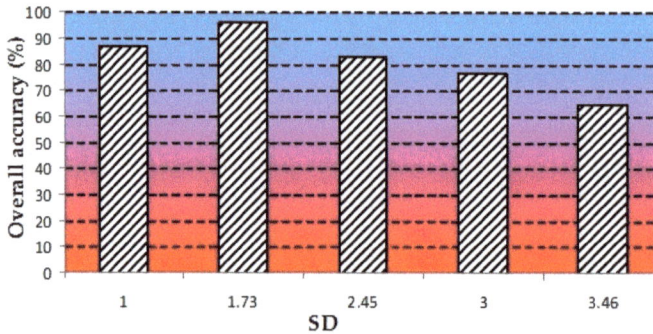

**Figure 8.** Overall accuracy for five setups of the *SD* parameter.

Based on the optimal *maxDist* and *SD* values previously discussed, we studied the effect of the *ScaleRF* and *ScaleOverlap* parameters on the network training and overall accuracy obtained in the classification of the images.

As mentioned above, the *ScaleRF* and *ScaleOverlap* parameters are related to the scale or the resolution of the images that are presented to the network; thus, by changing these parameters, we can vary the number of different scales of the image that are presented to the nodes and the overlap among them. This change is critical because changes in the image resolutions allow the network to extract patterns of the same image in different levels to create invariant representations (or models of stored patterns) used to classify new images.

The basic network starts from intermediate values of *ScaleOverlap* and *ScaleRF* (1 and 1, respectively). Figure 9 presents a bar chart in which the *ScaleOverlap* and the *ScaleRF* parameters are related to the overall accuracy for each case. The highest overall accuracy (97.1%) was obtained for a value of 4 for the *ScaleRF* parameter and 1 for the *ScaleOverlap*. The worst results were obtained for a *ScaleOverlap* parameter value of 0; this value creates no spatial overlap among the input patterns, worsening the training stage in the temporal module and thereby reducing the number of temporal groups formed and their time sequence.

In general, it is observed in this study that a value of 4 for the *ScaleRF* parameter optimizes the capacity of the network to extract patterns from images at different resolutions. From a value of 5, the overall classification accuracy starts to fall again.

After the analysis of the videos, the abilities of the model to learn the invariant representation of the visual pattern, to store these patterns in the hierarchy and to automatically retrieve them associatively, was verified.

For this experience, the maximum overall accuracy obtained among the different classifications made was 97.1% (Figure 10), avoiding problems related to the use of images with high spatial resolution, as in the salt-and-pepper noise effect. The salt-and-pepper effect makes it difficult to obtain and cleanly classify images, resulting in different cases for a plot where there should only be a single case.

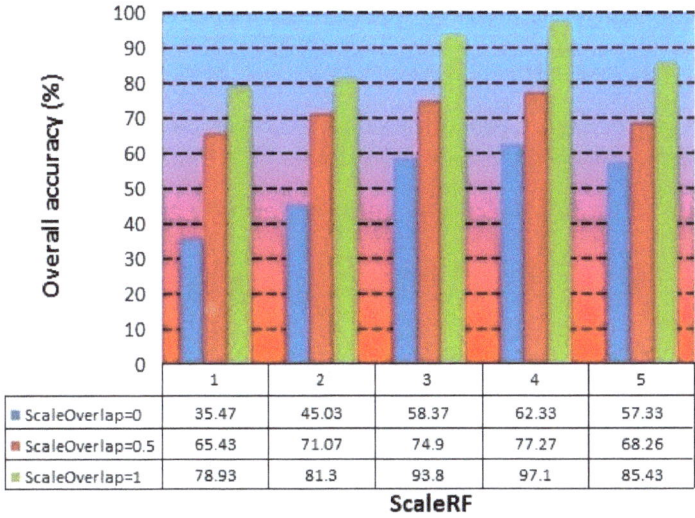

| | 1 | 2 | 3 | 4 | 5 |
|---|---|---|---|---|---|
| ScaleOverlap=0 | 35.47 | 45.03 | 58.37 | 62.33 | 57.33 |
| ScaleOverlap=0.5 | 65.43 | 71.07 | 74.9 | 77.27 | 68.26 |
| ScaleOverlap=1 | 78.93 | 81.3 | 93.8 | 97.1 | 85.43 |

**ScaleRF**

**Figure 9.** Overall accuracy for different values of *ScaleRF* and *ScaleOverlap*.

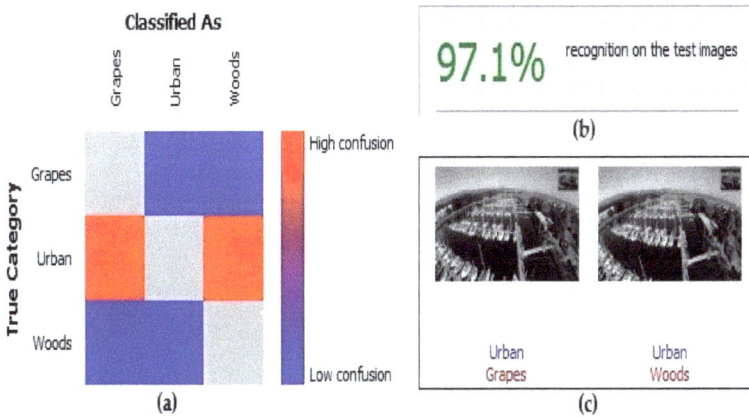

**Figure 10.** Classification results for the best HTM configuration presented by the HTM Camera Toolkit API (Application Programming Interface). (**a**) Confusion matrix; (**b**) overall accuracy; (**c**) clicking on confusion matrix the user can display the misclassified frames (for example, Urban class classified as Grapes or Urban class classified as Woods).

Comparing the results of the Confusion matrix (Table 5), lower accuracy in the Urban class is observed; there were a few misclassified frames because in the same image, two different classes could coexist, such as buildings and parks (Table 5). In 59 frames, the Urban class was classified as the Woods class, and in 11 frames, it was classified as the Grapes class. The higher accuracy obtained was for the Grapes class, where one frame was classified as Urban class and five frames as Woods class.

**Table 5.** Confusion matrix of the best performing system.

| Classes | Grapes | Urban | Woods |
|---|---|---|---|
| Grapes | 994 | 1 | 5 |
| Urban | 11 | 930 | 59 |
| Woods | 4 | 7 | 989 |
| Overall accuracy | 97.10% | | |

*Case Study: Golf Course*

The analyzed patterns to check the accuracy of this case study were turfgrass (see Figure 11) and other uses, namely urban, water, bunker and wood areas.

**Figure 11.** Sample image of the video sequences studied.

For each of these categories, 300 training videos and 150 testing videos with a total duration of 60 min were used. The videos were obtained in different areas of a golf course in Pilar, Buenos Aires (34°29'52.62'' S; 58°56'11.68'' O; Figure 12).

**Figure 12.** Golf course location: case study.

Based on the optimal parameter values previously discussed, we studied the effect of and the overall accuracy obtained in the classification of the images.

For this case study, the overall accuracy obtained, using the optimal values parameters studied above, was 98.28% (Table 6).

**Table 6.** Confusion matrix of the best performing system.

| Classes | Turfgrass | Urban | Water | Bunker | Woods |
|---|---|---|---|---|---|
| Turfgrass | 986 | 1 | 3 | 4 | 6 |
| Urban | 0 | 980 | 7 | 3 | 10 |
| Water | 2 | 7 | 974 | 15 | 2 |
| Bunker | 3 | 3 | 7 | 986 | 1 |
| Woods | 3 | 4 | 0 | 5 | 988 |
| Overall accuracy | 98.28% | | | | |

We compared our results to those of other works. For example, Revollo et al. [49] develop an autonomous application for geographic feature extraction and recognition in coastal videos and obtained an overall accuracy of 95%; Duro et al. [50] used object-oriented classification and decision trees in Spot images to identify vegetal coverings and obtained an overall accuracy of 95%; Karakizi et al. [51] developed and evaluated an object-based classification framework towards the detection of vineyards reaching an overall accuracy rate of 96%.

Therefore, the accuracy value obtained from the classification using the algorithm based on HTM is similar or superior to values obtained by other authors using object-oriented classification and neural networks, which demonstrates that the methodology is appropriate for discriminating agricultural covers in real-time.

Furthermore, as an added benefit, HTM and the methodology developed in this study enable the classification and decision making to be performed in real-time. As we commented before, the operator can control the camera using Wi-Fi. The Wi-Fi computer camera system allows for real-time viewing of everything being seen by the camera, even without taking an image or video. Once the network has been trained and tested, the algorithm classifies the videos, which are received in real-time from the Wi-Fi computer camera system of the DJI Phantom 2 Vision+ (Figure 13).

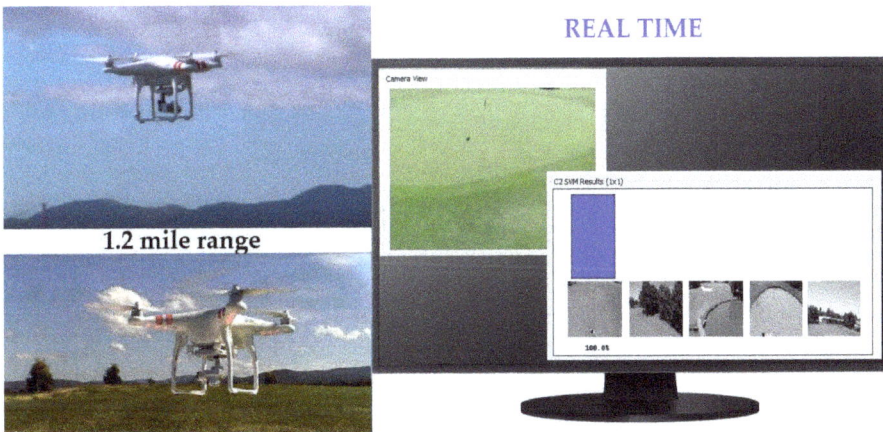

**Figure 13.** Display of the classification in real-time.

In contrast, in the works [1,2,51] of classical classification, post-processing work was required.

## 4. Conclusions

Pattern recognition is an important step in remote sensing applications for precision agriculture. Unmanned aerial vehicles (UAVs) are currently a valuable source of aerial photographs and video images for inspection, surveillance and mapping in precision agriculture purposes. This is because UAVs can be considered in many applications as a low-cost alternative to classical remote sensing. New applications in the real-time domain are expected. The problem of video image analysis taken from an UAV is approached in this paper. A new recognition methodology based on the Hierarchical Temporal Memory (HTM) algorithm for classifying video imagery was proposed and tested for agricultural areas.

As a case study of precision agriculture, golf courses have been considered, namely precision turfgrass. The analyzed patterns to check the accuracy of this case study were turfgrass (see Figure 11) and other uses, namely urban, water, bunker and wood areas.

In the classification process, based on the optimal parameter values obtained during the first stage, a maximum overall accuracy of 98.28% was obtained with a minimum number of misclassified frames. In this case study, a rapid classification of turfgrass, among others, can play an important role to determine water requirements of the different areas in order to plan water use.

Additionally, these results provide evidence that the analysis of UAV-based video images through HTM technology represents a first step for video imagery classification. As a final conclusion, the use of HTM has shown that it is possible to perform, in real-time, pattern recognition of video data images taken from an UAV. This opens new perspectives for precision irrigation methods in order to save water, increase yields and improve water, as well as indicating many possible future research topics.

**Acknowledgments:** The authors would like to thank all of the researchers who kindly shared the codes used in our studies. This research was supported by the Universities of Cordoba and Almeria.

**Author Contributions:** Alberto-Jesús Perea-Moreno and José-Emilio Meroño-De Larriva conceived and designed the experiments; Alberto-Jesús Perea-Moreno, Francisco Manzano-Aguglirao, José-Emilio Meroño-De Larriva, and María-Jesús Aguilera-Ureña performed the experiments; Alberto-Jesús Perea-Moreno, José Emilio Meroño-De Larriva analyzed the data; Alberto-Jesús Perea-Moreno, Francisco Manzano-Agugliaro and María-Jesús Aguilera-Ureña contributed equipment and analysis tools; Alberto-Jesús Perea-Moreno and Francisco Manzano-agugliaro wrote the paper.

**Conflicts of Interest:** The authors declare no conflict of interest.

### Abbreviations

The following abbreviations are used in this manuscript:

| | |
|---|---|
| DTM | Digital terrain model |
| GCP | Ground control point |
| HTM | Hierarchical temporal memory |
| LIDAR | Light detection and ranging |
| NDVI | Normalized difference vegetation index |
| OBIA | Object-based image analysis |
| PA | Precision agriculture |
| SD | Standard deviation |
| SVM | Support vector machine |
| UAV | Unmanned aerial vehicle |

### References

1. Carrow, R.N.; Krum, J.M.; Flitcroft, I.; Cline, V. Precision turfgrass management: Challenges and field applications for mapping turfgrass soil and stress. *Precis. Agric.* **2010**, *11*, 115–134. [CrossRef]
2. Beard, J.B.; Kenna, M.P. *Water Quality and Quantity Issues for Turfgrasses in Urban Landscapes*; CAST Special Publication 27; Council for Agricultural Science and Technology: Ames, IA, USA, 2008.

3. Baghzouz, M.; Devitt, D.A.; Morris, R.L. Evaluating temporal variability in the spatial reflectance response of annual ryegrass to changes in nitrogen applications and leaching fractions. *Int. J. Remote Sens.* **2006**, *27*, 4137–4158. [CrossRef]

4. López-Bellido, R.J.; López-Bellido, L.; Fernández-García, P.; López-Bellido, J.M.; Muñoz-Romero, V.; López-Bellido, P.J.; Calvache, S. Nitrogen remote diagnosis in a creeping bentgrass golf green. *Eur. J. Agron.* **2012**, *37*, 23–30. [CrossRef]

5. Devitt, D.A.; Morris, R.L.; Kopec, D.; Henry, M. Golf course superintendents' attitudes and perceptions toward using reuse water for irrigation in the southwestern United States. *Hortechnology* **2004**, *14*, 1–7.

6. Devitt, D.A. Irrigation management considerations when using reuse water on golfcourses in the arid southwest. In Proceedings of the World Environmental and Water Resources Congress 2008, Ahupua'a, HI, USA, 12–16 May 2008; p. 316.

7. Manzano-Agugliaro, F. *Study for Economic and Environmental Sustainability in the Planning, Construction and Maintenance of Golf Courses in Mediterranean Climate*; National Report for Ministry of the Presidency of the Government of Spain 2012; Grant Project Number 190/UPB10/12; Ministry of Education, Culture and Sport: Madrid, Spain, 2012.

8. Watts, A.C.; Ambrosia, V.G.; Hinkley, E.A. Unmanned Aircraft Systems in Remote Sensing and Scientific Research: Classificacion and Considerations of Use. *Remote Sens.* **2012**, *4*, 1671–1692. [CrossRef]

9. Borra-Serrano, I.; Peña, J.M.; Torres-Sánchez, J.; Mesas-Carrascosa, F.J.; López-Granados, F. Spatial Quality Evaluation of Resampled Unmanned Aerial Vehicle-Imagery for Weed Mapping. *Sensors* **2015**, *15*, 19688–19708. [CrossRef] [PubMed]

10. Feng, Q.; Liu, J.; Gong, J. Urban Flood Mapping Based on Unmanned Aerial Vehicle Remote Sensing and Random Forest Classifier—A Case of Yuyao, China. *Water* **2015**, *7*, 1437–1455. [CrossRef]

11. Vallet, J.; Panissod, F.; Strecha, C.; Tracol, M. Photogrammetric performance of an ultra-light weight swinglet UAV. In Proceedings of the International Archives of the Photogrammetry, Remote Sensing and Spatial Information Sciences, ISPRS Archives, Zurich, Switzerland, 14–16 September 2011; Volume 38 (1C22), pp. 253–258.

12. Candiago, S.; Remondino, F.; De Giglio, M.; Dubbini, M.; Gattelli, M. Evaluating Multispectral Images and Vegetation Indices for Precision Farming Applications from UAV Images. *Remote Sens.* **2015**, *7*, 4026–4047. [CrossRef]

13. Torres-Sánchez, J.; López-Granados, F.; Serrano, N.; Arquero, O.; Pena, J.M. High-Throughput 3-D Monitoring of Agricultural-Tree Plantations with Unmanned Aerial Vehicle (UAV) Technology. *PLoS ONE* **2015**, *10*, e0130479. [CrossRef] [PubMed]

14. Hassan, F.M.; Lim, H.S.; MatJafri, M.Z. Cropcam UAV images for land use/land cover over Penang Island, Malaysia using neural network approach. In Proceedings of the SPIE: Earth Observing Missions and Sensors: Development, Implementation, and Characterization, Incheon, Korea, 13–14 October 2010; Volume 7862, p. 78620.

15. Wang, J.; Wang, L.; Yue, X.; Liu, Y.; Quan, D.; Qu, X.; Gan, H.; Wang, J. Design and test of unmanned aerial vehicle video transfer system based on WiFi. *Trans. Chin. Soc. Agric. Eng.* **2015**, *31*, 47–51.

16. Yang, G.; Li, C.; Yu, H.; Xu, B.; Feng, H.; Gao, L.; Zhu, D. UAV based multi-load remote sensing technologies for wheat breeding information acquirement. *Trans. Chin. Soc. Agric. Eng.* **2015**, *31*, 184–190.

17. Kusnierek, K.; Korsaeth, A. Challenges in using an analog uncooled microbolometer thermal camera to measure crop temperature. *Int. J. Agric. Biol. Eng.* **2014**, *7*, 60–74.

18. Peña-Barragán, J.M.; Ngugi, M.K.; Plant, R.E.; Six, J. Object-based crop identification using multiple vegetation indices, textural features and crop phenology. *Remote Sens. Environ.* **2011**, *115*, 1301–1316. [CrossRef]

19. Berni, J.A.J.; Zarco-Tejada, P.J.; Suarez, L.; Fereres, E. Thermal and Narrow-band Multispectral Remote Sensing for Vegetation Monitoring from an Unmanned Aerial Vehicle. *IEEE Trans. Geosci. Remote Sens.* **2009**, *47*, 722–738. [CrossRef]

20. Zarco-Tejada, P.J.; Berni, J.A.J.; Suárez, L.; Sepulcre-Cantó, G.; Morales, F.; Miller, J.R. Imaging Chlorophyll Fluorescence from an Airborne Narrow-Band Multispectral Camera for Vegetation Stress Detection. *Remote Sens. Environ.* **2009**, *113*, 1262–1275. [CrossRef]

21. Suárez, L.; Zarco-Tejada, P.J.; González-Dugo, V.; Berni, J.A.J.; Sagardoy, R.; Morales, F.; Fereres, E. Detecting water stress effects on fruit quality in orchards with time-series PRI airborne imagery. *Remote Sens. Environ.* **2010**, *114*, 286–298. [CrossRef]

22. Yang, B.; Chen, C. Automatic registration of UAV-borne sequent images and LiDAR data. *ISPRS J. Photogramm.* **2015**, *101*, 262–274. [CrossRef]

23. Zhang, L.Y.; Peng, Z.R.; Li, L.; Wang, H. Road boundary estimation to improve vehicle detection and tracking in UAV video. *J. Cent. South Univ. Technol.* **2014**, *12*, 4732–4741. [CrossRef]

24. Caturegli, L.; Corniglia, M.; Gaetani, M.; Grossi, N.; Magni, S.; Migliazzi, M.; Angelini, L.; Mazzoncini, M.; Silvestri, N.; Fontanelli, M.; et al. Unmanned aerial vehicle to estimate nitrogen status of turfgrasses. *PLoS ONE* **2016**, *11*, e0158268. [CrossRef] [PubMed]

25. Perea-Moreno, A.J.; Meroño-De Larriva, J.E.; Aguilera-Ureña, M.J. Comparison between pixel based and object based methods for analysing historical building façades. *Dyna* **2016**, *91*, 681–687.

26. Ma, L.; Cheng, L.; Han, W.Q.; Zhong, L.S.; Li, M.C. Cultivated land information extraction from high-resolution unmanned aerial vehicle image data. *J. Appl. Remote Sens.* **2014**, *8*, 083673. [CrossRef]

27. Díaz-Varela, R.A.; Zarco-Tejada, P.J.; Angileri, V.; Loudjani, P. Automatic identification of agricultural terraces through object-oriented analysis of very high resolution DSMs and multispectral imagery obtained from an unmanned aerial vehicle. *J. Environ. Manag.* **2014**, *134*, 117–126. [CrossRef] [PubMed]

28. Chung, M.; Detweiler, C.; Hamilton, M.; Higgins, J.; Ore, J.-P.; Thompson, S. Obtaining the Thermal Structure of Lakes from the Air. *Water* **2015**, *7*, 6467–6482. [CrossRef]

29. Feng, Q.; Liu, J.; Gong, J. UAV Remote Sensing for Urban Vegetation Mapping Using Random Forest and Texture Analysis. *Remote Sens.* **2015**, *7*, 1074–1094. [CrossRef]

30. DJI Company. Phantom 2 Vision+ User Manual, V1.6. Available online: http://download.dji-innovations.com/downloads/phantom_2_vision_plus/en/Phantom_2_Vision_Plus_User_Manual_v1.6_en.pdf (accessed on 19 October 2015).

31. Andreopoulos, A.; Tsotsos, J.K. 50 Years of object recognition: Directions forward. *Comput. Vis. Image Underst.* **2013**, *117*, 827–891. [CrossRef]

32. Li, Y.; Wang, S.; Tian, Q.; Ding, X. Feature representation for statistical-learning-based object detection: A review. *Pattern Recogn.* **2015**, *48*, 3542–3559. [CrossRef]

33. Battiato, S.; Farinella, G.M.; Furnari, A.; Puglisi, G.; Snijders, A.; Spiekstra, J. An Integrated System for Vehicle Tracking and Classification. *Expert Syst. Appl.* **2015**, *42*, 7263–7275. [CrossRef]

34. Siddiqi, M.H.; Ali, R.; Khan, A.M.; Kim, E.S.; Kim, G.J.; Lee, S. Facial expression recognition using active contour-based face detection, facial movement-based feature extraction, and non-linear feature selection. *Multimed. Syst.* **2015**, *21*, 541–555. [CrossRef]

35. Hare, S.; Saffari, A.; Torr, P.H. Efficient online structured output learning for keypoint-based object tracking. In Proceedings of the IEEE Conference on Computer Vision and Pattern Recognition (CVPR), Providence, RI, USA, 16–21 June 2012; IEEE: San Diego, CA, USA, 2012; pp. 1894–1901.

36. Li, Y.; Gong, J.; Wang, D.; An, L.; Li, R. Sloping farmland identification using hierarchical classification in the Xi-He region of China. *Int. J. Remote Sens.* **2013**, *34*, 545–562. [CrossRef]

37. Hawkins, J.; Blakeslee, S. *On Intelligence*, 1st ed.; Henry Holt: New York, NY, USA, 2004; p. 296.

38. Hawkins, J.; Ahmad, S.; Dubinsky, D. Hierarchical Temporal Memory including HTM Cortical Learning Algorithms. Available online: http://numenta.org/resources/HTM_CorticalLearningAlgorithms.pdf (accessed on 19 October 2015).

39. Rozado, D.; Rodriguez, F.B.; Varona, P. Extending the bioinspired hierarchical temporal memory paradigm for sign language recognition. *Neurocomputing* **2012**, *79*, 75–86. [CrossRef]

40. Boone, A.; Karnowski, T.P.; Chaum, E.; Giancardo, L.; Li, Y.; Tobin, K.W. Image Processing and Hierarchical Temporal Memories for Automated Retina Analysis. In Proceedings of the Biomedical Sciences and Engineering Conference (BSEC), Oak Ridge, TN, USA, 25–26 May 2010; IEEE: San Diego, CA, USA, 2010.

41. Perea, A.J.; Meroño, J.E.; Aguilera, M.J. Application of Numenta® Hierarchical Temporal Memory for land-use classification. *S. Afr. J. Sci.* **2009**, *105*, 370–375.

42. Perea, A.J.; Meroño, J.E.; Crespo, R.; Aguilera, M.J. Automatic detection of urban areas using the Hierarchical Temporal Memory of Numenta®. *Sci. Res. Essays* **2012**, *7*, 1662–1673.

43. Numenta Inc. *HTM Cortical Learning Algorithms*; Numenta, Inc.: Redwood City, CA, USA, 2010.

44. Hawkins, J.; George, D. *Hierarchical Temporal Memory, Concepts, Theory, and Terminology*; Numenta, Inc.: Redwood City, CA, USA, 2007.

45. Perea, A.J.; Meroño, J.E. Comparison between New Digital Image Classification Methods and Traditional Methods for Land-Cover Mapping. In *Remote Sensing of Land Cover: Principles and Applications*, 1st ed.; Giri, C., Ed.; CRC Press Taylor and Francis Group: Boca Raton, FL, USA, 2012; pp. 1662–1673.

46. George, D.; Jaros, B. *The HTM Learning Algorithms*; Numenta, Inc.: Redwood City, CA, USA, 2007.

47. Weldon, T.P.; Higgins, W.E.; Dunn, D.F. Efficient Gabor filter design for texture segmentation. *Pattern Recogn.* **1996**, *29*, 2005–2015. [CrossRef]

48. Cortes, C.; Vapnik, V. Support-vector networks. *Mach. Learn.* **1995**, *20*, 273–297. [CrossRef]

49. Revollo, N.V.; Delrieux, C.A.; Perillo, G.M.E. Automatic methodology for mapping of coastal zones in video sequences. *Mar. Geol.* **2016**, *381*, 87–101. [CrossRef]

50. Duro, D.C.; Franklin, S.E.; Dubé, M.G. A comparison of pixel-based and object-based image analysis with selected machine learning algorithms for the classification of agricultural landscapes using SPOT-5 HRG imagery. *Remote Sens. Environ.* **2012**, *118*, 259–272. [CrossRef]

51. Karakizi, C.; Oikonomou, M.; Karantzalos, K. Vineyards Detection and Vine Variety Discrimination from very hight Resolution Satellite Data. *Remote Sens.* **2016**, *8*, 235. [CrossRef]

*water*

MDPI

Article

# Comparison of IMERG Level-3 and TMPA 3B42V7 in Estimating Typhoon-Related Heavy Rain

Ren Wang [1,2], Jianyao Chen [1,2,*] and Xianwei Wang [2,*]

[1] Department of Water Resources and Environment, School of Geography and Planning, Sun Yat-sen University, Guangzhou 510275, China; rwang91@foxmail.com

[2] Guangdong Key Laboratory for Urbanization and Geo-simulation, Sun Yat-sen University, Guangzhou 510275, China

\* Correspondence: chenjyao@mail.sysu.edu.cn (J.C.); wangxw8@mail.sysu.edu.cn (X.W.); Tel.: +86-20-8411-5930 (J.C.)

Academic Editor: Ataur Rahman

Received: 17 January 2017; Accepted: 10 April 2017; Published: 22 April 2017

**Abstract:** Typhoon-related heavy rain has unique structures in both time and space, and use of satellite-retrieved products to delineate the structure of heavy rain is especially meaningful for early warning systems and disaster management. This study compares two newly-released satellite products from the Integrated Multi-satellitE Retrievals for Global Precipitation Measurement (IMERG final run) and the Tropical Rainfall Measuring Mission (TRMM) Multi-satellite Precipitation Analysis (TMPA 3B42V7) with daily rainfall observed by ground rain gauges. The comparison is implemented for eight typhoons over the coastal region of China for a two-year period from 2014 to 2015. The results show that all correlation coefficients (CCs) of both IMERG and TMPA for the investigated typhoon events are significant at the 0.01 level, but they tend to underestimate the heavy rainfall, especially around the storm center. The IMERG final run exhibits an overall better performance than TMPA 3B42V7. It is also shown that both products have a better applicability (i.e., a smaller absolute relative bias) when rain intensities are within 20–40 and 80–100 mm/day than those of 40–80 mm/day and larger than 100 mm/day. In space, they generally have the best applicability within the range of 50–100 km away from typhoon tracks, and have the worst applicability beyond the 300-km range. The results are beneficial to understand the errors of satellite data in operational applications, such as storm monitoring and hydrological modeling.

**Keywords:** IMERG final run; TMPA 3B42V7; typhoon; heavy rain; coastal region of China

## 1. Introduction

Heavy rain events have profound impacts on human society, hydrological processes, and natural ecosystems [1,2]. They can adjust river regimes, flood peak, and waterlogging patterns rapidly, and even cause significant losses in human life and social economy [3–5]. A typhoon is a type of cyclone formed in the tropical ocean and often brings heavy rainfall to coastal territory. Typhoon-related heavy rain has unique patterns in both time and space, e.g., it can last from one day to several days, and is dominated by typhoon track, translation speed, atmospheric environment, etc. Therefore, reliable measurements of the heavy rainfall provide essential information to monitor and forecast its changing patterns, which are crucial for early warning systems and disaster management strategies [6–9]. Moreover, detailed regularity of heavy rainfall across different spatiotemporal scales leads to insights about the variability of runoff, which can further contribute to reduce inundation of urban regions [10]. However, rainfall is highly variable in both space and time during a typhoon event, creating significant challenges in its accurate monitoring.

Radar and satellite precipitation measurements provide more homogeneous datasets than ground gauge observations [11,12]. Radar precipitation estimates are constrained by the monitoring scope of radar, while satellite precipitation products have advantages in global coverage and fine resolution [13,14]. There are currently various open access satellite-based precipitation products that could bring valuable scientific and societal benefits. Meanwhile, those products often contain large uncertainties and inevitable errors in different aspects, such as the variability of the precipitation fields and systematic errors [15–17]. These various errors and differential resolutions influence the accuracy of hydrological modeling [18,19]. Evaluation of these products is, therefore, necessary for further understanding of their error characteristics, and is vital to algorithm improvement and subsequent applications.

The Global Precipitation Measurement (GPM) mission, which was launched on 27 February 2014, provides the next generation satellite-based global observations of rainfall and snow. GPM is built upon the success of the Tropical Rainfall Measuring Mission (TRMM). Currently, the accessible Integrated Multi-satellitE Retrievals for GPM (IMERG) Level-3 products have finer resolutions ($0.1° \times 0.1°$, 30 min) than the TRMM precipitation series ($0.25° \times 0.25°$, 3 h), and are valuable for applications over the band of 60° N to 60° S [20]. Meanwhile, the TRMM Multi-satellitE Precipitation Analysis (TMPA) products have provided abundant precipitation information since 1997 [12]. The recently-updated version, 3B42 version 7 (3B42V7), comprises near-real-time and research-grade products with a resolution of $0.25° \times 0.25°$ in space and 3 h in time [21,22].

Scientists, worldwide, have been investigating the error characteristics of the satellite precipitation series at different spatial and temporal scales. Some studies [6,17,23] demonstrated that TMPA 3B42V7 performs better than TMPA 3B42V6, while both products have larger errors in mountainous regions [6,8]. Some other studies [24–30] made comparisons between IMERG and TMPA products, and reported that IMERG generally exhibits an overall better performance than TMPA, especially for estimating heavy and light precipitation. However, IMERG still has room to improve, such as in arid and high-latitude zones [26,29], and mountainous areas [28,31]. These studies provide a large amount of information to understand the applicability of IMERG and TMPA, but tend to focus on annual, seasonal, and monthly scales, not sufficient for short-term heavy rainfall events, especially when associated with the tropical cyclone rainfall system [4,32]. Evaluation of the products for heavy rainfall is a high standard to verify their performance, and is important in practical applications, such as flood forecasting and urban stormwater collection.

The coastal region of China has experienced frequent typhoons and encountered severe socio-economic losses. In general, typhoons strike the coastal region frequently during the period of July to September every year and generate a large number of rainstorm events. There were more than $5.3 billion economic losses per year since 2001, and approximately 34 typhoons landed in this region, which caused 422 losses to life during 2011–2015 [33]. It is of great significance to focus on typhoon heavy rains over the coastal region of China.

Therefore, our motivations are: (1) to evaluate the performance of two recently-released products, i.e., the IMERG final run and TMPA 3B42V7, in estimating typhoon-related heavy rain over the coastal region of China; and (2) to analyze their applicability with respect to different rainfall intensities and ranges away from the typhoon track. This paper is organized as follows: Section 2 describes the study area and datasets; Section 3 describes the statistical methods used in this study; Section 4 presents and analyzes the results; Section 5 discusses the causes of the error characteristics and the comparative results; and Section 6 provides a summary and some concluding remarks.

## 2. Study Area and Datasets

### 2.1. Study Area

The coastal region of China (18°–38.5° N; 104°5′–123° E), which is located at the leading edge of Eurasia and Pacific Ocean, comprises seven provinces and one city, i.e., Shandong, Jiangsu, Zhejiang,

Fujian, Guangdong, Hainan, and Guangxi provinces and Shanghai city (Figure 1). The region is a typical monsoon climate zone, with annual average precipitation ranging from 550 to 2600 mm, and a decreasing pattern from the south to the north. The temporal distribution of precipitation is also uneven, with more than 70% of the annual rainfall concentrating in summer and autumn, when rainfall is primarily controlled by summer wet monsoons and typhoons. Owing to its rapid urbanization and topographic feature, i.e., the eastern areas are mainly plains and rivers downstream where most of the large cities are located, therefore, the risk of flooding is very high over the developed areas. In addition, the coastal region of China has a well-developed economy and dense distribution of cities and people. According to the 2015 China Statistical Yearbook, the region possessed a total gross domestic product (GDP) of $4.48 trillion, accounting for 48% of the national GDP, and had a total resident population of 459 million, accounting for 34% of the country's total population.

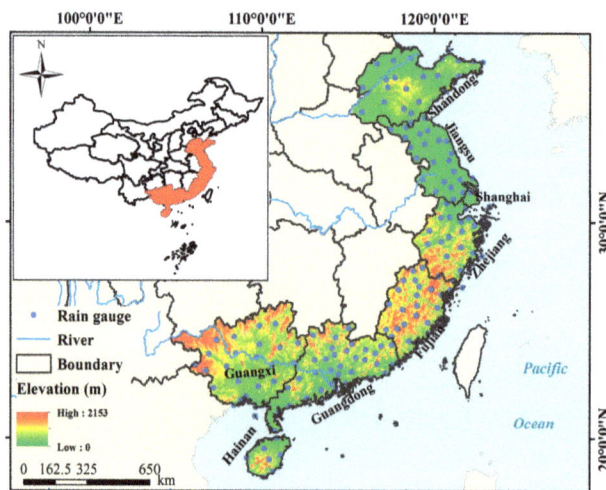

**Figure 1.** Study area and the spatial distribution of rain gauges.

*2.2. Typhoon Events*

Eight typhoon events, which landed in the coastal region of China during the two-year period of 2014–2015, are investigated in this study. The study period of 2014 to 2015 is constrained by the availability of the IMERG final run data. Additionally, the investigated typhoons are different in magnitude, typhoon track, duration and the affected geographical areas. The basic information of typhoon events and their storm tracks are obtained from China Typhoon Online [34] and National Meteorological Center [35]. For the convenience of making comparisons, the typhoon events are further divided into two groups according to the geographical areas where they made landfall and their moving directions. Rammasun, Mujigae, Kalmaegi, and Linfa, which made landfall in the southern areas (Guangdong or Hainan province) and moved to the south, are divided into Group I. Chon-hom, Matmo, Soudelor, and Dujuan, which made landfall in the eastern areas (Fujian or Zhejiang province) and moved to the north, are divided into group II. Matmo and Soudelor are two stronger typhoons among them and have almost impacted the entire coastal region. The basic information of the eight investigated typhoons is listed in Table 1.

**Table 1.** Basic information of the investigated typhoon events over the coastal region of China.

| Group | Typhoon Event | Period | Mainly Affected Province (City) | Number of Investigated Station | Maximum Daily Rainfall |
|---|---|---|---|---|---|
| Group I | Rammasun | 18–19 July 2014 | Guangdong, Guangxi, Hainan | 55 | 303.6 mm |
| | Mujigae | 4–5 October 2015 | Hainan, Guangdong, Guangxi | 58 | 192.9 mm |
| | Kalmaegi | 16–17 September 2014 | Hainan, Guangdong, Guangxi | 67 | 296.5 mm |
| | Linfa | 9–10 July 2015 | Guangdong, Fujian | 39 | 158.8 mm |
| Group II | Chon-hom | 11–12 July 2015 | Zhejiang, Jiangsu, Fujian, Shanghai | 37 | 267.7 mm |
| | Matmo | 23–25 July 2014 | Fujian, Guangdong, Jiangsu, Shandong | 98 | 238.3 mm |
| | Soudelor | 8–10 August 2015 | Fujian, Zhejiang, Jiangsu, Guangdong | 104 | 232.1 mm |
| | Dujuan | 28–30 September 2015 | Fujian, Zhejiang, Jiangsu | 80 | 170.9 mm |

## 2.3. Gauge Observations

Daily rain gauge observations are collected from the National Meteorological Information Center of the China Meteorological Administration (CMA). There are 165 observation stations in total over the study area (Figure 1) with the gauge density of 1.734 stations per $10^4$ km$^2$. Regarding the data quality control, the dataset has passed homogeneity assessments through the Standard Normal Homogeneity Test method [36]. Non-uniform stations, such as "Shaoguang" "Qingyuan", and "Shangchuandao", have been corrected using the ratio method by CMA. Few missing records, which were amended by the CMA, have been replaced by the mean value of adjacent dates. This study mainly analyzes these stations which had total rainfall larger than 10 mm during a typhoon event and have substantial spatial relation with the typhoon track. Thus, the 10 mm rainfall depth threshold [37], which refers to accumulated rainfall from gauge observations, is applied to screen out the light rainfall since the focus of this study is to investigate the performance of both latest satellite precipitation products for heavy rainfall. In addition, some stations with total rainfall less than 10 mm are still retained if they are within a 150-km range of the typhoon track. One and two stations with rainfall <10 mm have been retained for the Rammasun and Linfa events, respectively.

## 2.4. TMPA 3B42V7

The latest post real-time 3B42V7 precipitation product of TMPA is used in this study. It integrates various satellite microwave radar data, including that from TRMM Microwave Image (TMI), Special Sensor Microwave Image (SSMI), Special Sensor Microwave Image/Sounder (SSMIS), Advanced Microwave Scanning Radiometer-EOS (AM-SR-E), Advanced Microwave Sounding Unit-B (AMSU-B), and Microwave Humidity Sounder (MHS) [20]. In addition, the 3B42V7 version combines the ground rain gauge products of the Global Precipitation Climatology Center (GPCC) [21]. The improved 3B42V7 data (0.25° × 0.25°, 3 h) are collected from the Precipitation Measurement Missions website [38]. This study utilized all of the TMPA 3B42V7 data, in HDF format, during the period of the typhoons. The unit of the precipitation field is mm per hour, which refers to the precipitation rate. The three-hour 3B42V7 precipitation is further accumulated into daily and event-total rainfall during the period of each typhoon event, based on ENVI version 5.1 which is developed by Exelis Visual Information Solutions company in the United States, and MATLAB R2015a which is developed by MathWorks company in Natick, Massachusetts, USA. The precipitation in a grid that corresponds to the ground gauges can be extracted by ArcMap 10.1 which is provided by Environmental Systems Research Institute in RedLands, California, USA.

## 2.5. IMERG Final Run

The IMERG final run-calibrated precipitation data are analyzed in this study. The geophysical parameters of IMERG Level-3 have been spatially or temporally re-sampled from Level-1 or Level-2 data, and the Level-3 products include early run, late run, and final run versions. Currently, IMERG employs the 2014 version of the Goddard Profiling Algorithm (GPROF2014) to compute precipitation estimates from all passive microwave (PMW) sensors onboard GPM satellites, and is an improvement compared to TMPA's retrieval algorithm (GPROF2010) [21,39]. The IMERG final run data [38], with a latency of four months, are available from March 2014 to present, so that the investigated typhoon of this study is constrained in the period of March 2014 to September 2015. All of the IMERG final run datasets in HDF5 format, are collected for the periods of typhoon events. The unit of the precipitation field is mm per half hour. Similar to TMPA 3B42V7, the half-hour IMERG precipitation is also accumulated into daily and event-total rainfall.

## 3. Methods

A series of common statistical metrics, which include relative bias (*RB*), mean error (*ME*), mean absolute error (*MAE*), root-mean-squared error (*RMSE*), and Pearson linear correlation coefficient (*CC*), are used to perform the comparative evaluation. *RB* is used to evaluate the errors in a gauge-grid pair while *ME*, *MAE*, and *RMSE* are for regional-scale evaluations [40].

*RB* is the ratio of underestimating or overestimating, in percentage, and it is applicable to reflect errors between the satellite estimates and the corresponding gauge observations. *RB* is calculated for a typhoon event at individual sites by Equation (1):

$$RB_i = \frac{S_i - G_i}{G_i} \times 100 \tag{1}$$

where $i$ is a rain gauge number; $S_i$ represents the satellite precipitation estimates; and $G_i$ represents the gauge observations.

The other metrics (*ME*, *MAE*, and *RMSE*) are used to measure the magnitude of errors for a whole region or sub-region in this study. *MAE* is a statistical metric with absolute value, while *ME* is a metric having positive and negative value, so it can be used to reflect the direction of accumulated errors, i.e., overestimation or underestimation at all stations. Furthermore, *RMSE* is the squared root of errors emphasizing extremes [8]. These statistical metrics can be calculated by Equations (2)–(4):

$$ME = \frac{1}{n} \sum_{i=1}^{n} (S_i - G_i) \tag{2}$$

$$MAE = \frac{1}{n} \sum_{i=1}^{n} |S_i - G_i| \tag{3}$$

$$RMSE = \sqrt{\frac{\sum_{i=1}^{n} (S_i - G_i)^2}{n}} \tag{4}$$

where $n$ represents the number of samples.

In addition, Pearson linear correlation analysis is used to examine the linear agreement of satellite precipitation estimates and rain gauge observations. The *CC* can be obtained by Equation (5) [32].

$$CC = \frac{\sum (x_i - \bar{x})(y_i - \bar{y})}{\sqrt{\sum (x_i - \bar{x})^2 \sum (y_i - \bar{y})^2}} \tag{5}$$

where *CC* is the correlation coefficient, ranging from $-1$ to 1; $\bar{x} = \frac{1}{n} \sum_{i=1}^{n} x_i$ and $\bar{y} = \frac{1}{n} \sum_{i=1}^{n} y_i$ $n$ represents the number of gauge-grid samples; and $x_i$ and $y_i$ represent the grid-scale satellite measurements and rain gauge observations, respectively.

## 4. Results

### 4.1. Characteristics of the Metrics

The statistical metrics (*ME*, *MAE*, RSME, and *CC*), which can reflect the error characteristics of satellite rainfall data over the region during typhoon events, were computed for each gauge-grid pair. The results are presented in Figure 2. With the exception of the products for Mujigae and Rammasun and the IMERG final run for Soudelor, all *ME*s have a negative value, highlighting that both IMERG and TMPA tend to underestimate typhoon heavy rainfall at the regional scale. Moreover, except for the typhoon Dujuan, the absolute *ME* of IMERG is smaller than that of TMPA. Since some samples have large values of positive *RB* (more than 100%), the positive *ME* for Mujigae and Rammasun is highly possible. Meanwhile, there are six typhoon events where the *MAE* of IMERG is slightly larger than that of TMPA. This is also likely to be influenced by some large values of positive *RB* in IMERG. The RSME presents a similar pattern with *MAE*, i.e., RSMEs of IMERG are larger than that of TMPA during the periods of Rammasun, Mujigae, Linfa, Soudelor, and Dujuan. Regarding the correlations between the satellite products and gauge observations, all *CC*s for the investigated typhoon events are significant at the 0.01 confidence level, and there are five typhoon events, i.e., Mujigae, Kalmaegi, Linfa, Soudelor, and Dujuan, that the *CC* of TMPA 3B42V7 is higher than that of the IMERG final run. These larger *CC*s are partially attributed to the smoothing effect of the larger grid size of TMPA (25 km) than IMERG (10 km).

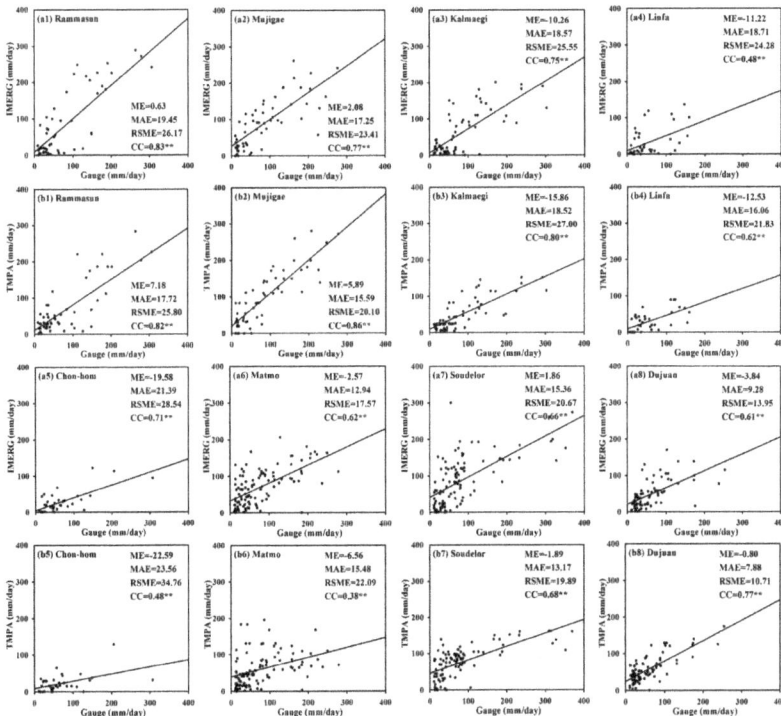

**Figure 2.** Statistical metrics (*ME*, *MAE*, RSME, and *CC*) of (**a1–a8**) the IMERG final run and (**b1–b8**) TMPA 3B42V7 against gauge observations for each typhoon event over the coastal region of China. The units of *ME*, *MAE*, and RSME is mm/day, and the range of *CC* is −1 to 1. ** The correlation is significant at the 0.01 level.

Overall, both satellite products underestimate the heavy rainfall at the regional scale. The IMERG final run can provide a slightly better performance than TMPA 3B42V7 in estimating typhoon heavy rainfall. Taking one of the largest typhoon events, Matmo, for example, the Kriging interpolation method [41] was used to map the spatial patterns of total rainfall from gauge observations, as shown in Figure 3. It is also shown that the IMERG final run and TMPA 3B42V7 have captured similar spatial patterns of total rainfall, but they all tend to underestimate the extreme values of total rainfall in the storm center.

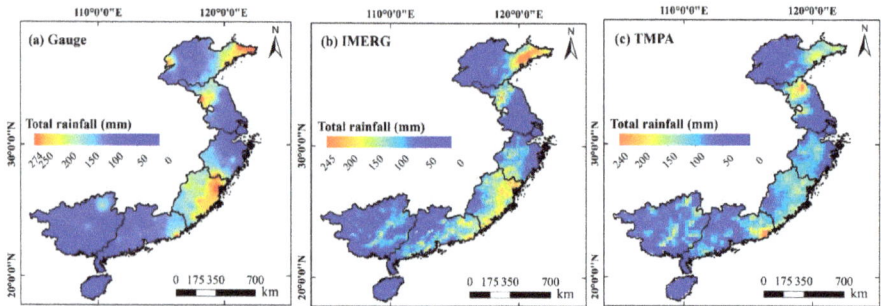

**Figure 3.** Spatial distribution of total rainfall plotted for (**a**) gauge observations, (**b**) the IMERG final run, and (**c**) TMPA 3B42V7 for typhoon Matmo during the period of 23–25 July 2014.

The characteristics of gauge-grid *RB* of over- and underestimated amounts are illustrated in Figures 4 and 5 for the typhoon events of Group I and Group II, respectively. Table 2 summarizes the percentages of overestimate and underestimate that are illustrated in Figures 4 and 5.

**Table 2.** The percentage (%) of gauge-grid pairs of over- and underestimation of IMERG and TMPA against gauge observations illustrated in Figure 2.

| Typhoon Events | | IMERG | | TMPA | |
|---|---|---|---|---|---|
| | | Overestimate | Underestimate | Overestimate | Underestimate |
| Group I | Rammasun | 47.27 | 52.73 | 38.18 | 61.82 |
| | Mujigae | 58.62 | 41.38 | 67.24 | 32.76 |
| | Kalmaegi | 26.87 | 73.13 | 23.88 | 76.12 |
| | Linfa | 25.51 | 74.49 | 17.95 | 82.05 |
| Group II | Chon-hom | 8.11 | 91.89 | 8.11 | 91.89 |
| | Matmo | 39.80 | 60.20 | 35.71 | 64.29 |
| | Soudelor | 55.77 | 44.23 | 50.96 | 49.04 |
| | Dujuan | 38.75 | 61.25 | 46.25 | 53.75 |

The typhoons of group I made landfall in Southern China, such as in Guangdong, Guangxi, or Hainan province. Both IMERG and TMPA underestimated the total rainfall in the storm centers and along the typhoon tracks (Figure 4). For instance, the percentages of IMERG and TMPA's underestimated samples are 74.49% and 82.05% for typhoon Linfa, and 73.13% and 76.12% for typhoon Kalmaegi, respectively (Table 2). However, only one exception exists, for typhoon Mujigae, when both products overestimated the total rainfall in most regions and their percent of overestimated samples are 58.62% (IMERG) and 67.24% (TMPA).

The typhoons of group II made landfall in Eastern China (Fujian and Zhejiang provinces) and moved toward northern areas. Similar to Group I, both the IMERG final run and TMPA 3B42V7 show underestimation for the total rainfall in the storm centers (Figure 5). Both products underestimated the total rainfall of most samples for typhoon Chon-hom, and their underestimated percentages are all 91.89% (Table 2), with different magnitudes of underestimation (Figure 5). This relates to the rain intensity of Chon-hom, as it was a powerful typhoon with a maximum daily rainfall of 267.7 mm (Table 1). Moreover, its track through the coastal areas should also contribute to the largest underestimation. In contrast, for typhoon Soudelor, IMERG and TMPA both displayed larger values than the gauge observations, especially in the east of the typhoon track. Their overestimated samples were 55.77% (IMERG) and 50.96% (TMPA).

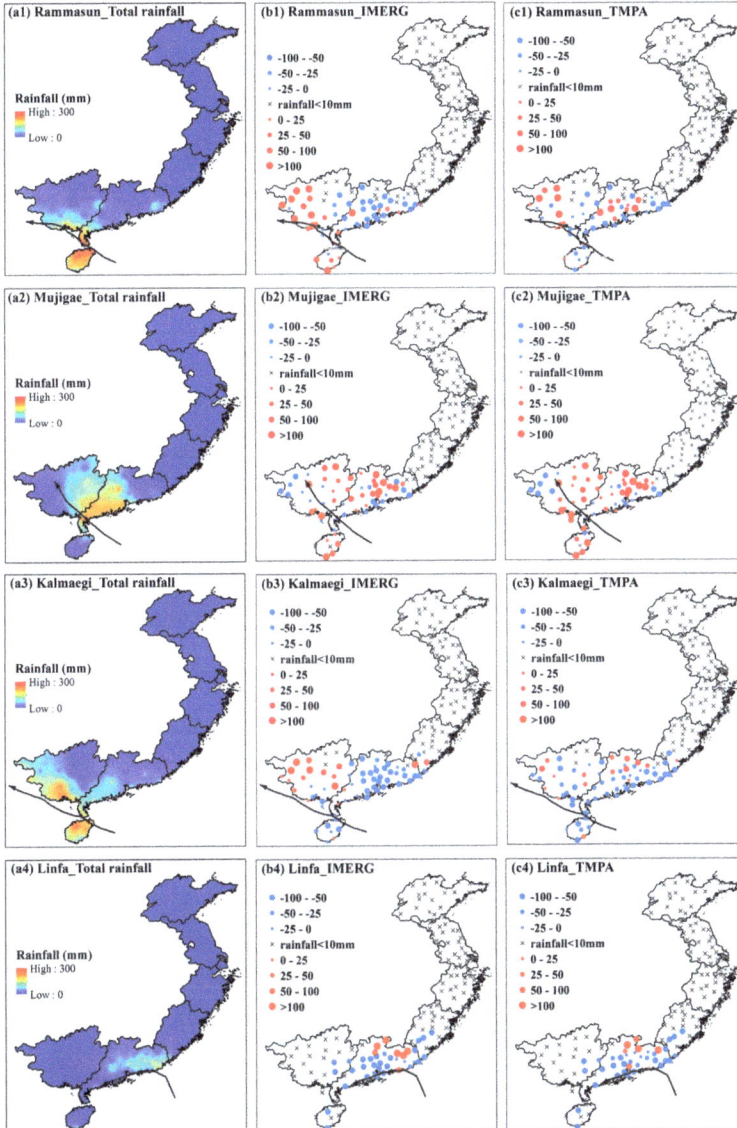

**Figure 4.** Spatial distribution of total rainfall (**a1–a4**), *RB* (%) in (**b1–b4**) IMERG and (**c1–c4**) TMPA for the typhoon events of Group I (Rammasun, Kalmaegi, Linfa, and Mujigae). Dots are scaled according to the magnitude of the overestimation or underestimation. The arrowed lines represent typhoon tracks.

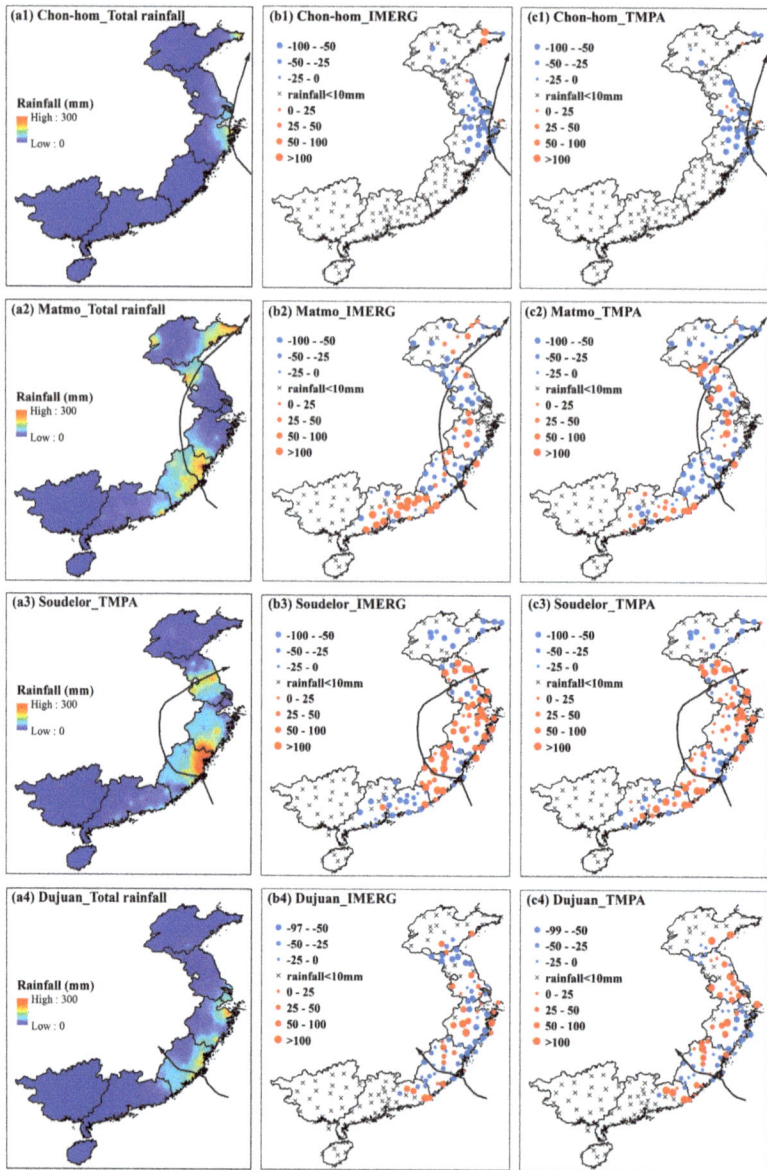

**Figure 5.** Spatial distribution of total rainfall (**a1–a4**), *RB* (%) in (**b1–b4**) IMERG and (**c1–c4**) TMPA for the typhoons of Group II (Chon-hom, Matmo, Soudelor, and Dujuan). Dots are scaled according to the magnitude of the overestimation or underestimation. The arrowed lines represent typhoon tracks.

*4.2. Applicability Associated with Rain Intensity and Typhoon Track*

The above analysis indicates that the performance of IMERG and TMPA are associated with the storm center or rain intensity. Therefore, the applicability of the products associated with rain intensity and typhoon track are further investigated. Figure 6 shows that both IMERG and TMPA have many large overestimate samples (*RB* > 100%) when rain intensity is less than 20 mm/day, and the percentages of IMERG and TMPA's overestimated samples are 44.31% and 54.90%, with mean an *RB* of 31.51% and 45.23% (Table 3). When the rain intensity is larger than 20 mm/day, both IMERG and TMPA capture smaller rainfall than gauge observations, and the magnitude of the underestimation is generally increased with the increased rain intensity. When rain intensities are 20–40 and 80–100 mm/day, the absolute *RB* ( |*RB*| ) are smaller than those of 40–80 and larger than 100 mm/day. Meanwhile, the |*RB*| of IMERG are much smaller than those of TMPA in all rain intensity ranges, which again suggest the better performance of IMERG than TMPA.

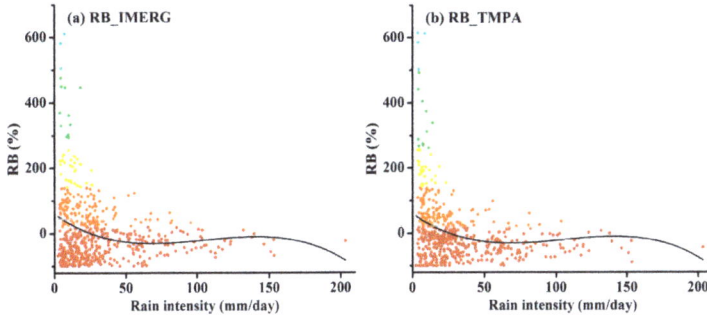

**Figure 6.** Scatter diagram and fitted curve of the rain intensity and *RB* in (**a**) IMERG final run and (**b**) TMPA 3B42V7 for the eight investigated typhoon events. The colors of the dots represent different magnitudes of *RB*.

**Table 3.** The percentage (%) of overestimate (over-per) and underestimate (under-per) and mean *RB* (%) in IMERG and TMPA at different rain intensity (mm/day) for the eight typhoon events.

| Rain Intensity | IMERG | | | TMPA | | |
|---|---|---|---|---|---|---|
| | Over-Per | Under-Per | RB | Over-Per | Under-Per | RB |
| 0–20 | 44.31 | 55.69 | 31.51 | 54.90 | 45.10 | 45.23 |
| 20–40 | 39.86 | 60.14 | −9.23 | 33.11 | 66.89 | −19.58 |
| 40–60 | 33.33 | 66.66 | −12.14 | 21.57 | 78.43 | −23.77 |
| 60–80 | 18.18 | 81.82 | −27.92 | 18.18 | 81.82 | −41.13 |
| 80–100 | 50.00 | 50.00 | −7.87 | 25.00 | 75.00 | −22.50 |
| >100 | 12.50 | 87.50 | −30.17 | 12.50 | 87.50 | −34.81 |

Furthermore, the distance of stations away from the typhoon central track also has an influence on the rain intensity, as well as the performance of satellite rainfall products. As shown in Figure 7, the rain intensity mainly decreases with the increase of the distance from typhoon tracks, which are consistent with storm centers that are around the typhoon tracks (Figures 4 and 5). Meanwhile, the averaged *RB* of both IMERG and TMPA are mainly larger for those stations with a range >300 km than those within smaller ranges. Table 4 shows that IMERG has the smallest absolute mean *RB* in the range of 50–100 km, and has the largest *RB* in a range within 300 km. TMPA also has the smallest absolute mean *RB* in the range of 50–100 km, but its largest value is within the range of 50 km, although *RB* has fluctuations for the eight typhoon events when the range is larger than 300 km. Therefore, IMERG and TMPA mainly have the best applicability within the range of 50–100 km away from the typhoon tracks, and the worst applicability beyond a range of 300 km. This is also illustrated in Figures 4 and 5.

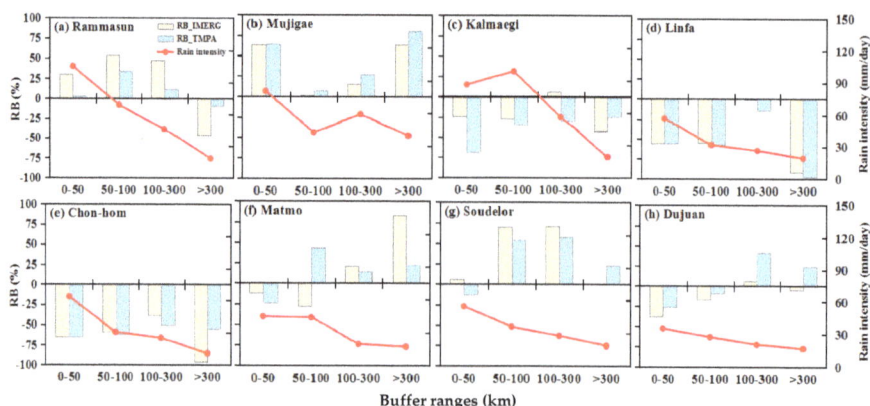

**Figure 7.** Variations of mean *RB* (%) and mean rain intensity (mm/day) within different buffer ranges (km) away from typhoon tracks during the period of (**a**) Rammasun, (**b**) Mujigae, (**c**) Kalmaegi, (**d**) Linfa, (**e**) Chon-hom, (**f**) Matmo, (**g**) Soudelor, and (**h**) Dujuan. The yellow histograms represent IMERG's mean *RBs*, and the blue histograms represent TMPA's mean *RBs*. The red polylines are the changes of mean rain intensity.

**Table 4.** Mean *RB* (%) of IMERG and TMPA within the different buffer ranges away from typhoon tracks.

| Buffer Ranges (km) | IMERG | | | | | | | | Mean |
|---|---|---|---|---|---|---|---|---|---|
| | Group I | | | | Group II | | | | |
| | Rammasun | Mujigae | Kalmaegi | Linfa | Chon-hom | Matmo | Soudelor | Dujuan | |
| <50 | 30.18 | 63.66 | −24.11 | −55.77 | −65.57 | −11.76 | 6.32 | −37.47 | −11.82 |
| 50–100 | 54.36 | 1.31 | −17.26 | −55.49 | −58.78 | −28.33 | 69.95 | −17.47 | −6.46 |
| 100–300 | 47.02 | 15.00 | 6.19 | −1.13 | −38.93 | 21.10 | 70.48 | 6.37 | 15.76 |
| >300 | −46.94 | 63.79 | −42.82 | −91.89 | −96.40 | 82.81 | 0.84 | −4.92 | −16.94 |

| Buffer Ranges (km) | TMPA | | | | | | | | Mean |
|---|---|---|---|---|---|---|---|---|---|
| | Group I | | | | Group II | | | | |
| | Rammasun | Mujigae | Kalmaegi | Linfa | Chon-hom | Matmo | Soudelor | Dujuan | |
| <50 | 2.65 | 63.83 | −69.00 | −55.57 | −65.57 | −23.96 | −13.21 | −26.76 | −23.45 |
| 50–100 | 34.03 | 7.64 | −35.35 | −57.20 | −62.21 | 43.31 | 54.52 | −9.42 | −3.09 |
| 100–300 | 11.50 | 27.12 | −30.00 | −13.73 | −50.51 | 14.50 | 58.09 | 40.88 | 7.23 |
| >300 | −9.30 | 79.40 | −24.80 | −98.11 | −55.34 | 22.23 | 22.23 | 23.90 | −4.97 |

## 5. Discussion

Why do both IMERG and TMPA generally perform better along the typhoon track than farther away from it? One possible explanation is its association with the physical structure of the typhoon and the underlying surface topography of typhoon tracks, both of which can influence the spatial distribution of rainfall intensity [42]. As shown in Figure 7, the mean rain intensity decreases with the increase of buffer ranges. Most storm centers are within the range of 50 km, where both IMERG and TMPA tend to underestimate the heavy rainfall. Meanwhile, when the range is larger than 300 km away from the typhoon track, there is light rain, for which the satellite products show large *RB* and have large uncertainties. Moreover, the impact range of typhoon-related heavy rain is also associated with the magnitude of each typhoon. This applicability range suggested in this study is just a simplified analysis and indicator of the error characteristics of satellite products in estimating typhoon heavy rain.

It has been reported that the current satellite rainfall products have limitations for monitoring typhoon heavy rain [43]. In particular, Chen et al. [44] found that TMPA 3B42V7 is least capable in the coastal region and significantly underestimates the heavy rainfall, which is the primary motivation of this study to investigate the performance of the latest released IMERG in the typhoon-affected coastal region and to compare its performance with TMPA 3B42V7. Our results also confirm this error pattern of both IMERG and TMPA in the coastal region of China. In addition, the daily rain gauge observations are used to validate the total rainfall estimated by both IMERG and TMPA. This is constrained by the availability of the hourly precipitation data in a large region in a timely way. Moreover, the density of the observation gauges in the studied area have impacts on the satellite-based rainfall errors. Thus, higher spatiotemporal density of gauges or gauge-satellite merged products could, potentially, be better in evaluating the errors and to monitor the evolution of heavy rainfall [32,45].

IMERG Level-3 and TMPA 3B42V7 are the emerging satellite precipitation products with relatively high resolutions in time and space, and have the potential to provide more reliable information for flood/drought monitoring, hydrologic modeling, and global climate change study. Previous studies [17,24,25] have demonstrated that the latest products have better performance, but few studies focused on typhoon-related heavy rain events that occur in a short time. That is another objective of this study to evaluate the performance of both products for typhoon rainfall. Overall, IMERG shows a better performance than TMPA.

## 6. Summary

This study compared the performance of the IMERG final run and TMPA 3B42V7 for typhoon heavy rain using ground gauge observations for reference, with focus on eight typhoon events that made landfall in the coastal region of China from July 2014 to October 2015. The main conclusive remarks are as follows:

1. All correlation coefficients (*CCs*) both of IMERG and TMPA for the investigated typhoon events are significant at the 0.01 level, but they tend to underestimate a total amount of heavy rainfall, especially around the storm center.
2. The IMERG final run shows an overall better performance than TMPA 3B42V7.
3. Both IMERG and TMPA exhibit a better performance (i.e., smaller absolute *RB*) when rain intensities are within 20–40 and 80–100 mm/day than those of 40–80 mm/day and larger than 100 mm/day. Meanwhile, both products generally have the best applicability in the range of 50–100 km away from typhoon tracks, and have the worst applicability beyond a 300-km range.
4. It needs to be emphasized that the study lacks physical insights to strengthen the statistical analysis. Future works, which will be devoted to further understand the limits of the applicability and accuracy of such satellite products in monitoring typhoon rainfall, should be focused on the physical process of typhoon rainfall, with consideration for the moving speed and direction of the typhoon, and the underlying topography.

**Acknowledgments:** This research was supported by the China National Key Technology R&D Program of 2015 (2015BAK11B02), the National Natural Science Foundation of China (41371055, 41611140112, and 41371404), and the Water Resource Science and Technology Innovation Program of Guangdong Province (#2016–19). We would like to express our appreciation to the Water Editorial Office, and two anonymous reviewers for their valuable comments and constructive suggestions.

**Author Contributions:** Ren Wang performed the experiments, processed the data, analyzed the results, and wrote the manuscript; Jianyao Chen conceived the experiments and contributed to the design of the manuscript; and Xingwei Wang improved the design of the experiments and the analysis of the results, and contributed to the writing of the manuscript.

**Conflicts of Interest:** The authors declare no conflict of interest.

# References

1. Easterling, D.R.; Meehl, G.A.; Parmesan, C.; Changnon, S.A.; Karl, T.R.; Mearns, L.Q. Climate extremes: Observations, modeling, and impacts. *Science* **2000**, *289*, 2068–2074. [CrossRef] [PubMed]
2. Ban, N.; Schmidli, J.; Schar, C. Heavy precipitation in a changing climate: Does short-term summer precipitation increase faster? *Geophys. Res. Lett.* **2015**, *42*, 1165–1172. [CrossRef]
3. Yang, T.H.; Yang, S.C.; Ho, J.Y.; Lin, G.F.; Hwang, G.D.; Lee, C.S. Flash flood warnings using the ensemble precipitation forecasting technique: A case study on forecasting floods in Taiwan caused by typhoons. *J. Hydrol.* **2015**, *520*, 367–378. [CrossRef]
4. Habib, E.; Henschke, A.; Adler, R.F. Evaluation of TMPA satellite-based research and real-time rainfall estimates during six tropical-related heavy rainfall events over Louisiana, USA. *Atmos. Res.* **2009**, *94*, 373–388. [CrossRef]
5. Wang, X.; Xie, H.; Mazari, N.; Zeitler, J.; Sharif, H.; Hammond, W. Evaluation of a near-real time NEXRAD DSP product in evolution of heavy rain events on the Upper Guadalupe River Basin, Texas. *J. Hydroinform.* **2013**, *15*, 464–485. [CrossRef]
6. Prakash, S.; Mitra, A.K.; AghaKouchak, A.; Pai, D.S. Error characterization of TRMM Multisatellite Precipitation Analysis (TMPA-3B42) products over India for different seasons. *J. Hydrol.* **2015**, *529*, 1302–1312. [CrossRef]
7. Mehran, A.; AghaKouchak, A. Capabilities of satellite precipitation datasets to estimate heavy precipitation rates at different temporal accumulations. *Hydrol. Process.* **2014**, *28*, 2262–2270. [CrossRef]
8. Bharti, V.; Singh, C. Evaluation of error in TRMM 3B42V7 precipitation estimates over the Himalayan region. *J. Geophys. Res.* **2015**, *120*, 12458–12473. [CrossRef]
9. Collischonn, B.; Collischonn, W.; Tucci, C.E.M. Daily hydrological modeling in the Amazon basin using TRMM rainfall estimates. *J. Hydrol.* **2008**, *360*, 207–216. [CrossRef]
10. Hyun, J.Y.; Rockaway, T.D.; French, M.N. Ground-level rainfall variation in Jefferson County, Kentucky. *J. Hydrol. Eng.* **2016**, *21*, 05016029. [CrossRef]
11. Kitzmiller, D.; Miller, D.; Fulton, R.; Feng, D. Radar and multisensor precipitation estimation techniques in National Weather Service hydrologic operations. *J. Hydrol. Eng.* **2013**, *18*, 133–142. [CrossRef]
12. Huffman, G.J.; Adler, R.F.; Bolvin, D.T.; Gu, G.; Nelkin, E.J.; Bowman, K.P.; Hong, Y.; Stocker, E.F.; Wolff, D.B. The TRMM multisatellite precipitation analysis (TMPA): Quasi-global, multiyear, combined–sensor precipitation estimates at fine scales. *J. Hydrometeorol.* **2007**, *8*, 38–55. [CrossRef]
13. Kidd, C.; Huffman, G. Global precipitation measurement. *Meteor. Appl.* **2011**, *18*, 334–353. [CrossRef]
14. Hou, A.Y.; Skofronick-Jackson, G.; Kummerow, C.D.; Shepherd, J.M. Global precipitation measurement. In *Precipitation: Advances in Measurement, Estimation and Prediction*; Silas, M., Ed.; Springer: Belin/Heidelberg, Germany, 2008; pp. 131–169.
15. AghaKouchak, A.; Mehran, A.; Norouzi, H.; Behrangi, A. Systematic and random error components in satellite precipitation data sets. *Geophys. Res. Lett.* **2012**, *39*, L09406. [CrossRef]
16. AghaKouchak, A.; Nasrollahi, N.; Habib, E. Accounting for uncertainties of the TRMM satellite estimates. *Remote Sens.* **2009**, *1*, 606–619. [CrossRef]
17. Chen, S.; Hong, Y.; Gourley, J.J.; Huffman, G.J.; Tian, Y.; Cao, Q.; Yong, B.; Kirstetter, P.E.; Hu, J.; Hardy, J.; et al. Evaluation of the successive V6 and V7 TRMM multisatellite precipitation analysis over the Continental United States. *Water Resour. Res.* **2013**, *49*, 8174–8186. [CrossRef]
18. Goldstein, A.; Foti, R.; Montalto, F. Effect of spatial resolution in modeling stormwater runoff for an urban block. *J. Hydrol. Eng.* **2016**, *21*, 06016009. [CrossRef]
19. Liao, S.L.; Li, G.; Sun, Q.Y.; Li, Z.F. Real-time correction of antecedent precipitation for the Xinanjiang model using the genetic algorithm. *J. Hydroinform.* **2016**, *18*, 803–815. [CrossRef]
20. Huffman, G.J.; Bolvin, D.T.; Braithwaite, D.; Hsu, K.; Joyce, R.; Kidd, C.; Nelkin, E.J.; Xie, P. *Algorithm Theoretical Basis Document (ATBD) Version 4.4 for the NASA Global Precipitation Measurement (GPM) Integrated Multi-Satellite Retrievals for GPM (IMERG)*; NASA: Greenbelt, MD, USA, 2014; pp. 1–30.
21. Huffman, G.J.; Bolvin, D.T. TRMM and Other Data Precipitation Data Set Documentation. Available online: ftp://precip.gsfc.nasa.gov/pub/trmmdocs/3B42_3B43_doc.pdf (accessed on 13 April 2017).

22. Mitra, A.K.; Momin, I.M.; Rajagopal, E.N.; Basu, S.; Rajeevan, M.N.; Krishnamurti, T.N. Gridded daily Indian monsoon rainfall for 14 seasons: Merged TRMM and IMD gauge analyzed values. *J. Earth Syst. Sci.* **2013**, *122*, 1173–1182. [CrossRef]

23. Chen, S.; Hong, Y.; Cao, Q.; Gourley, J.J.; Kirstetter, P.E.; Yong, B.; Tian, Y.; Zhang, Z.; Shen, Y.; Hu, J.; et al. Similarity and difference of the two successive V6 and V7 TRMM multisatellite precipitation analysis performance over China. *J. Geophys. Res.* **2013**, *118*, 13060–13074. [CrossRef]

24. Tang, G.; Ma, Y.; Long, D.; Zhong, L.; Hong, Y. Evaluation of GPM Day-1 IMERG and TMPA Version-7 legacy products over Mainland China at multiple spatiotemporal scales. *J. Hydrol.* **2016**, *533*, 152–167. [CrossRef]

25. Tang, G.; Zeng, Z.; Long, D.; Guo, X. Statistical and hydrological comparisons between TRMM and GPM Level-3 products over a midlatitude basin: Is Day-1 IMERG a good successor for TMPA 3B42V7? *J. Hydrometeorol.* **2016**, *17*, 121–137. [CrossRef]

26. Sharifi, E.; Steinacker, R.; Saghafian, B. Assessment of GPM-IMERG and other precipitation products against gauge data under different topographic and climatic conditions in Iran: Preliminary results. *Remote Sens.* **2016**, *8*, 135. [CrossRef]

27. Guo, H.; Chen, S.; Bao, A.; Behrangi, A.; Hong, Y.; Ndayisaba, F.; Hu, J.; Stepanian, P.M. Early assessment of integrated multi-satellite retrievals for global precipitation measurement over China. *Atmos. Res.* **2016**, *176–177*, 121–133. [CrossRef]

28. Prakash, S.; Mitra, A.K.; Pai, D.S.; AghaKouchak, A. From TRMM to GPM: How well can heavy rainfall be detected from space? *Adv. Water Resour.* **2016**, *88*, 1–7. [CrossRef]

29. Chen, F.; Li, X. Evaluation of IMERG and TRMM 3B43 monthly precipitation products over Mainland China. *Remote Sens.* **2016**, *8*, 472. [CrossRef]

30. Liu, Z. Comparison of Integrated Multisatellite Retrievals for GPM (IMERG) and TRMM Multisatellite Precipitation Analysis (TMPA) monthly precipitation products: Initial results. *J. Hydrometeorol.* **2016**, *17*, 777–790. [CrossRef]

31. Sahlu, D.; Nikolopoulos, E.; Moges, S.; Anagnostou, E.; Hailu, D. First evaluation of the Day-1 IMERG over the upper Blue Nile Basin. *J. Hydrometeorol.* **2016**, *17*, 2875–2882. [CrossRef]

32. Wang, D.; Wang, X.; Liu, L.; Wang, D.; Huang, H.; Pan, C. Evaluation of CMPA precipitation estimate in the evolution of typhoon-related storm rainfall in Guangdong, China. *J. Hydroinform.* **2016**, *18*, 1055–1068. [CrossRef]

33. Kang, B. Statistical analysis of typhoon events in China. *China Flood Drought Manag.* **2016**, *26*, 36–40. (In Chinese).

34. China Typhoon Online. Available online: http://typhoon.weather.com.cn/ (accessed on 13 April 2017).

35. National Meteorological Center: Typhoon and Marine Weather Monitoring and Warning. Available online: http://typhoon.nmc.cn/web.html (accessed on 13 April 2017).

36. Toreti, A.; Kuglitsch, F.; Xoplaki, E.; Della-Marta, P.; Aguilar, E.; Prohom, M.; Luterbacher, J. A note on the use of the standard normal homogeneity test to detect inhomogeneities in climatic time series. *Int. J. Climatol.* **2011**, *31*, 630–632. [CrossRef]

37. Hirpa, F.A.; Gebremichael, M.; Hopson, T. Evaluation of high-resolution satellite precipitation products over very complex terrain in Ethiopia. *J. Appl. Meteorol. Clim.* **2010**, *49*, 1044–1051. [CrossRef]

38. NASA's Precipitation Measurement Missions. Available online: http://pmm.nasa.gov/data-access (accessed on 13 April 2017).

39. Skofronick-Jackson, G.; Huffman, G.; Stocker, E.; Walter, P. Successes with the Global Precipitation Measurement (GPM) Mission. In Proceedings of the Geoscience and Remote Sensing Symposium (IGARSS), Beijing, China, 10–15 July 2016; pp. 3910–3912.

40. Nastos, P.T.; Kapsomenakis, J.; Philandras, K.M. Evaluation of the TRMM 3B43 gridded precipitation estimates over Greece. *Atmos. Res.* **2016**, *169*, 497–514. [CrossRef]

41. Oliver, M.A.; Webster, R. Kriging: A method of interpolation for geographical information systems. *Inter. J. Geogr. Inf. Syst.* **1990**, *4*, 313–332. [CrossRef]

42. Huang, J.C.; Yu, C.K.; Lee, J.Y.; Cheng, L.W.; Lee, T.Y.; Kao, S.J. Linking typhoon tracks and spatial rainfall patterns for improving flood lead time predictions over a mesoscale mountainous watershed. *Water Resour. Res.* **2012**, *48*, W09540. [CrossRef]

43. Chen, Y.; Ebert, E.E.; Walsh, K.J.E.; Davidson, N.E. Evaluation of TRMM 3B42 precipitation estimates of tropical cyclone rainfall using PACRAIN data. *J. Geophys. Res.* **2013**, *118*, 2184–2196. [CrossRef]

44. Chen, S.; Hong, Y.; Cao, Q.; Kirstetter, P.E.; Gourley, J.J.; Qi, Y.; Zhang, J.; Howard, K.; Hu, J.; Wang, J. Performance evaluation of radar and satellite rainfalls for Typhoon Morakot over Taiwan: Are remote-sensing products ready for gauge denial scenario of extreme events? *J. Hydrol.* **2013**, *506*, 4–13. [CrossRef]
45. Shen, Y.; Zhao, P.; Pan, Y.; Yu, J. A high spatiotemporal gauge-satellite merged precipitation analysis over China. *J. Geophys. Res.* **2014**, *119*, 3063–3075. [CrossRef]

*water*

MDPI

*Article*

# Improvement to the Huff Curve for Design Storms and Urban Flooding Simulations in Guangzhou, China

Cuilin Pan [1], Xianwei Wang [1,*], Lin Liu [1,2,*], Huabing Huang [1,*] and Dashan Wang [1]

[1]   Center of Integrated Geographic Information Analysis, School of Geography and Planning, and Guangdong
     Key Laboratory for Urbanization and Geo-simulation, Sun Yat-sen University, Guangzhou 510275, China;
     pancuil@mail2.sysu.edu.cn (C.P.); wangdash@mail2.sysu.edu.cn (D.W.)
[2]   Department of Geography, University of Cincinnati, Cincinnati, OH 45221, USA
*   Correspondence: xianweiw@vip.qq.com (X.W.); liulin2@mail.sysu.edu.cn (L.L.);
     huanghb7@mail.sysu.edu.cn (H.H.); Tel.: +86-20-84114623 (X.W., L.L. & H.H.)

Academic Editor: Hongjie Xie
Received: 24 March 2017; Accepted: 5 June 2017; Published: 8 June 2017

**Abstract:** The storm hyetograph is critical in drainage design since it determines the peak flooding volume in a catchment and the corresponding drainage capacity demand for a return period. This study firstly compares the common design storms such as the Chicago, Huff, and Triangular curves employed to represent the storm hyetographs in the metropolitan area of Guangzhou using minute-interval rainfall data during 2008–2012. These common design storms cannot satisfactorily represent the storm hyetographs in sub-tropic areas of Guangzhou. The normalized time of peak rainfall is at $33 \pm 5\%$ for all storms in the Tianhe and Panyu districts, and most storms (84%) are in the 1st and 2nd quartiles. The Huff curves are further improved by separately describing the rising and falling limbs instead of classifying all storms into four quartiles. The optimal time intervals are 1–5 min for deriving a practical urban design storm, especially for short-duration and intense storms in Guangzhou. Compared to the 71 observed storm hyetographs, the Improved Huff curves have smaller RMSE and higher NSE values (6.43, 0.66) than those of the original Huff (6.62, 0.63), Triangular (7.38, 0.55), and Chicago (7.57, 0.54) curves. The mean relative difference of peak flooding volume simulated with SWMM using the Improved Huff curve as the input is only 2%, −6%, and 8% of those simulated by observed rainfall at the three catchments, respectively. In contrast, those simulated by the original Huff (−12%, −43%, −16%), Triangular (−22%, −62%, −38%), and Chicago curves (−17%, −19%, −21%) are much smaller and greatly underestimate the peak flooding volume. The Improved Huff curve has great potential in storm water management such as flooding risk mapping and drainage facility design, after further validation.

**Keywords:** Huff curve; design storm; urban flooding; SWMM

---

## 1. Introduction

   The storm hyetograph is crucial not only for urban storm water management, but also for the catchment hydrology in general [1–4]. Given a total rainfall depth and duration for a certain return period, the storm hyetograph determines the peak flow/time and the drainage capability demand in a catchment [5,6]. Therefore, the accurate representation of the storm hyetograph is significant for designing suitable drainage facilities and reducing the flooding risk in an urban catchment.

   Urban flooding events have frequently occurred and increased in many cities worldwide in recent years in the context of global warming [7–11]. China faces even more severe challenges in urban flooding due to its dramatic urbanization and relatively poor storm water management [12]. In order to mitigate the impacts of urban flooding, the Chinese government has issued a series of regulations on urban storm water management [13–15]. Several design storms are recommended for drainage facility design in those regulations, including the Triangular curve [16], the Chicago curve [17], and the Soil Conservation Service (SCS) curve [18].

The Triangular curve was developed by Yen & Chow [16] for drainage design in a small catchment and is widely used in natural watersheds and small urban catchments [19,20]. It is a one-parameter model that is estimated by preserving the first moment of the rainfall depth [19]. The Chicago curve is constructed by fitting the equations of the intensity-duration-frequency curves given the total rainfall depth and duration for a return period [17–21]. It is often applied in sewer and flooding drainage design [22,23]. The SCS curve is a dimensionless hyetograph/hydrograph with a single parameter [24]. It was originally developed for designing safe water storage facilities in agricultural applications and has been widely applied in various situations, especially for long duration storm events of 6, 12, 24 h, and even longer [4,25–27].

The Huff curve is another popular design hyetograph for characterizing the temporal distributions of rainfall depth in an area [19]. Like the SCS curve, it is also a dimensionless cumulative hyetograph with specified probabilities of occurrence [28] and is widely utilized as a design storm, downscaling analysis of rainfall depth data, and inputs to rainfall-runoff models for drainage design [5,29]. In practice, historic storm data are first classified into four quartiles according to the normalized time of peak rainfall, and a series of Huff curves are then developed at different probabilities within each quartile [29].

The above design storm curves are derived for the entire storms using historic storm data with hourly rainfall accumulation in most cases, except for the Chicago curve. The time of peak rainfall has a critical influence on the classification of the hyetograph [30]. Separating a storm into the rising and falling limbs could better represent the rainfall hyetograph [31]. The Chicago curve uses two formulas to represent the rising and falling limbs, where the rainfall intensity exponentially decreases on both sides of the peak rainfall [32].

Most storms are less than three hours in the Guangzhou Metropolitan areas in South China, identified according to the criteria reported in the following Section 3.1. The partial reason for this frequent flooding is that the pipe system underestimates the peak runoff, which is simulated by using design storms like Chicago and Triangular curves for a given return period. Those design storms are usually derived from hourly rainfall data and thus underestimate the rainfall intensity. As a result of this, they do not satisfactorily represent the real storm hyetograph.

Therefore, the primary objective of this study is to develop a suitable storm hyetograph by improving the Huff model through studying the rising and falling limbs separately based on minute-interval rainfall depth data in the Guangzhou metropolitan areas. A secondary objective is to investigate the sensitivity of the design storm to time intervals of rainfall depth and to offer a suggestion for selecting optimal time intervals of rainfall depth data with urban design storm research. The Improved Huff curve is then validated and compared to the Huff curve, Triangular curve, and Chicago curve by using in situ measurements of storm events and by applying for flooding volume simulation as inputs to the Storm Water Management Model (SWMM) in three small urban catchments.

## 2. Study Area and Data

### 2.1. Study Area

The study area is located in the Guangzhou Metropolitan areas in South China (Figure 1). It has a sub-tropic climate controlled by the East Asian Monsoon, more specifically the South China Sea Monsoon. It has warm and wet summers and dry winters, with a mean annual air temperature of 22 °C and annual precipitation of 1700 mm [33]. Over 80% of the annual precipitation falls during the rainy season from April to September [12,34]. This area is well known for its dense interlocking river network and has gone through dramatic urbanization over the past 20 years; the impervious land changed from 12,998 ha in 1990 to 59,911 ha in 2009 [35,36]. In the Tianhe (Site/Rain gauge 2) and Panyu (Site/Rain gauges 1, 3–6) Districts, the imperious land ratio (based on Landsat images) increased from 16% to 71% and from 2% to 40% from 1990 to 2013, respectively. However, most (83%) of the drainage pipes adopted the design standards for storms of a one-year return period, and only 9% of the pipes adopted these for a two-year return period [37]. It is a very low return period. In many countries, a 25-year return period is adopted. Hence, the streets in the city of Guangzhou were frequently inundated.

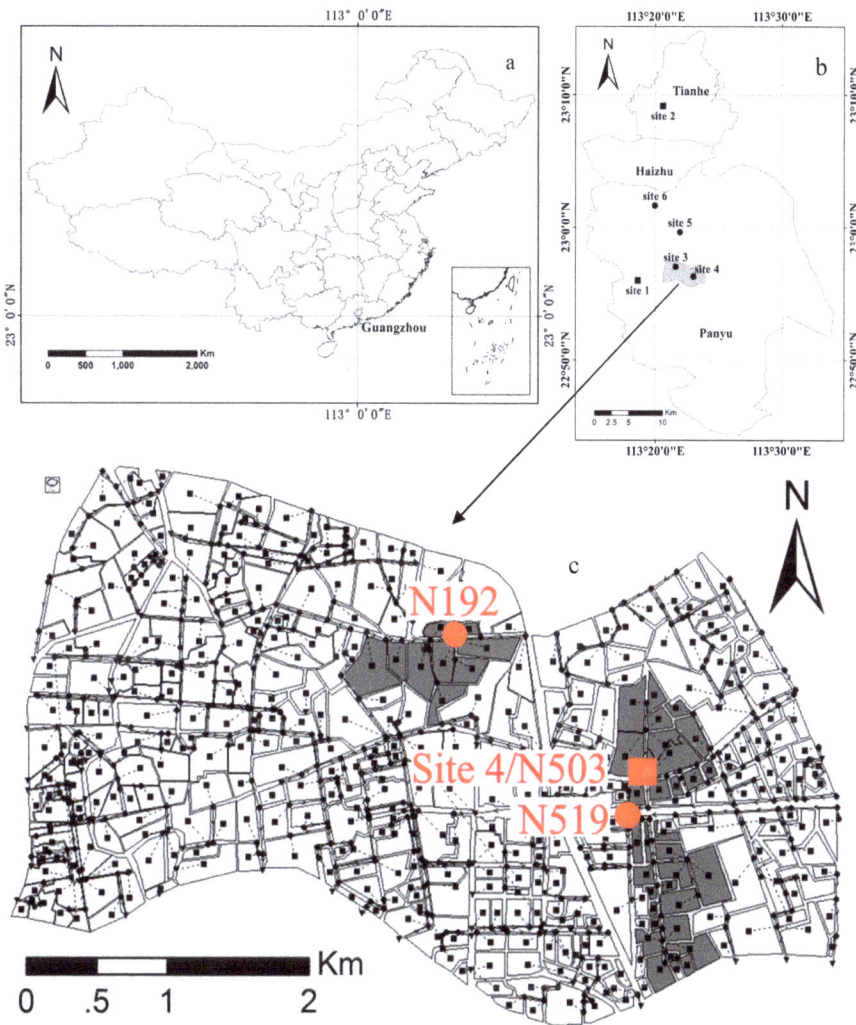

**Figure 1.** The locations of Guangzhou (**a**); meteorological sites (**b**); and the three selected nodes and their catchment area/ modeling areas by SWMM in Panyu District (**c**). Street flooding water depth is recorded near Site 4/Node 503 by a wireless electronic water depth meter.

*2.2. Rainfall Depth Data*

This study uses two types of rainfall data from six automatic gauges. The first type is from national standard meteorological sites (Sites 1 and 2) of China, where rainfall depth data are automatically recorded at one-minute intervals with a precision of 0.1 mm. Site 2 is within the downtown area of the Tianhe District, whereas Site 1 is in the Panyu District, a sub-urban area. The two sites are 25 km apart. Five-year rainfall data from 2008 to 2012 are obtained to develop and validate the coefficients of the design storms at Sites 2 and 1, respectively. The other four sites (Sites 3–6) were set up at the Panyu District in the summer of 2014 by our research team. Rainfall data are recorded at one-minute intervals with a precision of 1 mm. Meanwhile, an electronic water depth meter was also set up to record the street water depth at Site 4 (Node 503), where flooding inundation has occurred several times each year recently. The storm rainfall and water depth data recorded at Site 4/Node503 is used to optimize the SWMM model and validate the design storms.

## 3. Methodology

The rainfall data at Site 2 are used to develop the design storms, including the Huff curve, Improved Huff curve, and the Triangular curve. The storm data at Site 1 and Sites 3–6 are used to validate the design storms. These design storms are further applied as inputs to the SWMM model to simulate the flooding volume, which is compared with that from in situ measurements at Node 503. The detailed procedure and methods are arranged below.

*3.1. Storm Events*

In this study, storm events are identified based on the following criteria: (a) rainfall duration > 20 min [31]; (b) rainfall depth in a one-hour moving window > 20 mm [13]; (c) storm event separation, hourly rainfall depth < 1 mm [29]. According to these criteria, 175 storms at Sites 1 (71) and 2 (104) during the five years from 2008 to 2012 were extracted and are summarized in Table 1.

**Table 1.** The number of storms in different durations at two sites for the period 2008–2012.

| Rain Gauges | Duration (h) | | | | | | | |
|---|---|---|---|---|---|---|---|---|
| | <1 | 1–2 | 2–3 | 3–4 | 4–5 | 5–6 | >6 | Total |
| Site 1 (Sub-urban) | 12 | 27 | 15 | 3 | 6 | 5 | 3 | 71 |
| Site 2 (Urban) | 25 | 37 | 16 | 11 | 6 | 5 | 4 | 104 |
| Total | 37(21%) | 64(36%) | 31(18%) | 14(8%) | 12(7%) | 10(6%) | 7(4%) | 175 |

*3.2. Design Storms*

Four design storms are developed and validated for comparison in this study, including the Huff curve, Improved Huff curve, Triangular curve, and Chicago curve.

3.2.1. Huff Curve

The Huff curve was initially developed by Huff [28] for characterizing temporal rainfall distributions in an area and has been widely applied to describe the hyetograph and to predict the runoff in a watershed [28,38–43]. The Huff curve is a dimensionless hyetograph. First, the storm durations ($X$ axis) of different storms are normalized by dividing the total storm duration. The 10% interval of time is normally applied. Next, the cumulative rainfall depth ($Y$ axis) within each time interval from 0–10% to 90–100% is normalized by dividing the storm-total rainfall depth. When developing the Huff curve from historic storm data, the percent of the cumulative rainfall depth within a time interval (e.g., 0–10%) is sorted into a descending order for all storm events, and the rank of each storm is then normalized into a probability from 0 to 100% by the storm count [29]. The Huff curve is an isopleth, i.e., the percent of cumulative rainfall depth within each time interval at a certain probability. These isopleths are usually developed by the probability in a 10% increment from 10% up to 90%. The 50% (median) curve is the most representative curve [44] and is developed for comparison in this study using the storm data from 2008 to 2012 at Site 2, while the 10% and 90% curves represent the two extreme cases, which are the highest and lowest ranks in percent of the cumulative rainfall depth within a time interval.

The aforementioned curve is a general Huff curve that is derived using all historic storm rainfall data. In practice, the number of storms in each quartile is defined according to the occurrence of peak rainfall in a normalized rainfall duration, i.e., the 1st (0–25%), 2nd (25–50%), 3rd (50–75%), and 4th (75–100%) quartiles. Then, a series of Huff curves are developed at different probabilities within each quartile [28]. All Huff curves are derived at the probability of 50% within a quartile in this study. The storm count within each quartile at Sites 1 and 2 is summarized in Table 2. Most of the storms (84%) are in the first two quartiles.

**Table 2.** The number of storms in each quartile defined according to the occurrence of peak rainfall considering a normalized time at two sites for the period 2008–2012.

| Quartiles | 1st | 2nd | 3rd | 4th | Total |
| --- | --- | --- | --- | --- | --- |
| Site 1 | 31(44%) | 27(38%) | 9(13%) | 4(6%) | 71 |
| Site 2 | 33(32%) | 56(54%) | 8(8%) | 7(7%) | 104 |
| Total | 64(37%) | 83(47%) | 17(10%) | 11(6%) | 175 |

### 3.2.2. Improved Huff Curve

The Huff curve model is applied to describe the hyetograph of a storm event within a quartile of its normalized time of peak rainfall intensity. Instead of separating the storms into different quartiles, all storms at Site 2 from 2008 to 2012 are first separated into the rising and falling limbs. Then, a series of Huff curves are derived separately from both limbs at different probabilities, and finally form an Improved Huff curve by combining both limbs. Figure 2a,b are the developed dimensionless hyetographs using the percent of rainfall intensity at the probabilities of 10%, 50%, and 90%. Accordingly, Figure 2c,d are the curves using the percent of the cumulative rainfall depth. The mean normalized time of peak rainfall is 33 ± 5% at Site 2 (Table 3). The hyetographs of the percent of rainfall intensity at 50% are further fitted into Equations (1) and (2) by regression models for the rising and falling limbs, respectively. The fitting coefficients ($R^2$ and RMSE) are 0.985 and 0.008 in the rising limb, and 0.993 and 0.009 in the falling limb. Both equations can be easily applied to compute the rainfall depth distribution with time once the total rainfall depth and duration are given for a drainage facility design and other purposes.

$$i(t_b) = 0.007 + 0.406t_b - 0.927t_b^2 + 0.785t_b^3 \tag{1}$$

$$i(t_a) = 0.017 + 0.040/t_a \tag{2}$$

where $i(t_b)$ and $i(t_a)$ are the time series of rainfall intensity in the rising and falling limbs, respectively, and $t_b$ and $t_a$ represent the normalized time prior to and post the peak rainfall intensity, respectively.

The Improved Huff curves are also derived at the probability of 50% in both rising and falling limbs using the storm data within each quartile (Figure 3a,b). The first and second quartiles play a dominant role in forming the rainfall hyetograph and have a similar shape to that from all rainfall. In contrast, the hyetographs in the third and fourth quartiles show some discrepancy, especially in the falling limb.

**Table 3.** Statistics of the rising and falling limbs for storm events at Sites 1 and 2 from 2008 to 2012.

| Title | Rainfall Depth (%) | | Rainfall Duration (%) | | Intensity (mm/min) | |
|---|---|---|---|---|---|---|
| | Rising | Falling | Rising | Falling | Rising | Falling |
| Site 1 | 45 ± 5% | 55 ± 5% | 33 ± 5% | 67 ± 5% | 0.66 ± 0.26 | 0.36 ± 0.03 |
| Site 2 | 41 ± 4% | 59 ± 4% | 33 ± 5% | 67 ± 5% | 0.62 ± 0.23 | 0.32 ± 0.07 |
| Mean | 43 ± 5% | 57 ± 5% | 33 ± 5% | 67 ± 5% | 0.64 ± 0.24 | 0.34 ± 0.05 |

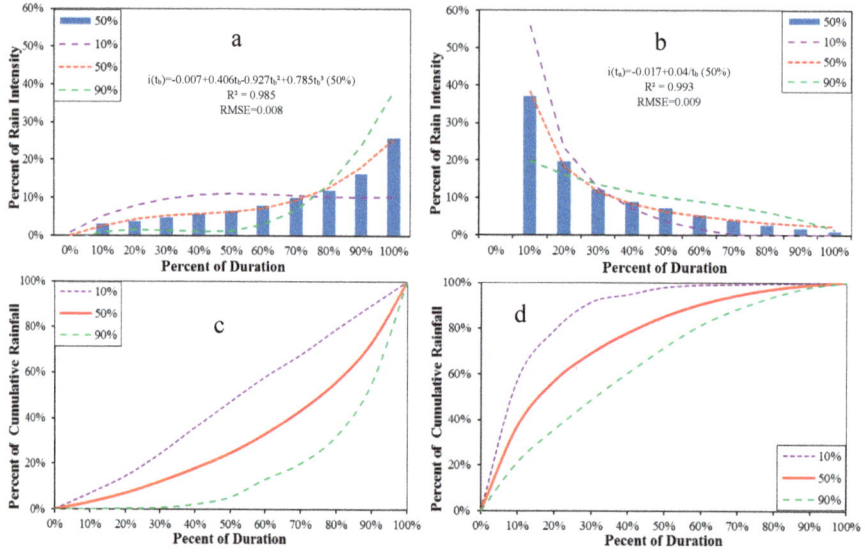

**Figure 2.** Comparisons of rainfall intensity and cumulative rainfall percent derived by the Improved Huff curve at three probabilities (10%, 50%, and 90%) for the rising limb (**a,c**), and the falling limb (**b,d**), which are separated by the time of peak rainfall (33%) at Site 2.

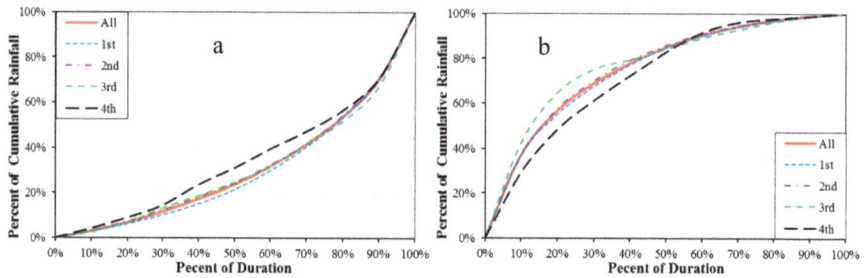

**Figure 3.** The Huff curves for the rising (**a**) and falling (**b**) limbs at a probability of 50% by all storm events and by those within the four quartiles at Site 2.

### 3.2.3. Triangular Curve

The triangular curve was developed by Yen & Chow [16] for drainage design in small areas and has been widely used in watershed and urban drainage designs [16,19]. The establishment of a triangular curve is used to determine the three vertexes of the triangular hyetograph, denoted by

(0, 0), (a, h), and (t_d, 0). The height (h) of the triangle is calculated by Equation (3) according to the area computation of a triangle.

$$h = \frac{2D}{t_d} \tag{3}$$

where $D$ is the storm total rainfall depth and $t_d$ is the storm duration. Both are given values in storm or drainage facility designs.

Then, a critical step is to determine $a$, the time of the peak rainfall intensity, which is estimated by preserving the first moment of the rainfall depth in Equations (4) and (5) [19].

$$a = 3\bar{t} - t_d \tag{4}$$

$$\bar{t} = \frac{\Delta t \left[ \sum\limits_{j=1}^{n} (j - 0.5)d_j \right]}{D} \tag{5}$$

where $\bar{t}$ is the first moment of the rainfall depth or the geometric center of the triangle, $d_j$ is the rainfall depth corresponding to the $j$th time interval, $n$ is the number of time intervals for a storm, and $\Delta t$ is the time interval.

### 3.2.4. Chicago Curve

The Chicago curve is developed by intensity-duration-frequency curves for the design of sewers and drainage management [17–22]. The applied formats and constants in Guangzhou are presented in Equation (6) [14].

$$I = \frac{167 A_1 (1 + C \log P)}{(t_d + b)^n}; a = 167 A1 (1 + C \log P) \tag{6}$$

where $I$ is the mean rainfall intensity, $A_1$ is the rainfall depth with a one-year return period, $C$ is the parameter of rainfall depth variations, $P$ is a return period, $t$ is the rainfall duration, and $b$ and $n$ are constants. The values of $C$, $b$, and $n$ are 0.438, 11.259, and 0.750, respectively, which are adopted by the Department of Water Authority in the Guangzhou metropolitan area based on historic rainfall data from 1990 to 2010 [14]. The general equations of the rising and falling limbs are:

$$i(t_b) = \frac{a \left[ \frac{(1-n)t_b}{r} + b \right]}{\left[ \frac{t_b}{r} + b \right]^{1+n}}; i(t_a) = \frac{a \left[ \frac{(1-n)t_a}{1-r} + b \right]}{\left[ \frac{t_a}{1-r} + b \right]^{1+n}} \tag{7}$$

where $i(t_b)$ and $i(t_a)$ are the time series of rainfall intensity in the rising and falling limbs, respectively; $t_b$ and $t_a$ are the time before and after the peak rainfall intensity, respectively; and $r$ is the ratio of the peak rainfall intensity time to the total duration.

### 3.3. Validations and Applications

The derived hyetographs are first validated by using the real hyetographs from Sites 1, 3–6. Two indices, the Root Mean Squared Error (RMSE) (Equation (8)) and Nash-Sutcliffe Efficiency (NSE) (Equation (9)), are used to evaluate their agreement [45].

$$RMSE = \sqrt{\frac{\sum\limits_{i=1}^{n} (P_i - O_i)^2}{N}} \tag{8}$$

$$NSE = 1 - \frac{\sum\limits_{i=1}^{n} (P_i - O_i)^2}{\sum\limits_{i=1}^{n} (O_i - \overline{O})^2} \tag{9}$$

where $P_i$ is the model-predicted value, $O_i$ is the observed value, $\overline{O}$ is the mean of the observed value, and $N$ is the number of observations.

The developed design storms are further applied as inputs to the SWMM model to simulate the flooding volume in three small urban catchments. SWMM is developed by the American Environmental Protection Agency [46]. It has been widely applied in urban drainage management, flood-control facility design, water quality modeling, and so on [47–52]. In order to verify the Improved Huff curve at the rainfall depth-runoff calculation, SWMM is established in the Shiqiao Street, in the downtown area of Panyu District in the south of Guangzhou (Figure 1c). The total study area of SWMM modeling is 15.53 km², and three catchments are selected to test the design storms, represented by Node 192, Node 503, and Node 519 (Figure 1c). The drainage boundaries of each sub-catchment are derived from detailed pipe network and fine airborne LiDAR DEM data (0.5 m grid), plus repeated field visits and validation. The selected three catchments have similar total catchment areas, but different areas in terms of their direct drainage catchment and upstream catchment (Table 4).

**Table 4.** Direct Catchment Area (DCA), Upstream Catchment Area (UCA), Total Catchment Area (TCA), and Drainage Capacity (DC) of the three selected nodes.

| Node | DCA (ha) | UCA (ha) | TCA (ha) | DC (m³/s) |
|------|----------|----------|----------|-----------|
| 503  | 8.8      | 26.2     | 35.0     | 0.88      |
| 192  | 10.1     | 27.3     | 37.4     | 1.76      |
| 519  | 2.4      | 40.5     | 43.0     | 1.69      |

The SWMM model is firstly validated using the in situ measured storm rainfall and flooding volume at Site 4/Node 503. Next, the verified SWMM model is applied to simulate the flooding volume using design storms according to the storm-total rainfall depth recorded at Site 4. Finally, the simulated peak flooding volume and time at Nodes 503, 519, and 192 by all design storms and by the same storms recorded at Site 1 are compared.

## 4. Results

### 4.1. Characteristics of Historic Storms

Storm events frequently occur in the study area of a tropical climate setting. According to the given criteria, there were 71 (14/year) and 104 (21/year) storm events during the five years from 2008 to 2012 (Table 1). Site 2, which is located within the Downtown area of Tianhe District, had 46% more storm events than Site 1, especially concerning those of less than 2 h. Those storm events at Site 2 mainly (54%) concentrate in the 2nd quartile, while 44% of storm events are in the 1st quartile at Site 1 (Table 2). This large difference in short-duration storm events between the two sites is likely caused by the surrounding conditions of the urban center for Site 2 and the sub-urban area of Site 1. Similar phenomena are also found in the urban areas of Beijing [7]. However, after normalizing the rainfall duration and depth, both sites have a similar distribution with rainfall depth, time of peak intensity, and mean intensity in the rising and falling limbs, respectively (Table 3). The time of peak rainfall intensity is around 33 ± 5% of the storm duration. The rising limb displays 43 ± 5% of the total rainfall depth, with a stronger rainfall intensity than the falling limb. This suggests that the storm hyetograph is similar at both sites, which are located 25 km apart, and the design storm curve developed at one site is able to represent the overall rainfall temporal distribution at least within the study area and even in the entire Guangzhou metropolitan area.

## 4.2. Validations of Design Storms

The developed design storms are first validated by the storm events recorded at our own research Sites 3–6 in 2014 and 2015 (Figure 4, Table 5). Three storm events at each site are selected for a detailed comparison according to the different rainfall depths and durations of storm events. Overall, the Improved Huff curves have the best agreements with the observations, exhibiting smaller RMSE and higher NSE values than the other three curves (Table 5). The NSE of the Improved Huff curves varies from 0.94 to 0.99, except for one event (0.82) on 21 July 2015, when all design storms have a relatively lower NSE than the other events. In contrast, the Huff curves have the largest variations in terms of NSE, ranging from 0.65 to 0.99. Both Triangular and Chicago curves display a similar performance, with much larger RMSE and lower NSE values than the Improved Huff curves.

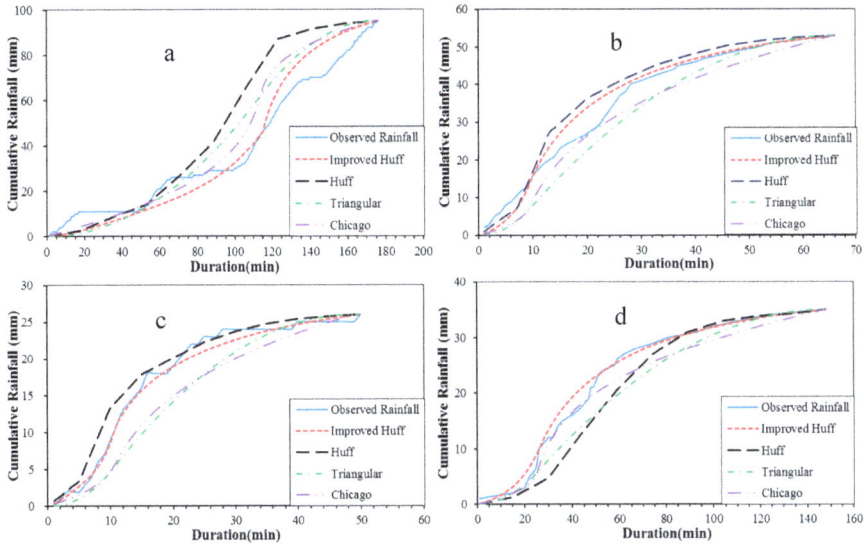

**Figure 4.** Comparisons of cumulative rainfall between ground observations and the design storms for the storm events at Site 3 on 21 September 2015 (**a**); at Site 4 on 25 May 2015 (**b**); at Site 5 on 3 October 2015 (**c**); and at Site 6 on 21 July 2015 (**d**).

**Table 5.** RMSE and NSE values computed between design storms and observed rainfall at Sites 3-6, where the observed storm data are not used to develop design storms. Both the Improved Huff and original Huff curves represent those at a probability of 50%.

| Site | Date | Rainfall Depth (mm) | Duration (min) | Intensity (mm/h) | Index | Improved Huff | Huff | Triangular | Chicago |
|------|------|------|------|------|------|------|------|------|------|
| | 19 August 2014 | 40 | 28 | 86 | RMSE | 1.08 | 1.22 | 4.91 | 4.33 |
| | | | | | NSE | 0.99 | 0.99 | 0.87 | 0.88 |
| 3 | 21 September 2015 | 95 | 175 | 33 | RMSE | 6.92 | 16.43 | 10.79 | 9.75 |
| | | | | | NSE | 0.94 | 0.65 | 0.85 | 0.88 |
| | 03 October 2015 | 55 | 136 | 24 | RMSE | 3.70 | 9.34 | 3.56 | 4.09 |
| | | | | | NSE | 0.96 | 0.75 | 0.96 | 0.95 |
| | 02 August 2014 | 42 | 84 | 30 | RMSE | 2.75 | 3.53 | 5.39 | 4.90 |
| | | | | | NSE | 0.97 | 0.94 | 0.88 | 0.90 |
| 4 | 11 May 2015 | 78 | 97 | 48 | RMSE | 5.58 | 8.54 | 14.25 | 12.25 |
| | | | | | NSE | 0.95 | 0.87 | 0.65 | 0.74 |
| | 25 May 2015 | 53 | 66 | 48 | RMSE | 2.74 | 3.91 | 4.60 | 3.87 |
| | | | | | NSE | 0.97 | 0.94 | 0.91 | 0.93 |
| | 16 May 2015 | 27 | 65 | 25 | RMSE | 1.74 | 2.80 | 2.05 | 1.22 |
| | | | | | NSE | 0.94 | 0.83 | 0.91 | 0.97 |
| 5 | 21 July 2015 | 22 | 59 | 22 | RMSE | 2.56 | 2.89 | 2.99 | 2.64 |
| | | | | | NSE | 0.82 | 0.76 | 0.75 | 0.80 |
| | 03 October 2015 | 26 | 49 | 32 | RMSE | 0.85 | 1.53 | 3.56 | 3.36 |
| | | | | | NSE | 0.99 | 0.96 | 0.81 | 0.83 |
| | 21 June 2014 | 30 | 105 | 17 | RMSE | 1.32 | 1.62 | 3.87 | 3.63 |
| | | | | | NSE | 0.95 | 0.94 | 0.55 | 0.61 |
| 6 | 20 August 2014 | 34 | 125 | 16 | RMSE | 1.28 | 3.38 | 2.88 | 1.81 |
| | | | | | NSE | 0.99 | 0.92 | 0.94 | 0.98 |
| | 21 July 2015 | 35 | 148 | 15 | RMSE | 1.17 | 2.95 | 2.84 | 1.78 |
| | | | | | NSE | 0.99 | 0.93 | 0.94 | 0.97 |

The cumulative hyetographs of one storm selected from Table 5 for each site is illustrated in Figure 4. Again, the Improved Huff curves successfully recover all of the rainfall processes recorded at the four sites. The Huff curves overestimate the observed rainfall depth at Sites 3–5, but greatly underestimate it at Site 6 prior to the peak rainfall intensity. Both the Triangular and Chicago curves tend to underestimate the rainfall prior to the peak rainfall intensity at Sites 4–6. A study in Reykjavik, the capital city of Iceland, has found that the Chicago curve underestimates the peak rainfall intensity [22]. Another study in Taiwan has shown that the Triangular curve does not satisfactorily simulate heavy storms because of its flat slope in the rising and falling limbs [53].

Besides the above events, all design storms are also validated using the 71 storm events at Site 1 from 2008 to 2012 (Table 6). The Improved Huff curves have the lowest RMSE and highest NSE values of 6.43 mm and 0.66, while they are 6.62, 7.38, and 7.57 mm and 0.63, 0.55, and 0.54 for the Huff, Triangular, and Chicago curves, respectively. Here, all design storms are derived using the parameters computed from 104 storms at Site 2. In contrast, the design storms illustrated in Figure 4 and Table 5 are derived using the parameters of each individual storm event. Therefore, the NSE values of design storms for a single storm event in Figure 4 and Table 5 are higher than those in Table 6. No matter which validation method is applied, the Improved Huff curve performs better than the others.

**Table 6.** Mean RMSE and NSE values between the 71 observed hyetographs at Site 1 and their design storms, and the mean relative difference of the simulated peak flooding volume (PV) and time (PT) at the three nodes by the design storms against those simulated by the 71 observed storms at Site 1. All design storms are computed using the parameters derived from the 104 storms at Site 2, together with the total rainfall depth and duration for each storm at Site 1.

| Index | | Improved Huff | Huff | Triangular | Chicago |
|---|---|---|---|---|---|
| RMSE | | 6.43 | 6.62 | 7.38 | 7.57 |
| NSE | | 0.66 | 0.63 | 0.55 | 0.54 |
| N503 | PV(%) | 2 | −12 | −22 | −17 |
| | PT(%) | 19 | 24 | 45 | 41 |
| N192 | PV(%) | −6 | −43 | −62 | −19 |
| | PT(%) | 17 | 19 | 15 | 24 |
| N519 | PV(%) | 8 | −16 | −38 | −21 |
| | PT(%) | 8 | 8 | 9 | 10 |

The above Huff curves are developed using the storm data at the original one-minute interval. In order to investigate the sensitivity of the Improved Huff curve to time intervals, we developed the curves by aggregating the one-minute interval data into time intervals of 5, 10, 30, 60, and 120 min and then compared the curves with the real storm data at Site 2. The storm events were divided into three groups by duration: <1 h, 1–3 h, and >3 h (Figure 5, Table 7). The results suggest that an optimal time interval for the original data is determined by the storm duration divided by 20, which is constrained by the 10% increment of time unit for computing the isopleths in the rising and falling limbs. For instances, for storms within 1 to 3 h, the percent of cumulative rainfall depth varies linearly with time for hourly and half-hourly data (Figure 5c,d). The mean storm duration for group 2 (1–3 h) is 100 min, and the optimal recording or computing time interval is 100/20 = 5 min. Similarly, the optimal time intervals for groups 1 (mean duration = 40 min) and 3 (mean duration = 300 min) are 2 and 15 min, respectively. The rising limb requires a smaller interval primarily due to its shorter time duration (33 ± 5%) than the falling limb. This further suggests that it is challenging to develop a storm hyetograph using hourly data in most conditions, and the rainfall depth data recorded at short-time intervals (1 min or 5 min) are required to derive a practical storm hyetograph, especially for short-duration intense storms in metropolitan areas like Guangzhou.

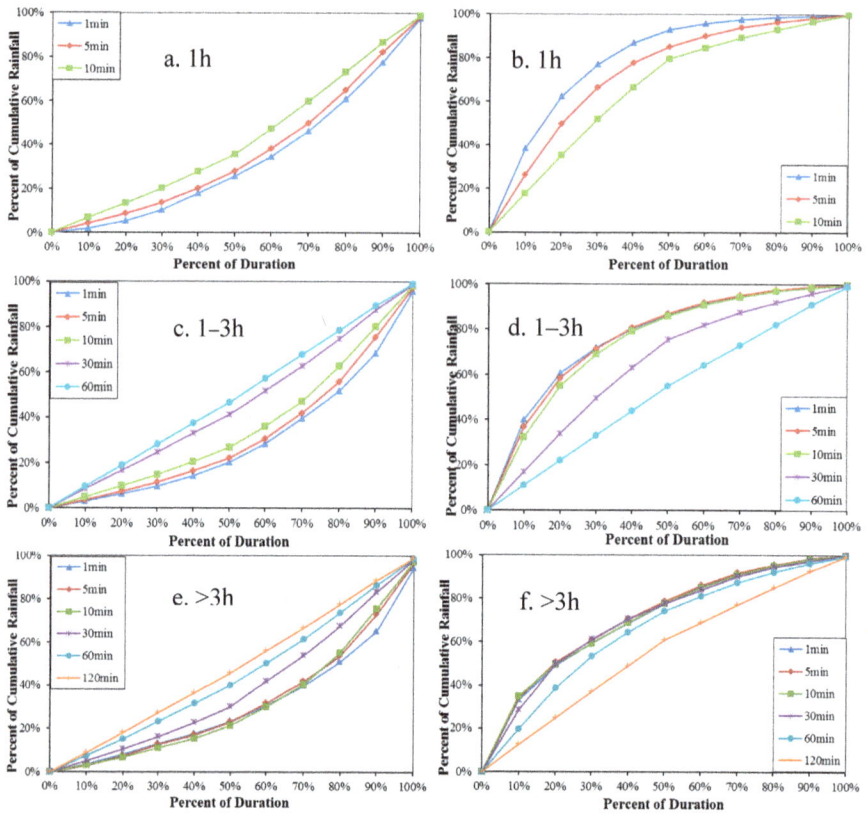

**Figure 5.** Percent of cumulative rainfall depth derived by the Improved Huff curves (probability 50%) using different time intervals for three groups (rainfall duration: 1 h, 1–3 h, and >3 h) of storms from 2008 to 2012 at Site 2. Plots (**a,c,e**) are for the rising limb, and plots (**b,d,f**) are for the falling limb.

**Table 7.** NSE, RMSE, (mm) and Relative Difference (RD) values between design storms derived by the Improved Huff curves (probability 50%) using different time intervals and observed rainfall depth for all storms at Site 2 from 2008 to 2012.

| Duration (h) | Title | Time Interval (min) | | | | | |
|---|---|---|---|---|---|---|---|
| | | 1 | 5 | 10 | 30 | 60 | 120 |
| | NSE | 0.97 | 0.94 | 0.81 | | | |
| 1 | RMSE | 1.32 | 2.12 | 3.80 | | | |
| | RD (%) | 2 | −5 | 13 | | | |
| | NSE | 0.94 | 0.94 | 0.94 | 0.89 | 0.75 | |
| 1–3 | RMSE | 2.22 | 2.24 | 2.36 | 3.61 | 5.27 | |
| | RD (%) | −2 | −3 | −3 | −5 | −13 | |
| | NSE | 0.92 | 0.92 | 0.92 | 0.92 | 0.90 | 0.82 |
| >3 | RMSE | 5.81 | 5.85 | 5.87 | 5.94 | 5.99 | 8.00 |
| | RD (%) | 2 | −3 | −3 | −4 | −4 | −10 |

*4.3. Applications in SWMM*

SWMM was first established and optimized using the observed rainfall and flooding volume measured at Site 4/Node503 on 2 August 2014 (Figure 6). The optimized SWMM was then utilized to simulate the flooding volume at the three nodes by using the observed rainfall and the according design storms (Figure 6b–d). At Node 503, the simulated flooding volume using the observed rainfall is in good agreement with that from the observation, especially for the four small crests (Figure 6b). The simulated peak flooding volume and time by the Improved Huff curve is also in good agreement with that from the observation. In contrast, the simulated peak flooding volumes by the Huff, Triangular, and Chicago curves all are lower than the observed values. Meanwhile, the simulated peak flooding time by the Huff, Triangular, and Chicago curves is 10, 14, and 27 min later than the real situation, respectively (Figure 6b, Table 8).

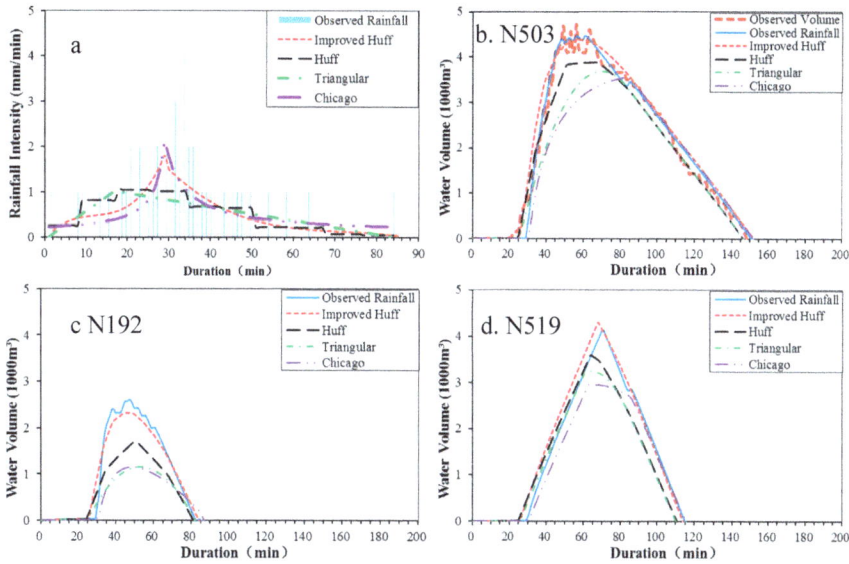

**Figure 6.** Comparisons of hyetographs (**a**) and the according ground measured and simulated flooding volume by SWMM using different hyetographs at Node 503 (**b**), Node 192 (**c**) and Node 519 (**d**) for the storm event recorded at Site 4 on 2 August 2014, with total rainfall of 42 mm and duration of 84 min.

**Table 8.** Peak flooding volume, time, and NSE simulated by SWMM using rainfall depths from gauge observations at Sites 4 and the considered design storms. The last four columns are the difference in the simulated peak flooding volume and time using design storms (DS) against those by gauge (G) rainfall depth. The water volume at Node 503 is computed according to the DEM and water depth recorded at Site 4/Node 503, while that at Node 192 and 519 is simulated by SWMM using gauge rainfall depth data at Site 4.

| Date Total P & T | Node | Flooding Volume & Time by Gauge Rain | Difference of Flooding Volume and Time (DS-G) | | | |
|---|---|---|---|---|---|---|
| | | | Improved Huff | Huff | Triangular | Chicago |
| | 503 | 4733 m$^3$ | −342 | −844 | −1044 | −1173 |
| | | 57 min | 0 | 10 | 14 | 27 |
| | | NSE | 0.97 | 0.93 | 0.90 | 0.86 |
| 2 August 2014 | 192 | 2596 | −281 | −703 | −1313 | −1319 |
| 42 mm | | 46 | −1 | 3 | 7 | 2 |
| 84 min | | NSE | 0.97 | 0.71 | 0.38 | 0.39 |
| | 519 | 4126 | 173 | −532 | −879 | −1171 |
| | | 70 | −2 | −7 | −6 | −5 |
| | | NSE | 0.98 | 0.92 | 0.93 | 0.92 |

At Nodes 192 and 519, there is no observed flooding volume. The simulated flooding volumes by the four design storms are compared to those simulated by the observed rainfall depth at Site 4. The simulated flooding volumes by gauge observed rainfall and the Improved Huff curve at Node 192 are similar and smaller than those at Node 503 (Figure 6c), which is consistent with the real situations that we learned about during our field survey. This node also has the largest drainage capacity/largest pipe radius among the three nodes (Table 4). The flooding volumes reported by the Huff, Triangular, and Chicago curves are much lower than seen for the observed rainfall. The overall patterns of the simulated flooding volumes by the four design storms are similar at Node 519, and the peak flooding volume reported by the Improved Huff curve displays the best agreement with that simulated by the observed rainfall (Figure 6d, Table 8).

All of the 71 observed storm events at Site 1 from 2008 to 2012 and their design storms are applied to drive SWMM. The simulated peak flooding volume and time of the four design storms are compared to those of the observed rainfall. The results show that the improved Huff curves exhibit the best performance, followed by the original Huff curves, and then the Chicago and Triangular curves (Table 6). Again, all design storms are derived using the parameters computed from the 101 storms at Site 2.

## 5. Discussion

### 5.1. Urban Flooding

There are generically two types of urban flooding. The first type of flooding is caused by a large stream flow due to long and continuous heavy rainfall in the upstream area of the cities, such as the typhoon-brought heavy rainfall in the Guangdong province. In such cases, the drainage system in the cities plays a small role. The other type of flooding is mainly caused by short and intense rainfall, which is the primary study target in this study. The large rainfall-runoff generated within a short time cannot be drained immediately by the drainage system and thus inundates the urban streets, which is also called waterlogging in some literature [54].

Why does waterlogging frequently occur in most cities in China? Global warming is often attributed as a scapegoat. The drainage system is mandatory in urban planning and community constructions, and there are all kinds of laws and regulations on drainage design in China [13,15,55]. In our preliminary study in the same area as in this study, we simulated the runoff by SWMM with real storm rainfall and found that the drainage capacity of the pipe was below the grade that is classified in construction. Then, we examined the design storms recommended in the pipe construction code [15], such as the Triangular and Chicago curves, and found similar results as demonstrated in this study (Figures 4 and 6, Table 8). These recommended curves underestimate the peak rainfall intensity, resulting in a lower peak flooding volume and thus a lower drainage capacity demand in drainage facility design for the same return period.

### 5.2. Improvement to the Huff Curve

It is quite challenging (if not impossible) to describe the storm hyetograph using a single design storm. A set of Huff curves are developed to describe the storm hyetographs for Peninsular Malaysia, where the majority (80%) of storms are between 3 and 6 h, belonging to the 2nd quartile [29]. Huff curves represent the temporal characteristics of rainfall quite well, although four types of Huff curves at different probabilities are required within the four quartiles [56]. Similarly, most storms (75%) are less than 3 h and dominate in the 2nd (47%) and 1st (37%) quartiles in Guangzhou (Tables 1 and 2). Thus, the Huff curves are investigated in this study to represent the storm hyetographs using minute-interval rainfall data. However, whilst the Huff curves work well in some cases, they work badly in other cases (Figures 4 and 6; Tables 5 and 8). How could we improve the Huff curve to be better representative of the temporal distribution of all storm rainfall depth?

The time of peak rainfall has a critical influence on the analysis of hyetographs [30]. The rainfall intensity exponentially decreases on both sides of the peak rainfall in the Chicago curve [32].

The Monte Carlo method could better simulate the storm hyetograph separately in the rising and falling limbs [31]. The normalized time of peak rainfall is similar at $33 \pm 5\%$ for both sites (Table 3). The rising limb ($0.64 \pm 0.24$ mm/min) has much stronger rainfall intensity than the falling limb ($0.34 \pm 0.05$ mm/min). Therefore, we separate the storms into the rising and falling limbs at the time of peak rainfall and then compute the Huff curves separately for both limbs in this study.

The storm events are firstly classified into the 1st, 2nd, 3rd, and 4th quartiles according to the time of peak rainfall. Each storm event is further divided into the rising and falling limb within each quartile, within which a set of Huff curves are derived at different probabilities for both limbs. The Huff curves for all storms in both limbs are quite similar with those for storms in the 1st and 2nd quartiles (Figure 3), which include 84% of all storm events (Table 2). The Huff curves in the rising limb for the 3rd and 4th quartiles are also similar to that for all storms, and they slightly deviate from the total Huff curve in the falling limb. Too many options for the different Huff curves in the four quartiles always generate challenges and even confusion to civil engineers in practice. Therefore, considering the limited storm events in the 3rd and 4th quartiles and their slight derivation, the total Huff curves derived separately for the rising and falling limbs from all storm events are recommended in this study. The Huff curves for both limbs are finally combined together to form a full storm hyetograph, which is known as the Improved Huff curve in this study.

The Improved Huff curve could better represent the storm hyetographs recorded at rain gauges than the original Huff curve, the Triangular curve, and the Chicago curve (Figure 4, Tables 5 and 6). This Improved Huff curve presents point-developed curves by using five-year data from one site and is verified at several neighboring sites. More sites and a longer period of time are needed to verify the curves in future studies.

The Triangular curve does not work well in our validation. One possible reason for this is that the Triangular curve is relatively flat on both sides of the peak rainfall and is often used in arid and semi-arid areas [19]. A double Triangular curve was tested to simulate the typhoon-related storms in Taiwan, China, and the central triangle of the double Triangular curve can better simulate the peak rainfall intensity than a single Triangular curve [53]. This is a possible option to improve the single Triangular curve to better represent the storm hyetographs in South China and is scheduled in our further study.

The Chicago curve is applied using constants recommended by the local water authority in this study [14]. These constants are derived from rainfall data during 1990–2010. The constants required in the Chicago curve equations may differ in different regions and climate settings and vary with the climate change, even in the same region [23]. This study does not try to apply or derive new constants by using the minute-interval rainfall depth data and this is another possible direction for our continuing study.

### 5.3. Application of the Improved Huff Curve

SWMM is established to verify the Improved Huff curve at the rainfall-runoff calculation (Figure 6; Tables 6 and 8). The optimal model parameters are obtained by comparing the flooding volume between the street observations at Node 503 and SWMM simulations using the storm data at Site 4. Thus, the flooding volumes simulated by SWMM using different design storm hyetographs could represent the real situations to some extent. Of course, more storm events data could provide better model parameters and simulations [47].

The Improved Huff curve displays a better performance in SWMM than the Huff, Triangular, and Chicago curves at the three small urban catchments in the Panyu District, Guangzhou. More studies in other districts are needed to verify the results obtained in this study. After further validation, the Improved Huff curve will have great applications in drainage design in the metropolitan area of Guangzhou, other urban areas in the Guangdong province, and even in Southern China, where there are similar climate settings. Of course, the fitting coefficients or equations must be derived from the local storm data in different cities with optimal time intervals of one to five minutes.

## 6. Conclusions

China faces severe challenges in urban flooding due to its dramatic urbanization and relatively poor storm water management. A partial reason for this phenomenon in China is that the recommended design storms underestimate the rainfall intensity before the peak rainfall, resulting in a lower peak flooding volume and thus a lower drainage capacity demand in drainage facility design for the return period. The design storms vary with different regions and climatic conditions. This study derives a storm hyetograph to represent the temporal distributions of rainfall depth in the metropolitan area of Guangzhou by improving the Huff curve. The results are summarized below.

The Huff curve is improved by separately describing the rising and falling limbs instead of classifying the storms into four quartiles. The time of peak rainfall is at 33 ± 5% for both sites and has a critical influence on the classification of hyetographs. The rising limb has a much stronger rainfall intensity (0.64 ± 0.24 mm/min) and slightly lower rainfall depth (43 ± 5% of total rainfall depth) than the falling limb (0.34 ± 0.05 mm/min). Most (84%) of the storm events are in the 1st and 2nd quartiles, whose Huff curves are dominant and similar to those for all storms in both limbs. The Huff curves for both limbs are combined together to form a full storm hyetograph, which is known as the Improved Huff curve in this study.

The optimal time intervals are one to five minutes to derive a practical storm hyetograph, especially for short-duration and intense storms in metropolitan areas like Guangzhou. It is challenging to develop urban storm hyetographs using hourly data in most conditions. All design storms except for the Chicago Curve are derived using the minute-interval rainfall data in this study.

The Improved Huff curve works best in simulations of hyetographs and hydrographs, followed by the Huff curves, and then the Chicago curves and Triangular curves. The peak flooding volumes simulated using the Huff, Triangular, and Chicago curves as inputs to SWMM are lower than that presented by the observed rainfall, i.e., underestimating the rainfall intensity and resulting in a lower peak flooding volume.

The Improved Huff curve has great potential in storm water management such as flooding risk mapping and drainage facility design after further validation. The Improved Huff curve presents point-developed curves by using five-year data from one site and is verified at several neighboring sites in this study. More site data with longer periods are needed to verify the Improved Huff curves in future study.

**Acknowledgments:** This study was partially supported by the Water Resource Science and Technology Innovation Program of Guangdong Province (#2016-19), and the National Natural Science Foundation of China (#41371404; #41301419), and the Science and Technology Program of Guangzhou (#1561000154). We thank all researchers and staff for providing and maintaining the meteorological data, SWMM model, and the model required data from all agencies. We would like to express our appreciation to the Water Editorial Office and the three anonymous reviewers for their valuable comments and suggestions.

**Author Contributions:** Cuilin Pan performed the experiments, processed the data, analyzed the results, and wrote the manuscript; Xianwei Wang designed experiments, analyzed the results, and wrote and revised the manuscript; Lin Liu, Huabing Huang, and Dashan Wang contributed to obtaining the in situ measurements, set up the SWMM model, improved the experiments, and revised the manuscript.

**Conflicts of Interest:** The authors declare no conflict of interest.

## References

1. Ahmed, S.I.; Rudra, R.P.; Gharabaghi, B.; Mackenzie, K.; Dickinson, W.T. Within-storm rainfall distribution effect on soil erosion rate. *ISRN Soil Sci.* **2012**, *2012*, 1–7. [CrossRef]

2. Bonta, J.V. Development and utility of Huff curves for disaggregating precipitation amounts. *Appl. Eng. Agric.* **2004**, *20*, 641–653. [CrossRef]

3. Dolšak, D.; Bezak, N.; Šraj, M. Temporal characteristics of rainfall events under three climate types in Slovenia. *J. Hydrol.* **2016**, *541*, 1395–1405. [CrossRef]

4. Park, D.; Jang, S.; Roesner, L.A. Evaluation of multi-use stormwater detention basins for improved urban watershed management. *Hydrol. Processes* **2014**, *28*, 1104–1113. [CrossRef]

5.  Kang, M.S.; Goo, J.H.; Song, I.; Chun, J.A.; Her, Y.G.; Hwang, S.W.; Park, S.W. Estimating design floods based on the critical storm duration for small watersheds. *J. Hydro-Environ. Res.* **2013**, *7*, 209–218. [CrossRef]

6.  Grimaldi, S.; Petroselli, A.; Serinaldi, F. Design hydrograph estimation in small and ungauged watersheds: Continuous simulation method versus event-based approach. *Hydrol. Process.* **2012**, *26*, 3124–3134. [CrossRef]

7.  Yang, L.; Tian, F.; Smith, J.A.; Hu, H. Urban signatures in the spatial clustering of summer heavy rainfall events over the Beijing metropolitan region. *J. Geophys. Res.* **2014**, *119*, 1203–1217. [CrossRef]

8.  Shastri, H.; Paul, S.; Ghosh, S.; Karmakar, S. Impacts of urbanization on Indian summer monsoon rainfall extremes. *J. Geophys. Res.* **2015**, *120*, 495–516. [CrossRef]

9.  Madsen, H.; Lawrence, D.; Lang, M.; Martinkova, M.; Kjeldsen, T.R. Review of trend analysis and climate change projections of extreme precipitation and floods in Europe. *J. Hydrol.* **2014**, *519*, 3634–3650. [CrossRef]

10. Jato-Espino, D.; Sillanpää, N.; Charlesworth, S.; Andrés-Doménech, I. Coupling GIS with stormwater modelling for the location prioritization and hydrological simulation of permeable pavements in urban catchments. *Water* **2016**, *8*, 451. [CrossRef]

11. Liu, L.; Liu, Y.; Wang, X.; Yu, D.; Liu, K.; Huang, H.; Hu, G. Developing an effective 2-D urban flood inundation model for city emergency management based on cellular automata. *Nat. Hazards Earth Syst. Sci.* **2015**, *15*, 381–391. [CrossRef]

12. Yang, L.; Scheffran, J.; Qin, H.; You, Q. Climate-related flood risks and urban responses in the Pearl River Delta, China. *Reg. Environ. Chang.* **2015**, *15*, 379–391. [CrossRef]

13. China Meteorological Administration (CMA). *Regulations on Short-Term and Near-Real Time Forecasting Operations, No. (2010)19*; China Meteorological Administration: Beijing, China, 2011. Available online: http://www.njqxj.gov.cn/xwxt/ztjd/zcqnwmh/gzzd/201308/t20130819_1089526.html (accessed on 18 April 2015).

14. Guangzhou Bureau of Water Authority (GBWA). *The Rainstorm Formula and Calculation Chart in Guangzhou Downtown, No. (2011)214*; Guangzhou Bureau of Water Authority: Guangzhou, China, 2011. Available online: http://www.cma.gov.cn/2011xwzx/2011xxdqxywtx/2011xzhgcxt/201110/t20111029_141510.html (accessed on 10 April 2015).

15. Ministry of Housing and Urban-Rural Development of the People's Republic of China (MOHURD). *Code for Design of Outdoor Wastewater Engineering, GB 50014-2006*; Ministry of Housing and Urban-Rural Development of the People's Republic of China: Beijing, China, 2006. Available online: http://www.mohurd.gov.cn/wjfb/200610/t20061031_155882.html (accessed on 22 April 2015).

16. Yen, B.C.; Chow, V.T. Design hyetographs for small drainage structures. *J. Hydraul. Div. ASCE* **1980**, *106*, 1055–1076.

17. Keifer, G.J.; Chu, H.H. Synthetic storm pattern for drainage design. *J. Hydraul. Div. ASCE* **1957**, *83*, 1–25.

18. Soil Conservation Service (SCS). Urban hydrology for small watersheds. In *Technical Release 55*; U.S. Department of Agriculture, Washington, D.C. Soil Conservation Service: Washington, DC, USA, 1986.

19. Ellouze, M.; Habib, A.; Riadh, S. A triangular model for the generation of synthetic hyetographs. *Hydrol. Sci. J.* **2009**, *54*, 287–299. [CrossRef]

20. Asquith, W.H.; Bumgarner, J.R.; Fahlquist, L.S. A triangular model of dimensionless runoff producing rainfall hyetographs in Texas. *J. Am. Water Resour. Assoc.* **2003**, *39*, 911–921. [CrossRef]

21. Palynchuk, B.A.; Guo, Y. A probabilistic description of rain storms incorporating peak intensities. *J. Hydrol.* **2011**, *409*, 71–80. [CrossRef]

22. Hlodversdottir, A.O.; Bjornsson, B.; Andradottir, H.O.; Eliasson, J.; Crochet, P. Assessment of flood hazard in a combined sewer system in Reykjavik city centre. *Water Sci. Technol.* **2015**, *71*, 1471–1477. [CrossRef] [PubMed]

23. Berggren, K.; Packman, J.; Ashley, R.; Viklander, M. Climate changed rainfalls for urban drainage capacity assessment. *Urban Water J.* **2014**, *11*, 543–556. [CrossRef]

24. Soil Conservation Service (SCS). *National Engineering Handbook, Section 4*; U.S. Department of Agriculture, Washington, D.C. Soil Conservation Service: Washington, DC, USA, 1972.

25. Yang, X.; Zhu, D.; Li, C.; Liu, Z. Establishment of design hyetographs based on risk probability models. *J. Hydraul. Eng.* **2013**, *44*, 542–548. (In Chinese)

26. Vieux, B.E.; Vieux, J.E. Continuous Distributed Modeling of LID/GI: Scaling from Site to Watershed. In Proceedings of the International Low Impact Development Conference, Huston, TX, USA, 17–21 January 2015; pp. 285–294.

27. HEC-Hydrologic Modeling System (HEC-HMS). *User's Manual. Version 4.0*; US Army Corps of Engineers Hydrologic Engineering Center: Davis, CA, USA, 2013.

28. Yin, S.Q.; Xie, Y.; Nearing, M.A.; Guo, W.L.; Zhu, Z.Y. Intra-Storm Temporal Patterns of Rainfall in China Using Huff Curves. *Trans. ASABE* **2016**, *59*, 1619–1632.

29. Azli, M.; Rao, A.R. Development of Huff curves for Peninsular Malaysia. *J. Hydrol.* **2010**, *388*, 77–84. [CrossRef]

30. Lin, G.; Chen, L.; Kao, S. Development of regional design hyetographs. *Hydrol. Process.* **2005**, *19*, 937–946. [CrossRef]

31. Kottegoda, N.T.; Natale, L.; Raiteri, E. Monte Carlo Simulation of rainfall hyetographs for analysis and design. *J. Hydrol.* **2014**, *519*, 1–11. [CrossRef]

32. Niemczynowicz, J. Impact of the greenhouse effect on sewerage systems—Lund case study. *Hydrol. Sci. J.* **1989**, *34*, 651–666. [CrossRef]

33. Liu, T.; Zhang, Y.H.; Xu, Y.J.; Lin, H.L.; Xu, X.J.; Luo, Y.; Xiao, J.; Zeng, W.L.; Zhang, W.F.; Chu, C.; et al. The effects of dust–haze on mortality are modified by seasons and individual characteristics in Guangzhou, China. *Environ. Pollut.* **2014**, *187*, 116–123. [CrossRef] [PubMed]

34. Xie, L.; Wei, G.; Deng, W.; Zhao, X. Daily δ18O and δD of precipitations from 2007 to 2009 in Guangzhou, South China: Implications for changes of moisture sources. *J. Hydrol.* **2011**, *400*, 477–489. [CrossRef]

35. Fan, F.; Fan, W. Understanding spatial-temporal urban expansion pattern (1990–2009) using impervious surface data and landscape indexes: A case study in Guangzhou (China). *J. Appl. Remote Sens.* **2014**, *8*, 083609. [CrossRef]

36. Wu, H.; Huang, G.; Meng, Q.; Zhang, M.; Li, L. Deep Tunnel for regulating combined sewer overflow pollution and flood disaster: A case study in Guangzhou City, China. *Water* **2016**, *8*, 329. [CrossRef]

37. Lu, W. Problems and suggestions on Guangzhou urban flood disaster management. *Chin. Public Adm.* **2014**, *343*, 106–108. (In Chinese).

38. Bonta, J.V.; Rao, A.R. Factors affecting development of Huff curves. *Trans. ASAE* **1987**, *30*, 1689–1693. [CrossRef]

39. Todisco, F. The internal structure of erosive and non-erosive storm events for interpretation of erosive processes and rainfall simulation. *J. Hydrol.* **2014**, *519*, 3651–3663. [CrossRef]

40. Bonta, J.V.; Rao, A.R. Fitting equations to families of dimensionless cumulative hyetographs. *Trans. ASAE* **1988**, *31*, 756–760. [CrossRef]

41. Bonta, J.V.; Rao, A.R. Regionalization of storm hyetographs. *Trans. ASAE* **1989**, *25*, 211–217. [CrossRef]

42. Terranova, O.G.; Iaquinta, P. Temporal properties of rainfall events in Calabria (southern Italy). *Nat. Hazards Earth Syst. Sci.* **2011**, *11*, 751–757. [CrossRef]

43. Terranova, O.G.; Gariano, S.L. Rainstorms able to induce flash floods in a Mediterranean-climate region (Calabria, southern Italy). *Nat. Hazards Earth Syst. Sci.* **2014**, *14*, 2423–2434. [CrossRef]

44. Huff, F.A. Time distributions of heavy rainstorms in Illinois. In *Illinois State Water Survey, Circular 173*; Illinois State Water Survey: Champaign, IL, USA, 1990.

45. Thapa, R.B.; Watanabe, M.; Motohka, T.; Shimada, M. Potential of high-resolution ALOS–PALSAR mosaic texture for aboveground forest carbon tracking in tropical region. *Remote Sens. Environ.* **2015**, *160*, 122–133. [CrossRef]

46. United States Environmental Protection Agency (EPA). *Storm Water Management Model Applications Manual*; United States Environmental Protection Agency: Washington, DC, USA, 2009.

47. Versini, P.A.; Ramier, D.; Berthier, E.; de Gouvello, B. Assessment of the hydrological impacts of green roof: From building scale to basin scale. *J. Hydrol.* **2015**, *524*, 562–575. [CrossRef]

48. Croci, S.; Paoletti, A.; Tabellini, P. URBFEP Model for Basin Scale Simulation of Urban Floods Constrained by Sewerage's Size Limitations. *Procedia Eng.* **2014**, *70*, 389–398. [CrossRef]

49. Li, F.; Duan, H.; Yan, H.; Tao, T. Multi-Objective Optimal Design of Detention Tanks in the Urban Stormwater Drainage System: Framework Development and Case Study. *Water Resour. Manag.* **2015**, *29*, 2125–2137. [CrossRef]

50. Palla, A.; Gnecco, I. Hydrologic modeling of Low Impact Development systems at the urban catchment scale. *J. Hydrol.* **2015**, *528*, 361–368. [CrossRef]

51. Martínez-Solano, F.; Iglesias-Rey, P.; Saldarriaga, J.; Vallejo, D. Creation of an SWMM toolkit for its application in urban drainage networks optimization. *Water* **2016**, *8*, 259. [CrossRef]

52. Ngo, T.; Yoo, D.; Lee, Y.; Kim, J. Optimization of upstream detention reservoir facilities for downstream flood mitigation in urban areas. *Water* **2016**, *8*, 290. [CrossRef]

53. Lee, K.T.; Ho, J.Y. Design hyetograph for typhoon rainstorms in Taiwan. *J. Hydrol. Eng.* **2008**, *7*, 647–651. [CrossRef]

54. Xie, Z.; Du, Q.; Ren, F.; Zhang, X.; Jamiesone, S. Improving the forecast precision of river stage spatial and temporal distribution using drain pipeline knowledge coupled with BP artificial neural networks: A case study of Panlong River, Kunming, China. *Nat. Hazards* **2015**, *77*, 1081–1102. [CrossRef]

55. Ministry of Housing and Urban-Rural Development of the People's Republic of China (MOHURD). *Code of Urban Wastewater Engineering Planning, GB 50318–2000*; Ministry of Housing and Urban-Rural Development of the People's Republic of China: Beijing, China, 2000. Available online: http://www.czs.gov.cn/ghj/zcfg/jsbzygf/content_195311.html (accessed on 15 April 2015).

56. Bonnin, G.M.; Martin, D.; Lin, B.; Paryzbok, T.; Yekta, M.; Riley, D. *Precipitation Frequency Atlas of the United States*, Version 3.0; NOAA Atlas 14; National Weather Service: Silver Springs, MD, USA, 2006; Volume 2.

*water*

MDPI

*Article*

# Physically, Fully-Distributed Hydrologic Simulations Driven by GPM Satellite Rainfall over an Urbanizing Arid Catchment in Saudi Arabia

**Hatim O. Sharif [1],\*, Muhammad Al-Zahrani [2] and Almoutaz El Hassan [1]**

[1]  Department of Civil and Environmental Engineering, University of Texas at San Antonio, One UTSA Circle, San Antonio, TX 78249, USA; almoutaz@gmail.com
[2]  Department of Civil and Environmental Engineering, King Fahd University of Petroleum and Minerals, Water Research Group, Dhahran 31261, Saudi Arabia; mzahrani@kfupm.edu.sa
\*  Correspondence: hatim.sharif@utsa.edu; Tel.: +1-210-458-6478

Academic Editors: Hongjie Xie and Y. Jun Xu
Received: 4 December 2016; Accepted: 22 February 2017; Published: 24 February 2017

**Abstract:** A physically-based, distributed-parameter hydrologic model was used to simulate a recent flood event in the city of Hafr Al Batin, Saudi Arabia to gain a better understanding of the runoff generation and spatial distribution of flooding. The city is located in a very arid catchment. Flooding of the city is influenced by the presence of three major tributaries that join the main channel in and around the heavily urbanized area. The Integrated Multi-satellite Retrievals for Global Precipitation Measurement Mission (IMERG) rainfall product was used due to lack of detailed ground observations. To overcome the heavy computational demand, the catchment was divided into three sub-catchments with a variable model grid resolution. The model was run on three sub-catchments separately, without losing hydrologic connectivity among the sub-catchments. Uncalibrated and calibrated satellite products were used producing different estimates of the predicted runoff. The runoff simulations demonstrated that 85% of the flooding was generated in the urbanized portion of the catchments for the simulated flood. Additional model simulations were performed to understand the roles of the unique channel network in the city flooding. The simulations provided insights into the best options for flood mitigation efforts. The variable model grid size approach allowed using physically-based, distributed models—such as the Gridded Surface Subsurface Hydrologic Analysis (GSSHA) model used in this study—on large basins that include urban centers that need to be modeled at very high resolutions.

**Keywords:** floods; hydrologic modeling; GSSHA; satellite rainfall; IMERG; GPM

## 1. Introduction

Storm events that result in catastrophic floods are rare but do occur in arid environments, especially in urban centers. With increasing population and urbanization, the public susceptibility and economic impact of flooding in these areas will be increasing [1]. Since these events do not occur frequently, there may not be enough pressure on decision makers to invest in the development of robust hydrometeorological observing systems or hydrologic/hydraulic flood control structures. The dramatic societal impacts of these events motivate researchers to perform studies aimed at developing science-based recommendations on best approaches to help decision makers address this issue [2]. Such information can lead to solutions that help save lives and resources and provide opportunities to harness floodwaters and turn them into a resource that can benefit the society. However, conducting hydrological studies of these events in ungauged areas is hampered by lack of adequate rainfall and physiographical data. Understanding the hydrometeorological conditions and processes that lead to destructive flood events is a first step in this process [3]. Physiographic data that enable the development of hydrologic models at reasonable resolutions over ungauged basins have become available globally in recent years. However, the accuracy of hydrologic model simulations is controlled by the accuracy of the model inputs, especially precipitation. A major challenge is the low quality and spatio-temporal resolutions of precipitation data for regions that do not have adequate ground observation networks.

High-resolution, better-quality satellite precipitation products that are increasingly becoming available can potentially lead to improvements in hydrologic modeling and forecasting, improvements as substantial as those witnessed in the United States in the 1990s when the NEXRAD radar network became operational. The potential of satellite precipitation products for various hydrometeorological applications has been reported in numerous studies [4–7]. Weather radars brought about the advantage of better spatial coverage, but satellites have even a better spatial coverage, though at lower temporal and spatial resolutions, and their field of view is not obstructed by topography. A major difference between radar and satellite precipitations products for hydrological applications is that when the NEXRAD network was deployed, the radar retrieval techniques were in place; however, it took many years to develop robust retrieval techniques after the launch of the Tropical Rainfall Measuring Mission (TRMM) satellite in late 1997. Also, systematic and random error of satellite precipitation products are typically higher than those associated with radar products [8,9]. Some of the satellite precipitation products that have been widely used in the past two decades include the Precipitation Estimation from Remotely Sensed Information using Artificial Neural Networks (PERSIANN) [10], the Climate Prediction Center (CPC) MORPHing (CMORPH) [11], the Tropical Rainfall Measuring Mission (TRMM) Multi-satellite Precipitation Analysis (TMPA) [12], and Global Satellite Mapping of Precipitation (GSMap) [13] products. These multi-sensor techniques that merge quality-controlled satellite products with higher resolution data from radars and rain gauges are helping improve the accuracy of satellite-based rainfall products over time [14]. In hydrological applications, however, satellite rainfall errors may either be amplified on dampened in simulated runoff at the catchment scale depending on the interaction between the spatio-temporal patterns of errors and catchment properties such as size, slope, and initial moisture conditions [15,16].

Simulation of flash floods, which are typically triggered by abrupt and intense bursts of rainfall, will benefit most from the fine resolution of the recent satellite rainfall products. For example, Anquetin et al. [17] reported that a higher resolution precipitation product was able to capture features of the precipitation system that caused the devastating 2002 flash flood in France. These features were missed by low resolution products. Several other studies demonstrated the need of high-resolution rainfall data for flash flood studies [18–20]. The number of hydrometeorological and climatological applications of satellite precipitation products will definitely increase with the emergence of the latest satellite product, the Integrated Multi-satellitE Retrievals for Global Precipitation Measurement (IMERG) product, with spatial and temporal resolutions of $0.1 \times 0.1°$ and 30 min, respectively. IMERG is based on the Global Precipitation Mission (GPM), which was deployed in 2014 to consolidate and enhance precipitation measurements from a constellation of research and operational microwave sensors [21]. GPM is composed of one Core Observatory satellite, deployed by NASA and the Japan Aerospace Exploration Agency (JAXA), and carries a dual frequency radar and a multi-channel microwave imager, and about 10 partner satellites [22]. IMERG integrates the intermittent precipitation estimates from all GPM microwave sensors (high quality, low temporal resolution) with infra-red-based observations from geosynchronous satellites (lower quality, higher temporal resolution) and precipitation gauge data to produce a uniformly gridded, global, multi-sensor precipitation product. The IMERG product is designed to incorporate strengths and avoid major weaknesses of the previous multi-satellite algorithms supported by NASA: CMORPH-KF, TRMM-TMPA, and PERSIANN. This high-resolution precipitation data will significantly advance hydrological modeling and predictions worldwide, especially in ungauged and poorly gauged basins. The near-real-time IMERG Early and Late products are available within 6 h and 18 h after observations are made, respectively. After the gauge analysis is incorporated, the final satellite-gauge IMERG product becomes available, typically three or more months after the month in which the observations were made.

Since the hydrological response of a basin is very sensitive to the spatio-temporal variability in various physical attributes of soil, land use, and topography, hydrological models that consider the spatial variability are better suited for accurate flood simulation and predictions [23–25]. Distributed hydrologic models can also provide a detailed description of the flood hazard areas, especially in urban catchments [26]. Physically based, distributed models employ a gridded nature, which allows parameters to be constrained within certain ranges that have clear physical meanings. Recent research demonstrated that physically-based, distributed hydrologic models can potentially perform as well as—or outperform—calibrated conceptual, lumped models [24–27]. The physically-based Gridded Surface Subsurface Hydrologic Analysis (GSSHA) model is an example of a grid-based fully-distributed hydrologic models [28]. GSSHA is capable of simulating flow generated from Hortonian runoff, saturated source areas, exfiltration, and ground water discharge to streams [28]. Sharif et al. [29] successfully applied the GSSHA model to evaluate the effect of flood control structures on stream discharge in urbanized watersheds. Furl et al. [23] used the model to describe the flood hydrology of a small urbanized basin in Austin, TX. Chintalapudi et al. [30] used the GSSHA model to study the effect of land cover changes on peak discharge and runoff volumes with simulations driven by satellite rainfall products. Ogden et al. [31] compared the GSSHA distributed model to the HEC-HMS (a lumped model). Results showed that HEC-HMS failed to simulate some extreme events using standard parameters, whereas the GSSHA performed fairly well. Elhassan et al. [32] compared the simulated stream generated by the GSSHA and HEC-HMS models for different storm events. They concluded that the GSSHA simulated streamflow matched the observations much better than HEC-HMS. In addition to these studies, the model has been validated over a densely urbanized catchment in Texas and the results demonstrated the benefit of the use of 30-m model grid over urban areas [25]. Different satellite products we used as input to GSSHA in simulation of several floods over a 3000 km$^2$ catchment in Texas [30] with satellite products of higher spatiotemporal resolutions producing the most reasonable runoff estimates. Another study over the Guadalupe River in Texas demonstrated that GSSHA was more successful in simulating events with multiple rainfall hiatuses than the HEC model [33]. An experiment using rainfall forecasts over a semi-arid urban catchment in Colorado demonstrated that GSSHA was able to produce reasonable forecasts of inundation and peak discharge for lag times of up to 70 min [34].

Recent extreme precipitation and flooding events in the Arab Peninsula led to several hydrometeorological studies. Furl et al. [35] analyzed rainfall in the southwestern region of Saudi Arabia and highlighted the lack of dense rain gauge networks. Almazroui [36] studied the TRMM rainfall data over Saudi Arabia throughout the period from 1998 to 2009. Although mixed results were obtained regarding accuracy of the TRMM estimates, the study recommended using the product to complement rainfall data from the extremely sparse rain gauge network in the country. A more recent study [37], evaluated the use of TRMM rainfall estimates for flood warning in urban areas of the country and concluded that TRMM satellite rainfall will provide some helpful information for preparation during extreme events but with low accuracy in terms of the spatiotemporal distribution of the rainfall storms. However, detailed fully-distributed hydrological analysis of runoff generation during these events is lacking, especially for urban areas that are most vulnerable to flash flooding events. A major reason for that was the lack of rainfall data at resolutions suitable for physically-based hydrologic modeling. The main objective of this study is to examine the flooding potential in the arid Wadi Al Batin catchment, Saudi Arabia, with focus on the rapidly urbanizing city of Hafr Al Batin, located near the outlet of the catchment. The city started to witness frequent flooding in the last two decades. A short-lived storm that hit the city in 2009 caused major ephemeral streams in the city to flow overbank, resulting in devastating flooding in the residential areas and significant damage to public and private properties. A flood event that occurred in October 2015 is employed as a case study in this paper. Detailed understanding of the runoff response in the urbanized part of the city of Hafr Al Batin will provide invaluable information that can help city officials to identify appropriate actions for reducing the probability of future flooding as well as to implement mitigation measures for severe storm events. GPM satellite precipitation data was used as input to a physically-based distributed hydrologic model. At model grid resolutions of 30 m to 700 m, computational requirements were reduced by dividing the basin into fully hydrologically interconnected sub-basins. The high-resolution simulations in the urbanized portion of the catchment helped identify the areas most susceptible to flooding. The relative contribution of the three streams that meet the main channel near the urban center was examined through model simulations.

## 2. Study Area

Hafr Al Batin is an old town in northeastern Saudi Arabia not very far from the borders with Kuwait and Iraq. The city has been witnessing a high rate of urbanization in the past few decades with a current population of over 300,000. Hafr Al Batin lies in the valley of the mostly-dry Wadi Al Batin (Figure 1) from which it takes its name and is the sole source of its groundwater supplies. Wadi Al Batin represents the now-disconnected upstream segment of an ancient large river, Wadi Al Rimmah, originates in western Saudi Arabia and empties in the Arabian Gulf [38,39]. The main channel is highly incised, a sign of a history of very frequent deluges [40], which is probably the reason for the name Hafr (incision in Arabic).

As the purpose of the hydrologic simulations of this study is to quantify the runoff generation and spatial distribution of flooding in the city of Hafr Al Batin (Figure 2), the catchment outlet was selected just northeast of the city limits such that all of the city is included in the simulation. Digital Elevation Models (DEMs) at 30-m resolution of the catchment were based on the Saudi General Directorate of Military Survey (GDMS) national DEM data. These DEMs were based on photogrammetry and ground control data [41]. ArcGIS 10.3 software was used for processing and resampling of the data to different resolutions (the grid cell size of the hydrologic model). The Watershed Modeling System (WMS) software [42] was used to delineate the stream network and sub-catchments based on the DEM at variable resolutions. The stream network based on aerial photography was superimposed on the network generated through WMS-based delineation to make adjustments when necessary.

The drainage area of the delineated catchment is 4273 km$^2$, as seen in Figure 1, with the city of Hafr Al Batin representing only about 7.2% of the delineated area. The densely urbanized part represents about 4% of the total catchment. The catchment has a generally mild slope with elevation ranging between around 660 m above mean sea level at the western edge and 203 m at the outlet. The catchment is dominated by a combination of sand and gravel soil while the alluvial fans of the ephemeral streams (wadis) consist of weathered and fractured limestone and sandstone and permeable sediments. Soil and land use data over the catchment are shown in Figure 3. Land use/cover data were obtained from Spot 5 and Landsat remote sensing imagery. Soil and geologic data were obtained from the Saudi Geological Survey maps. Soil properties and parameters were derived from the digital soil map of the world [43]. These datasets were compiled and then processed in ArcGIS 10.3 to create GSSHA input files that represent the physical characteristics of the watershed.

The delineated catchment includes three other wadis that meet Wadi Al Batin (Figure 2) in or around the city. The North Fleaj originates northwest of the city and joins the main channel just north of the city. The Northwest Fleaj that joins Wadi Al Batin near the center of the city passes through a heavily urbanized area. The South Fleaj that originates southeast of the city and flows through sparsely urbanized areas except near its confluence with the Wadi Al Batin's main channel in the northern part of the city.

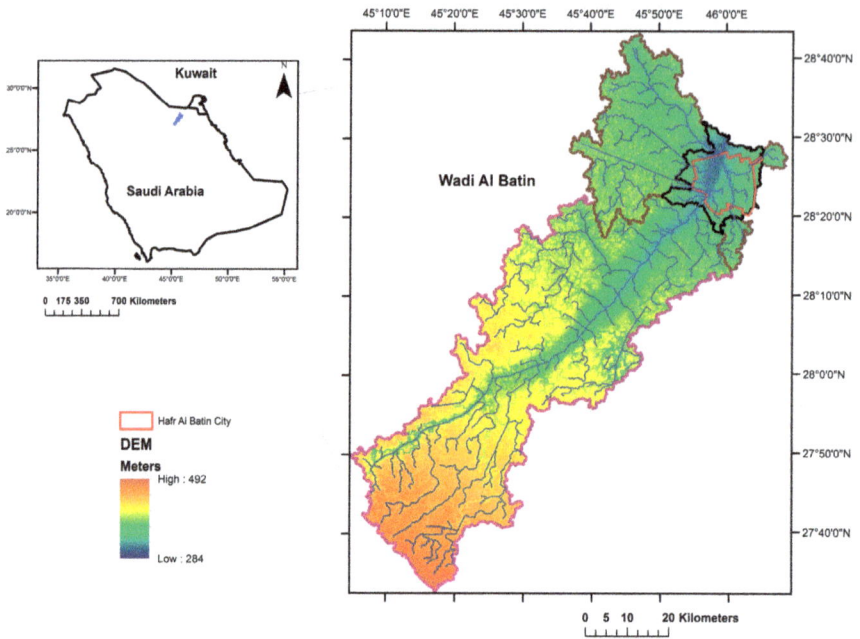

**Figure 1.** Location of the Wadi Al Batin catchment in Northeast Saudi Arabia.

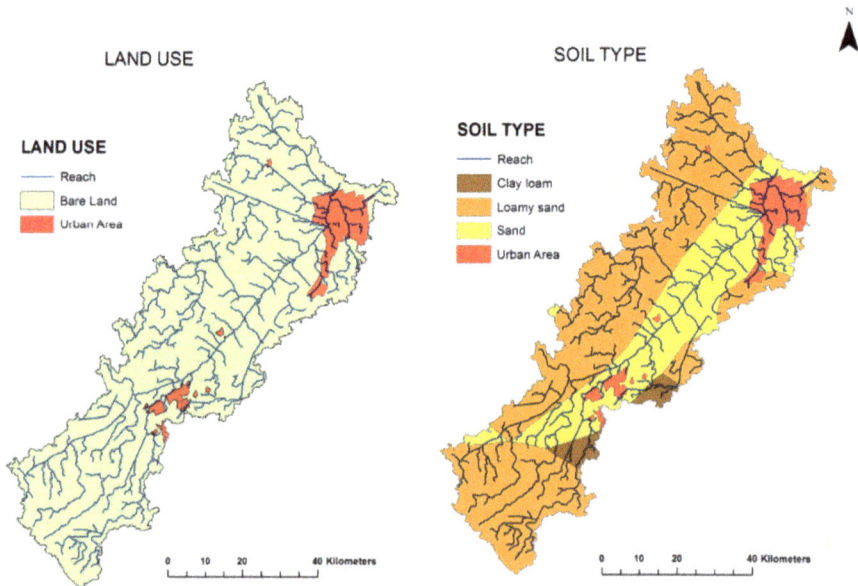

**Figure 2.** Soil and land use types of the Wadi Al Batin catchment in Northeast Saudi Arabia.

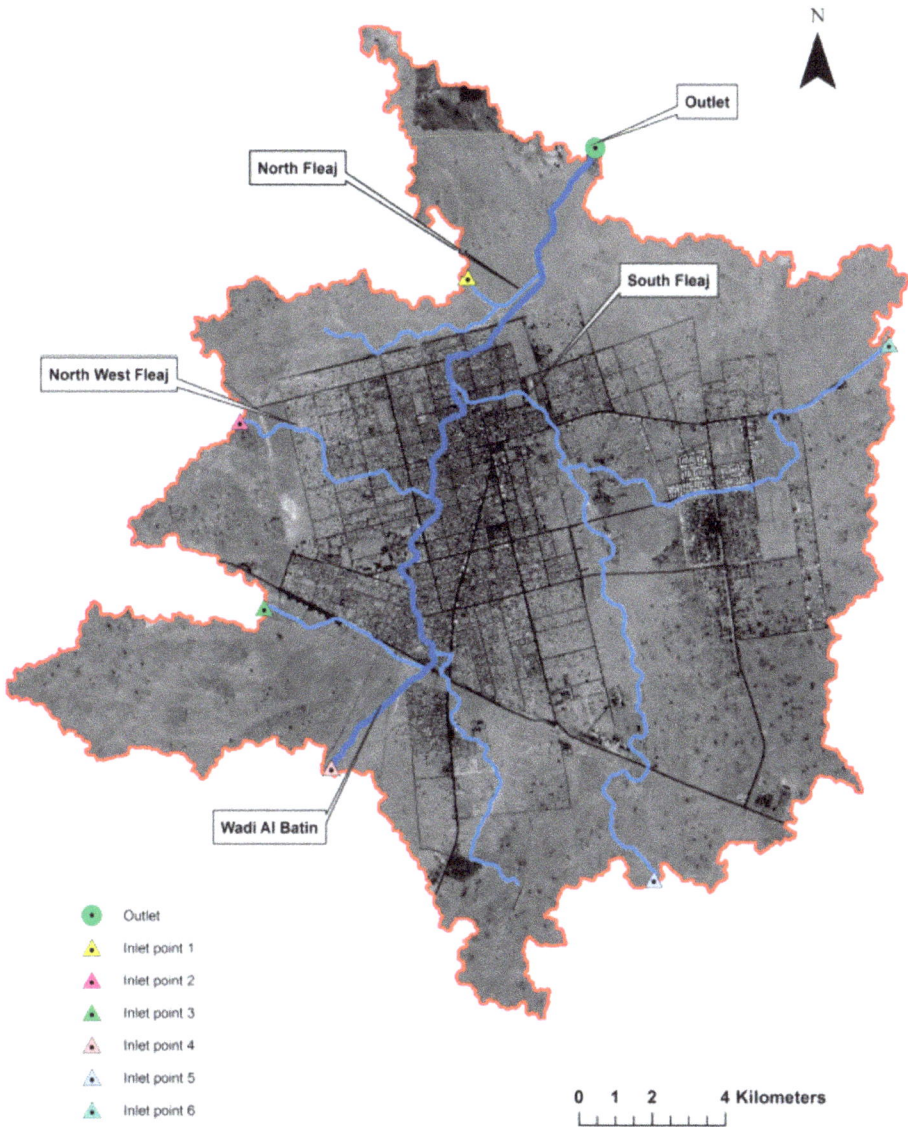

**Figure 3.** The city of Hafr Al Batin and Wadi Al Batin and its major tributaries.

## 3. Methods

### 3.1. Rainfall Data

The main hydrological simulations of this study were driven by satellite rainfall. The IMERG data were downloaded from the Precipitation Measurement Missions (PMM) website (http://pmm.nasa.gov/data-access/downloads/gpm). The three IMERG products, the early, late, and final were available for this event; all with spatial and temporal resolutions of $0.1 \times 0.1°$ (approximately $11 \times 11$ km) and 30 min, respectively. An R-based script was used to download GPM products, convert the rainfall data from HDF5 into a gridded ASCII format, and prepare the rainfall input files for the GSSHA model. Figure 4 shows the 3–5 November total rainfall I the region as estimated the IMERG product.

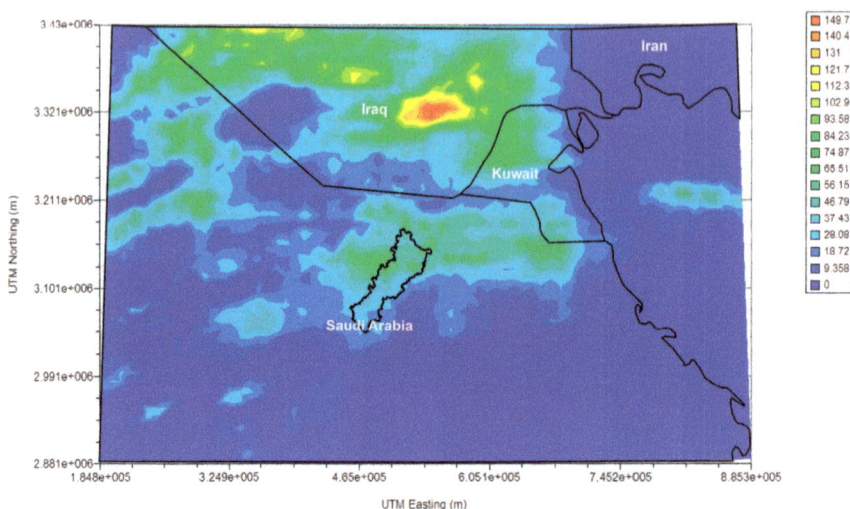

**Figure 4.** IMERG rainfall totals for the 2–3 November 2015 storm over the region.

*3.2. Hydrologic Model*

The physically-based, distributed-parameter Gridded Surface Subsurface Hydrologic Analysis (GSSHA) model is used to simulate recent flooding in Wadi Al Batin catchment. GSSHA is a process-based model for simulating all the hydrologic states and fluxes before, during and after storm events over each grid cell. GSSHA can accept spatially and temporally varying precipitation (e.g., from gauges, radar, satellite, or design storms). It can also ingest snowfall accumulation and simulate snow melting, abstractions due to interception, evapotranspiration, surface retention, and infiltration. The GSSHA model uses a simple two-parameter scheme to model interception of precipitation by plants. The user can select one of four infiltration methods: Green and Ampt (GA), Green and Ampt with Redistribution (GAR, [44]), multi-layered GA, and fully-implemented Richards's equation [45]. The last two methods are best suited for modeling continuous storm event with significant hiatuses. GSSHA can simulate overland runoff routing, unsaturated zone soil moisture dynamics, saturated groundwater flow, surface sediment erosion transport and deposition, in-stream sediment transport, simplified lake storage and routing, wetland peat layer hydraulics, and overland contaminant transport and uptake [45].

The GAR method was used in simulating the flooding event of November 2015. The method, which computes inter-storm redistribution of soil water and performs multiple ponding simulations using the GA methodology, is based on the following equation:

$$f(t) = K\left(\frac{S_f(\theta_s - \theta_i)}{F(t)} + 1\right) \tag{1}$$

where:

$f(t)$—potential Infiltration rate (cm/h)
$F(t)$—cumulative Infiltration (cm)
$S_f$—wetting front suction head (cm)
$K$—effective hydraulic conductivity (cm/h) = $K_s/2.0$

$K_s$—saturated hydraulic conductivity (cm/h)
$\theta_s$—water content of the soil at natural saturation.
$\theta_i$—initial soil water content.

The GAR methods uses the traditional rectangular wetting form assumption to execute variations of this equation to predict infiltration for multiple periods of ponding [44]. The configuration of GAR is illustrated in several figures and discussed in detail by Ogden and Saghafian [44]. After subtracting all abstraction, ponded flow over each grid cell is computed using the diffusive wave approximation of Saint-Venant's equation and routed into two orthogonal directions. For grid cells adjacent to the watershed divide, only inwards flow is allowed [46]. The GSSHA model uses three numerical schemes to solve the diffusive wave equation: the Explicit, Alternative Direction Explicit (ADE), and ADE-Prediction Correction (PC) schemes. The physiographic conditions of the catchment dictate the most appropriate method. The Explicit is the fastest, simplest, and least robust method and the ADE-PC is the slowest and most robust method [46]. The ADE-PC, which is more demanding computationally, is generally recommended for watersheds of complex terrain when minimal smoothing of the DEM is warranted. The ADE scheme is used in this study. The ADE method uses the following formulas to calculate the flows.

First, inter cell flows are calculated in the $x$-direction by using Equation (2).

$$p_{ij}^N = \frac{1}{n}\left(d_{ij}^N\right)^{\frac{5}{3}}\left(s_{fx}^N\right)^{\frac{1}{2}} \tag{2}$$

Based on the flows in the $x$-direction, depths in each cell are calculated at the $n+1$ time level by using Equation (3).

$$d_{ij}^{N+\frac{1}{2}} = d_{ij}^N + \frac{\Delta t}{\Delta x}\left(p_{i-1,j}^N - p_{ij}^N\right) \tag{3}$$

Equation (4) is used to calculate the interflows in the $y$-direction from each cell.

$$q_{ij}^{N+\frac{1}{2}} = \frac{1}{n}\left(d_{ij}^{N+\frac{1}{2}}\right)^{\frac{5}{3}}\left(s_{fy}^{N+\frac{1}{2}}\right)^{\frac{1}{2}} \tag{4}$$

Column depths are updated based on the interflows in the $y$-direction. Equation (5) is used to update the column depths.

$$d_{ij}^{N+1} = d_{ij}^{N+\frac{1}{2}} + \frac{\Delta t}{\Delta x}\left(q_{i,j-1}^{N+\frac{1}{2}} - q_{ij}^{N+\frac{1}{2}}\right) \tag{5}$$

where:

$p_{ij}$ and $q_{ij}$ are the overland flows from cell $ij$ in the x and y directions, respectively
$d_{ij}$ is the depth of water in cell $ij$ at the $N^{th}$ time level
$s_{fx}$ and $s_{fy}$ are the water surface slopes in the x and y directions, respectively.
$n$ is Manning's roughness coefficient

In this study, the Explicit solution scheme is used for solving the diffusive wave equation for 1-D channel routing. Overland flow that is entered into the stream is routed until it reaches the outlet. The volume of flow at each node is calculated using Equation (6).

$$V_i^{N+1} = V_i^N + \Delta t\left(q_{lat}^{N+1}\Delta x + q_{rec}^{N+1}\Delta x + Q_{i-1/2}^N - Q_{i+1/2}^N\right) \tag{6}$$

where:

$q_{lat}$—amount of lateral flow (m$^2$/s)
$q_{rec}$—amount of flow exchanged between the groundwater and channel (m$^2$/s)

$Q_{i-1/2}^N$ *and* $Q_{i+1/2}^N$ are the inter cell flows in the longitudinal direction ($x$) computed from depths $d$, at the $N^{th}$ time level. A simplified flowchart of GSSHA model is shown in Figure 5.

*3.3. Model Setup*

The GSSHA model step follows the flowchart shown in Figure 5. Al Batin catchment was divided into three sub-catchments: the upper sub-catchment of entirely barren desert with an area of 3117 km$^2$, the middle sub-catchment with increased density of small dry channels that join the main channel (an area of approximately 847 km$^2$), and the lower sub-catchment, which includes the city of Hafr Al Batin and its surroundings and covers an area of 309 km$^2$. To provide the level of detail needed to simulate runoff over the three sub-catchments, three sizes of GSSHA model grid were used, $270 \times 270$ m$^2$, $90 \times 90$ m$^2$, and $30 \times 30$ m$^2$ for the upper, middle, and lower sub-catchments, Figure 4. GSSHA model has to be run on each sub-catchment separately to allow for a variable grid size.

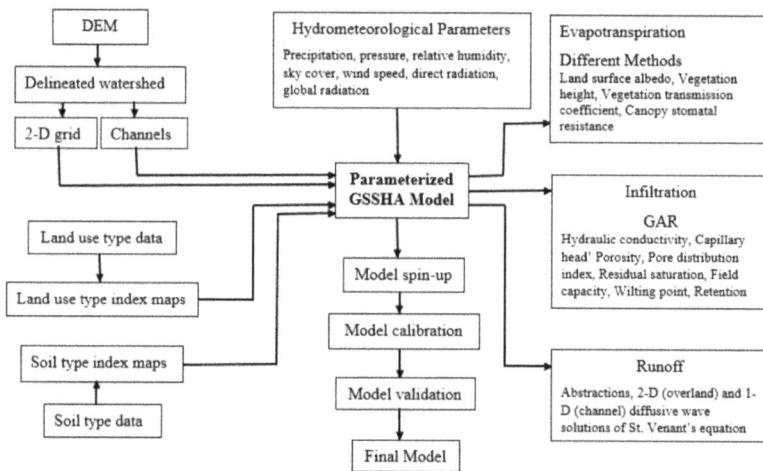

**Figure 5.** A simplified flowchart of GSSHA model.

The hydrologic connectivity between the three adjacent sub-catchments is maintained through the downstream channel flow from an upstream sub-catchment because there is no sub-surface flow interchange among them. For every model time step, outflow from the upstream sub-catchment is added as inflow to the most upstream cell of the corresponding channel in the downstream sub-catchment. This inflow is merged with the channel flow in the same manner lateral inflow is treated at every time step—i.e., it is not just added to the downstream channel flow—as is typically done in popular semi-distributed models. This method of hydrologic connectivity (not simple routing) will ensure that dividing the catchment into sub-catchments will have minimum effect on the hydrograph and the water balance at every simulation time step. The model has to be run on the three sub-catchments in sequence or in parallel separated by one time step. A script was written to run the simulations in sequence and transfer of output from one sub-catchment to the next. The three sub-catchments are shown in Figure 6 together with the location of the outflow from one sub-catchment into the next. Since the sub-catchment delineation was not based entirely on topography, overland flow across the divide between the sub-catchments is not included. The sub-catchments were delineated such that overland flow between sub-catchments is minimal and can be neglected in the simulation.

The $30 \times 30$ m DEMs were used to delineate the channel network. Minor adjustments were performed after comparing the delineated network with aerial photographs of the catchment. All stream channels within the urbanized sub-catchment were modeled with irregular cross sections as well as the main channel in the other two sub-catchments. Uniform trapezoidal channels were used for the tributaries in the un-urbanized sub-catchments. Manual adjustment of stream channels using the WMS 'smoothing' tools were used to adjust the profiles of the stream channels to remove several regions of adverse (negative) channel slope resulting from errors in the DEM. Infiltration is simulated using the Green and Ampt with redistribution and the ADE method was selected for overland flow routing. The infiltration model parameters' Manning roughness coefficients were taken from the GSSHA manual [46]. The initial soil moisture is estimated by running GSSHA model simulations over the few weeks prior to the event and extracting initial soil moisture for each model grid from the final map of spatially-distributed soil moisture values.

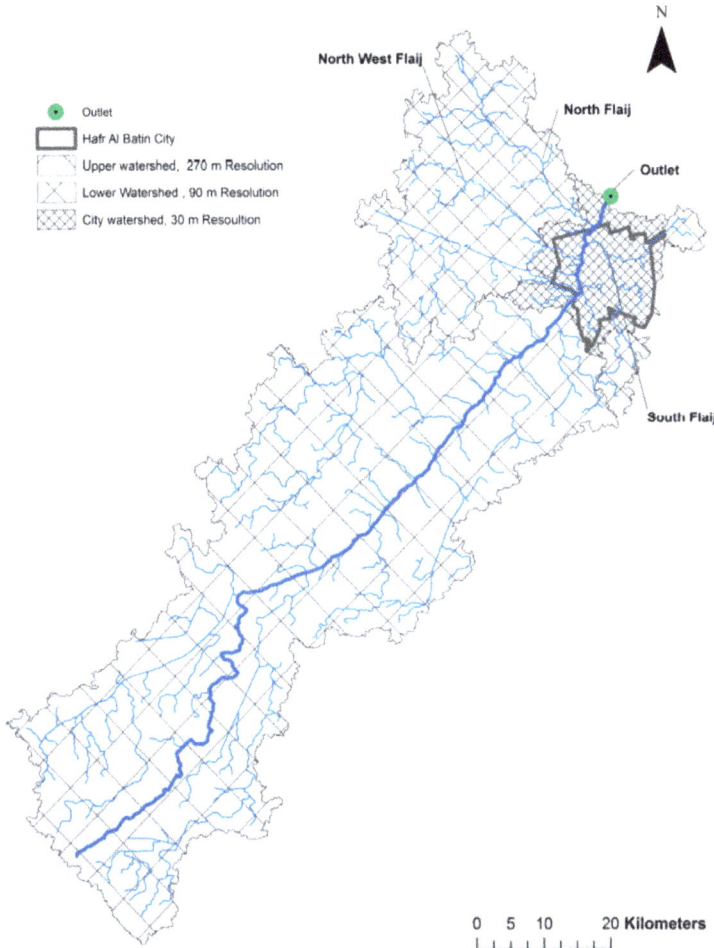

**Figure 6.** The three sub-catchments of the Wadi Al Batin with variable model grid sizes.

## 4. Results

The storm event started on 2 November 2015 at 9: 30 p.m. local time and ended at 11:30 a.m. on 3 November 2015. The spatial distribution of the total rainfall accumulations over the catchment estimated by the three GPM products, using Inverse Distance Weighted (IDW) interpolation, is shown in Figure 7. For the three products, the high amounts of rainfall fell on the urbanized portions of the catchment and immediately to the west of the city of Hafr Al Batin. Only two rain gauges were operational during the storm and they reported total accumulations (Figure 7). The Ministry of Environment, Water, and Agriculture rain gauge located inside the city recorded 63 mm while the Presidency of Meteorology and Environment rain gauge at the airport south of the city recorded 32 mm. The satellite grids collocated with the two gauges reported total accumulations of 47 mm and 23 mm for the Early IMERG product, respectively. The total storm precipitation averaged over the entire catchment for this product is 33.6–32 mm, 48 mm, and 47 mm over the upper, middle, and lower sub-catchments, respectively. The other two satellite products estimated slightly higher rainfall with the final IMERG product reporting the highest rainfall and a somewhat different spatial pattern of the storm. The total storm rainfall averaged over the entire catchment for the Late IMERG products are 37.2 (28.5, 44.7, and 43.1 mm for the three sub-catchments, respectively) and 46.2 mm for the final product (36.5, 52.9, and 51.9 mm for the three sub-catchments, respectively). The temporal distributions of rainfall the three IMERG products are shown in Figure 8.

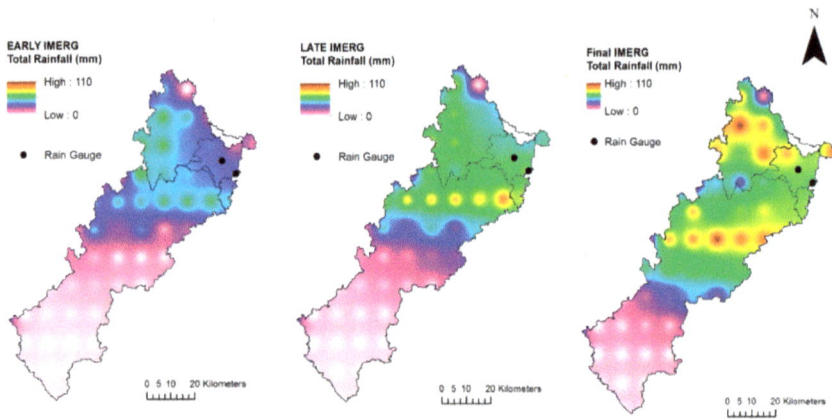

**Figure 7.** Rainfall totals for the 2–3 November 2015 storm as estimated by the three IMERG satellite products using IDW interpolation.

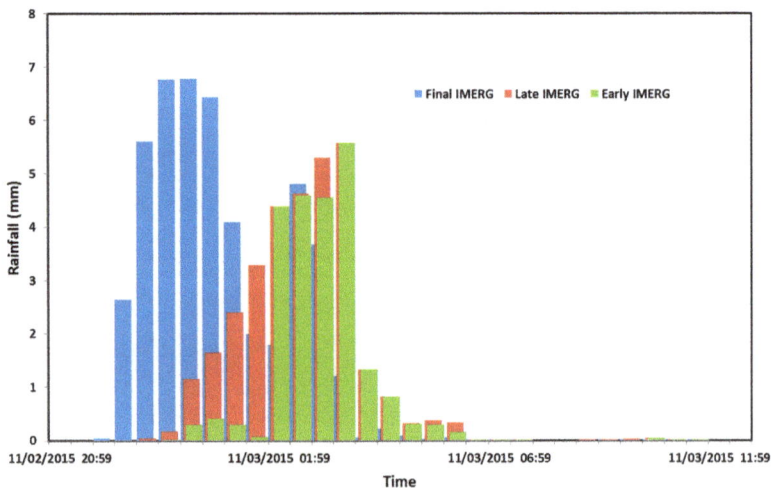

**Figure 8.** 30-min hyetographs of the the three IMERG products for the 2–3 November 2015 storm.

The GSSHA model was run for the period between 12:00 p.m. local time on 30 September 2015 and 10:30 a.m. on 5 November 2015 for the three catchments. The run was started one month in advance to spin up the model in order to obtain a reasonable estimate of the initial moisture content for each model grid (see Chintalapudi et al. [30,47] for details of model spin-up). The three GPM precipitation products were used to force the GSSHA model but the results forced by the final product will be discussed since this product is the one adjusted with climatology data and ground observations. The storm lasted for a period of 14 h which allowed for infiltration of most of the rainfall except in the urbanized areas. The discharge from the upper sub-catchment into the middle sub-catchment was very small due the small amount of rainfall and high infiltration in the barren dominantly sandy soils. The runoff ratio for the event was 1% and most of the water that reached the main channel infiltrated in the channel. The middle sub-catchment generated modest amounts of runoff with a runoff ratio of 5%. Most of the flow from the middle sub-catchment into the lower one occurred through the main channel. The discharge at the watershed was primarily due to runoff generated within the urbanized sub-catchment with inflow form the upstream sub-catchments representing more than 12% of the total discharge. The runoff ratio in the urbanized area was 15% due to higher rainfall, the significant impervious fraction, and efficient drainage by the natural channel network.

The peak discharge of 127 cms at the outlet occurred at 8:00 a.m. on 3 November 2015. Peak discharges were 10, 50, and 90 min earlier for the North Fleaj, the South Fleaj, and the Northwest Fleaj tributaries, respectively. The South Fleaj contributed most to the total discharge at the outlet (about 46%) and has the highest peak among the tributaries. The contribution from the North Fleaj and North West Fleaj was very small (each contributed just about 5%). Lack of development around the South Fleaj indicates that this tributary witnesses frequent flooding, even more than the main Al Batin Wadi. Underestimation of rainfall by the Early and Late IMERG products was amplified in runoff simulation, producing much smaller outlet peak discharges of 44.2 and 66.7 cms, respectively.

The predicted outlet hydrographs when the model was forced by the three IMERG products are shown in Figure 9. As described in the introduction section, the final IMERG product is the best rainfall product. There are no streamflow observations for the event, however, GSSHA simulations highlight the fact that rainfall errors will result in higher runoff errors and give an idea about the magnitude of runoff error (in this case, difference between using calibrated and raw satellite rainfall) to be expected when the real-time product is used for flood forecasting.

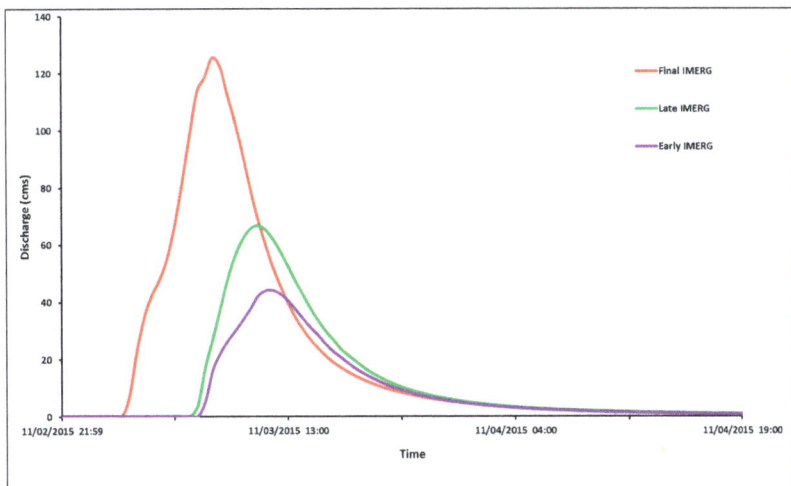

**Figure 9.** Outlet discharge predicted by GSSHA model for 2–3 November 2015 storm as estimated by the three GPM satellite products.

GSSHA outputs also include the overland and channel flow depths at every time step. The map shown in Figure 10 illustrates the maximum inundation over the city area as a result of the storm—only depth values above 3 cm are shown. The water depth values for each grid were estimated by GSSHA forced by the final IMERG product. The map indicates that most of the flooding occurred in the highly urbanized areas and near the center of the city. The map also shows significant flooding over some major streets. There is significant flooding also over the undeveloped areas adjacent to the South Fleaj. The storm did not result in any flooding at the confluences of the North West Fleaj and South Fleaj with Wad Al Batin. It is clear that the inundation is caused by topography and neither the Wadi Al Batin nor the South Fleaj witnessed overbank flooding. These results agree with media reports and photographs released after the event that described significant street flooding without a mention of channel overflow. The map shows that there is hardly any flooding outside the urban area, reinforcing the fact that the city was built over the ancient flood plain of Wadi Al Batin, which might be the reason for the city name, and the role of urbanization in increasing the flood hazard. When GSSHA was forced by the Early and Late GPM products, the inundation maps illustrate smaller flooding depths and extent.

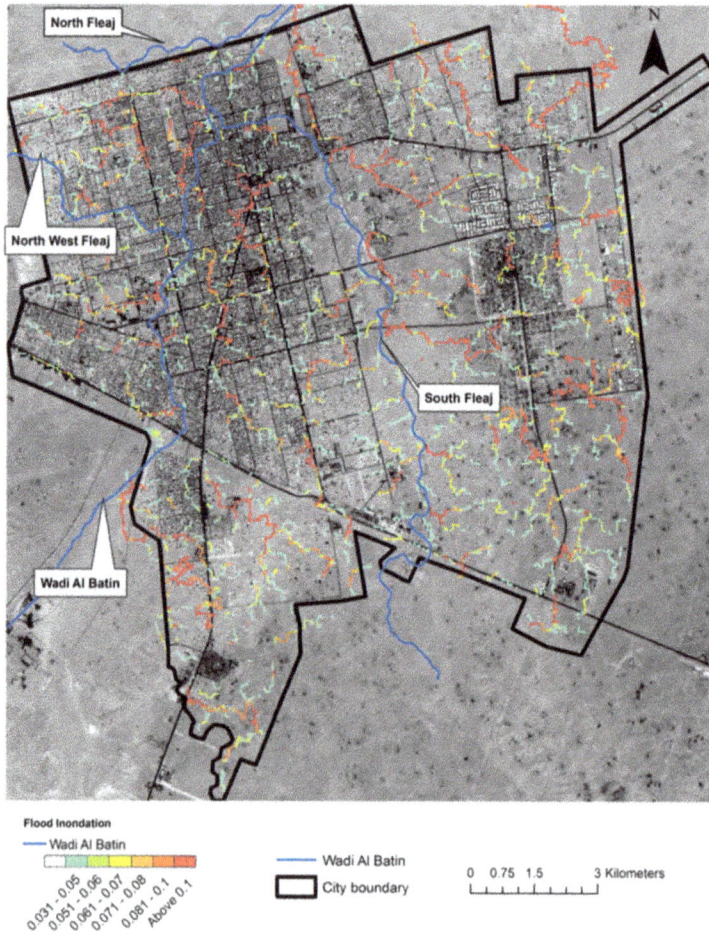

**Figure 10.** Flood inundation in the city of Hafr Al Batin caused by the 2–3 November 2015 storm as estimated by the GSSHA model.

To understand the role of tributaries and inflow from the upper portions of the catchment in the city of Hafr Al Batin flooding, two more GSSHA simulations were performed: a simulation representing the frequent flood events in the city and another simulation representing extreme events. GSSHA was run with the 5-year, 24-h, and 100-year, 24-h storms, which are considered spatially uniform. The two storms were based on statistical analysis of the historical rainfall data from the Ministry of Environment, Water, and Agriculture rain gauge with a record starting from 1979. The 24-h rainfall accumulations for the 5-year and 100-year, 24-h storms were found to be 31 mm and 73 mm, respectively, using Log-Pearson Type III distribution (LPT III) [48]. Due to the size of the catchment, an area-reduction factor of 0.8. Was used (see [49,50]). The temporal distributions of the designed storms, which are based on Type II storms [51], are shown in Figure 11. GSSHA simulation results indicate that the upper sub-catchment contributes less than 1% of the total discharge flooding in the city for the 5-year storm. Contribution from the middle sub-catchment is also small, about 7%, for this event. The 5-year event does not generate much discharge at the outlet. The South Fleaj, Northwest Fleaj, and North Fleaj tributaries contribute about 52.2, 13.6, and 6.9% of that discharge, respectively. For the 100-year event, the contributions from the middle sub-catchment is significant. The peak discharge at the outlet of the upper sub-catchment is 5.5 cms and it contributes less than 3% of the total discharge at the outlet. The middle sub-catchment contributes 21% of the total discharge with a peak of 50 cms at the main channel. Again, South Fleaj contributed most to the total discharge of the lower sub-catchment (53%) and has the highest peak among the tributaries while the other two tributaries contributed smaller amounts −6% from the Northwest Fleaj and 5% from the North Fleaj.

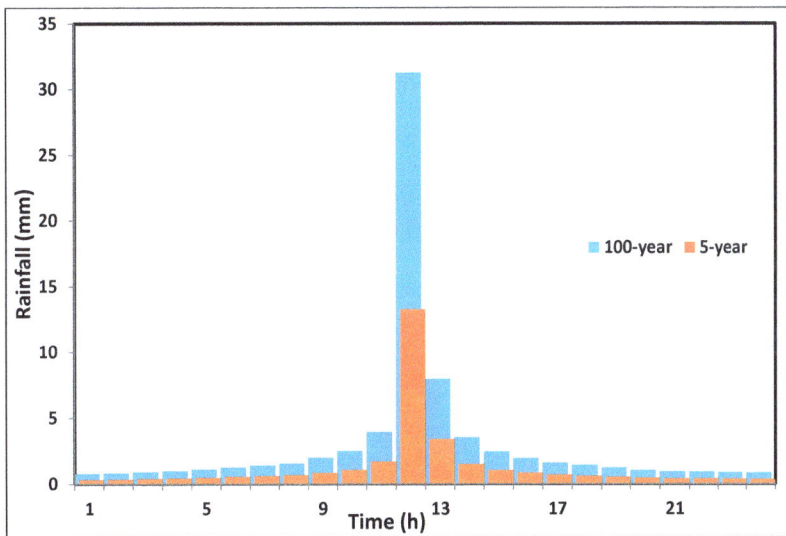

**Figure 11.** Hourly hyeographs of the design storms used for addition GSSHA model simulations.

## 5. Summary and Conclusions

In this study, the physically-based, distributed-parameter hydrologic model GSSHA, forced by Integrated Multi-satellite Retrievals for Global Precipitation Measurement Mission (IMERG) rainfall product, was used to study a recent flood event in the city of Hafr Al Batin, Saudi Arabia. Due to the large size of the catchment that encompasses the city, it was divided into three hydrologically connected sub-catchments: a lower sub-catchment that represents the urban area, an upper sub-catchment dominated by barren desert land, and a middle sub-catchment. A variable model grid size of $270 \times 270$ m$^2$, $90 \times 90$ m$^2$, and $30 \times 30$ m$^2$ was adopted for the upper, middle, and lower sub-catchments.

The semi-real time IMERG products underestimated the event rainfall with much more pronounced underestimation of the event runoff. The IMERG product is relatively news and is currently undergoing continuous enhancement. The references cited in the introduction section of this paper provide early assessments and discussions of the validity of the product. These products are expected to improve as the IMERG algorithms get refined through more ground validations. Nonetheless, they provide valuable information that can help improve flood prediction in ungauged basins. The authors are not aware of published hydrologic applications of the IMERG products. There was no streamflow data to validate the hydrographs predicted by the hydrologic. However, the model has been validated over similar environments in many previous studies as described in the introduction section. Notwithstanding, the GSSHA model simulations forced by the final IMERG product enabled quantifying the relative contribution of the sub-catchments and the major tributaries in the urbanized area to flooding of the city. For this event, the distributed model simulations demonstrated that most of the flooding (approximately 85%) was generated in the urbanized portion of the catchments (6.8% of the total area of the simulated catchment). The contribution of the upper portion of the catchment (68% of the area) was insignificant due to it sandy soils and the limited amounts of rainfall it received. The middle sub-catchment contributed about 13% of the discharge at the outlet.

One of the tributaries, the Northwest Fleaj, that meets the main channel inside the city, does not play an important role in flooding of the city center due to the size of its drainage area and its physiographic feature. For the same reasons, the North Fleaj, which meets the main channel just north of the city, does not contribute significant discharge. However, the South Fleaj that meets the main channel just north of the city center contributes significantly to the total discharge in the main channel and therefore has to be considered in any future flood control projects and urban development in the eastern part of the city. As this tributary originates well outside the city, different types of flood control measures can be applied to reduce the discharge it contributes. Distributed model simulations demonstrate that flooding in the city is driven primarily by topography rather that overbank flow in the main channel. A well designed urban drainage network might be needed to prevent flooding of residential areas and streets. This information is important if the city officials want to implement flood mitigation measures. For example, the presence of North Fleaj will limit expansion of the urban area to the north because of increasing flood hazards unless its flow is diverted before entering the urban area. Large detention basins outside the city on the main channel and Northwest Fleaj and South Fleaj can help mitigate flooding caused by extreme event. Also, diversion of the Northwest Fleaj and South Fleaj flow outside the city can be helpful.

Additional simulations were performed to understand the roles of the unique channel network in flooding in the city of Hafr Al Batin. The barren desert upper portions of the catchment contribute to flooding in the city only when it receives significant amounts of rainfall. It contributes about 3% of the total discharge when the 100-year storm covers the entire catchment, which should have a probability of much less than 1%. No significant discharge results from the 5-year storm. The middle sub-catchment contributes 7% of the total runoff for the 5-year storm and 23% for the 100-year storm. The results demonstrate that the upper portions of the catchment do not pose a significant flood threat unless they receive exceptional amounts of rainfall. In that case, the early or even the late satellite rainfall products can be invaluable since the flood peak will arrive many hours after the precipitation peak for such a large catchment. The best approach to control flooding in the city of Hafr Al Batin is to improve storm drainage to control runoff generated within and around the urbanized area as described above and use quantitative precipitation estimates, such as IMERG, and/forecast to prepare in the case of exceptional rainfall events in the upper portions of Wadi Al Batin catchment.

The use of variable grid size resulted in significant saving of computing time. It is necessary to use a $30 \times 30 \text{ m}^2$ grid size in the urbanized portion of the catchment (about 300 km$^2$) to make full use of the topographic resolution and include the major land surface in the simulations. However, there is no need to use such a resolution of the dominantly barren desert portion of the catchment. Employing this gridding scheme, the simulation run takes about 5% of what it would have taken had the $30 \times 30 \text{ m}^2$

grid been used for the entire catchment (more than 4000 km$^2$). We did not perform a synchronized run of the three simulations in parallel. The computing time for such a run is effectively not more than the time to run the model on the urbanized catchment alone. In addition to significant saving of computing time, the same approach used in this study will allow merging observed discharge from upstream portions of a catchment with the simulated runoff downstream.

Using this approach, physically-based, fully-distributed models like the GSSHA can be run on large basins that include urban centers that need to be modeled at very high resolutions. Previous studies demonstrated the validity of model predictions in urban settings, including semi-arid catchments. We believe that using high-resolution calibrated satellite products and land features would result in reasonable estimates of the flood inundation. However, several issues need to be taken into account. For example, the division of sub-catchments needs special treatment to make sure that no significant over land flow is lost. Also, the inflow into a sub-catchment must have the same time step of the model simulations (typically minutes or seconds). Disaggregation of observed discharge data represents a challenge as it may introduce significant errors. In addition, inclusion of sub-surface flow from one sub-catchment to the next, if needed, would add more complexity to the hydrological connectivity.

The main limitations of the study are that the IMERG product is relatively new and has not been validated extensively and there was no streamflow data to validate the hydrographs. Yet, the authors are confident that the main conclusions regarding the spatial distribution of the flood inundation and the relative contributions of different parts of the catchment to the flood generation will hold unless the hydrological model and IMERG are of very poor quality, which they do not believe to be the case.

**Acknowledgments:** The first author was funded in part by the U.S. Army Research Office (Grant W912HZ-14-P-0160). This support is cordially acknowledged. The second author would like to acknowledge the support provided by the Deanship of Scientific Research (DSR) at King Fahd University of Petroleum and Minerals (KFUPM) for funding his work through project No. RG1305-1&2.

**Author Contributions:** Hatim O. Sharif and Muhammad Al-Zahrani designed the overall study. Almoutaz El Hassan downloaded the remote sensing products and prepared and performed the hydrologic model simulations with input from Hatim O. Sharif and Muhammad Al-Zahrani. Hatim O. Sharif conducted post-analysis of the model outputs and prepared the first draft. Muhammad Al-Zahrani reviewed and revised the manuscript. Hatim O. Sharif did the final overall proofreading of the manuscript.

**Conflicts of Interest:** The authors declare no conflict of interest.

## References

1. De Moel, H.; Aerts, J.C.J.H. Effect of uncertainty in land use, damage models and inundation depth on flood damage estimates. *Nat. Hazards* **2011**, *58*, 407–425. [CrossRef]

2. Borga, M.; Stoffel, M.; Marchi, L.; Marra, F.; Jakob, M. Hydrogeomorphic response to extreme rainfall in headwater systems: Flash floods and debris flows. *J. Hydrol.* **2014**, *518*, 194–205. [CrossRef]

3. Piaget, N.; Froidevaux, P.; Giannakaki, P.; Gierth, F.; Martius, O.; Riemer, M.; Wolf, G.; Grams, C.M. Dynamics of a local Alpine flooding event in October 2011: Moisture source and large-scale circulation. *Q. J. R. Meteorol. Soc.* **2015**, *141*, 1922–1937. [CrossRef]

4. Prakash, S.; Mitra, A.K.; AghaKouchak, A.; Pai, D.S. Error characterization of TRMM Multisatellite Precipitation Analysis (TMPA-3B42) products over India for different seasons. *J. Hydrol.* **2015**, *529*, 1302–1312. [CrossRef]

5. Siddique-E-Akbor, A.H.M.; Hossain, F.; Sikder, S.; Shum, C.K.; Tseng, S.; Yi, Y.; Turk, F.J.; Limaye, A. Satellite Precipitation Data–Driven Hydrological Modeling for Water Resources Management in the Ganges, Brahmaputra, and Meghna Basins. *Earth Interact.* **2014**, *18*, 1–25. [CrossRef]

6. Wood, E.F.; Roundy, J.K.; Troy, T.J.; Van Beek, L.P.H.; Bierkens, M.F.; Blyth, E.; de Roo, A.; Döll, P.; Ek, M.; Famiglietti, J.; et al. Hyper-resolution global land surface modeling: Meeting a grand challenge for monitoring Earth's terrestrial water. *Water Resour. Res.* **2011**, *47*, W05301. [CrossRef]

7. Su, F.; Hong, Y.; Lettenmaier, D.P. Evaluation of TRMM Multisatellite Precipitation Analysis (TMPA) and its utility in hydrologic prediction in the La Plata basin. *J. Hydrometeorol.* **2008**, *9*, 622–640. [CrossRef]

8. Vergara, H.J.; Hong, Y.; Gourley, J.J.; Anagnostou, E.N.; Maggioni, V.; Stampoulis, D.; Kirstetter, P.-E. Effects of Resolution of Satellite-Based Rainfall Estimates on Hydrologic Modeling Skill at Different Scales. *J. Hydrometeorol.* **2013**, *15*, 593–613. [CrossRef]

9. Behrangi, A.; Khakbaz, B.; Jaw, T.C.; AghaKouchak, A.; Hsu, K.; Sorooshian, S. Hydrologic evaluation of satellite precipitation products over a mid-size basin. *J. Hydrol.* **2011**, *397*, 225–237. [CrossRef]

10. Sorooshian, S.; Kuo-Lin, H.; Xiaogang, G.; Gupta, H.V.; Imam, B.; Braithwaite, D. Evaluation of PERSIANN system satellite-based estimates of tropical rainfall. *Bull. Am. Meteor. Soc.* **2000**, *81*, 2035–2046. [CrossRef]

11. Joyce, R.J.; Janowiak, J.E.; Arkin, P.A.; Xie, P. CMORPH: A method that produces global precipitation estimates from passive microwave and infrared data at high spatial and temporal resolution. *J. Hydrometeorol.* **2004**, *5*, 487–503. [CrossRef]

12. Huffman, G.J.; Bolvin, D.T.; Nelkin, E.J.; Wolff, D.B.; Adler, R.F.; Gu, G.; Hong, Y.; Bowman, K.P.; Stocker, E.F. The TRMM multisatellite precipitation analysis (TMPA): Quasi-global, multiyear, combined-sensor precipitation estimates at fine scales. *J. Hydrometeorol.* **2007**, *8*, 38–55. [CrossRef]

13. Kubota, T.; Shige, S.; Hashizume, H.; Aonashi, K.; Takahashi, N.; Seto, S.; Takayabu, Y.N.; Ushio, T.; Nakagawa, K.; Iwanami, K.; et al. Global precipitation map using satellite-borne microwave radiometers by the GSMaP Project: Production and validation. *IEEE Trans. Geosci. Remote Sens.* **2007**, *45*, 2259–2275. [CrossRef]

14. Stisen, S.; Sandholt, I. Evaluation of remote-sensing-based rainfall products through predictive capability in hydrologic runoff modelling. *Hydrol. Process.* **2010**, *24*, 879–891. [CrossRef]

15. Nikolopoulos, E.I.; Anagnostou, E.N.; Borga, M. Using High-resolution satellite rainfall products to simulate a major flash flood event in Northern Italy. *J. Hydrometeorol.* **2013**, *14*, 171–185. [CrossRef]

16. Yong, B.; Hong, Y.; Ren, L.-L.; Gourley, J.J.; Huffman, G.J.; Chen, X.; Wang, W.; Khan, S.I. Assessment of evolving TRMM-based multisatellite real-time precipitation estimation methods and their impacts on hydrologic prediction in a high latitude basin. *J. Geophys. Res.* **2012**, *117*, D09108. [CrossRef]

17. Anquetin, S.; Yates, E.; Ducrocq, V.; Samouillan, S.; Chancibault, K.; Davolio, S.; Accadia, C.; Casaioli, M.; Mariani, S.; Ficca, G.; et al. The 8 and 9 September 2002 flash flood event in France: An intercomparison of operational and research meteorological models. *Nat. Hazards Earth Syst. Sci.* **2005**, *5*, 741–754. [CrossRef]

18. Deng, L.; McCabe, M.F.; Stenchikov, G.; Evans, J.P.; Kucera, P.A. Simulation of flash-flood-producing storm events in Saudi Arabia using the weather research and forecasting model. *J. Hydrometeor.* **2015**, *16*, 615–630. [CrossRef]

19. Gao, W.; Sui, C.-H. A modeling analysis of rainfall and water cycle by the cloud-resolving WRF Model over the western North Pacific. *Adv. Atmos. Sci.* **2013**, *30*, 1695–1711. [CrossRef]

20. Xie, B.; Zhang, F. Impacts of typhoon track and island topography on the heavy rainfalls in Taiwan associated with Morakot (2009). *Mon. Weather Rev.* **2012**, *140*, 3379–3394. [CrossRef]

21. Hou, A.Y.; Kakar, R.K.; Neeck, S.; Azarbarzin, A.A.; Kummerow, C.D.; Kojima, M.; Oki, R.; Nakamura, K.; Iguchi, T. The global precipitation measurement mission. *Bull. Am. Meteorol. Soc.* **2014**, *95*, 701–722. [CrossRef]

22. GPM, 2016: Precipitation Measurement Missions. Global Precipitation Measurement Program, NASA. Available online: http://gpm.nasa.gov/ (accessed on 15 November 2016).

23. Furl, C.; Sharif, H.O.; El Hassan, A.; Mazari, N.; Burtch, D.; Mullendore, G.L. Hydrometeorological Analysis of Tropical Storm Hermine and Central Texas Flash Flooding, September 2010. *J. Hydrometeor.* **2015**, *16*, 2311–2327. [CrossRef]

24. Smith, M.B.; Koren, V.; Zhang, Z.; Zhang, Y.; Reed, S.M.; Cui, Z.; Moreda, F.; Cosgrove, B.A.; Mizukami, N.; Anderson, E.A. DMIP Participants Results of the DMIP 2 Oklahoma experiments. *J. Hydrol.* **2012**, *418–419*, 17–48. [CrossRef]

25. Sharif, H.O.; Sparks, L.; Hassan, A.A.; Zeitler, X.W.J.; Xie, H. Application of a Distributed Hydrologic Model to the November 17, 2004 Flood of Bull Creek Watershed, Austin, Texas. *J. Hydrol. Eng.* **2010**, *15*, 651–657. [CrossRef]

26. Sharif, H.O.; Al-Juaidi, F.H.; Al-Othman, A.; Al-Dousary, I.; Fadda, E.; Jamal-Uddeen, S.; Elhassan, A. Flood Hazards in an Urbanizing Watershed in Riyadh, Saudi Arabia. *Geomat. Nat. Hazards Risk.* **2014**. [CrossRef]

27. Koren, V.I.; Reed, S.; Smith, M.; Zhang, Z.; Seo, D.-J. Hydrology laboratory research modeling system (HL-RMS) of the US National Weather Service. *J. Hydrol.* **2004**, *291*, 297–318. [CrossRef]

28. Downer, C.W.; Ogden, F.L. GSSHA: Model to simulate diverse stream flow producing processes. *J. Hydrol. Eng.* **2004**, *9*, 161–174. [CrossRef]

29. Sharif, H.O.; Chintalapudi, S.; Elhassan, A.; Xie, H.; Zeitler, J. Physically-based hydrological modeling of the 2002 floods in San Antonio, Texas. *J. Hydrol. Eng.* **2013**, *18*, 228–236. [CrossRef]

30. Chintalapudi, S.; Sharif, H.O.; Xie, H. Sensitivity of distributed hydrologic simulations to ground and satellite based rainfall. *Water* **2014**, *6*, 1221–1245. [CrossRef]

31. Ogden, F.L.; Downer, C.W.; Meselhe, E. U.S. Army Corps of Engineers Gridded Surface/Subsurface Hydrologic Analysis (GSSHA) Model: Distributed-Parameter, Physically Based Watershed Simulations. In Proceedings of the World Water and Environmental Resources Congress 2003, Philadelphia, PA, USA, 23–26 June 2003; Bizier, P., DeBarry, P., Eds.; American Society of Civil Engineers: Washington, DC, USA, 2003.

32. Elhassan, A.; Sharif, H.O.; Jackson, T.; Chintalapudi, S. Performance of a Conceptual and a Physically-Based Model in Simulating The Response of a Semi-Urbanized Watershed in San Antonio, Texas. *Hydrol. Process.* **2013**, *27*, 3394–3408. [CrossRef]

33. Sharif, H.O.; Hassan, A.A.; Bin-Shafique, S.; Xie, H.; Zeitler, J. Hydrologic Modeling of an Extreme Flood in the Guadalupe River in Texas. *J. Am. Water Resour. Assoc.* **2010**, *46*, 881–891. [CrossRef]

34. Sharif, H.O.; Yates, D.; Roberts, R.; Mueller, C. The use of an automated now-casting system to forecast flash floods in an urban watershed. *J. Hydrometeorol.* **2006**, *7*, 190–202. [CrossRef]

35. Furl, C.; Sharif, H.O.; Elhassan, M.A.A.; Mazari, N. Precipitation Amount and Intensity Trends Across Southwest Saudi Arabia. *J. Am. Water Resour. Assoc.* **2014**, *50*, 74–82. [CrossRef]

36. Almazroui, M.A.; Islam, M.N.; Jones, P.D.; Athar, H.; Rahman, M.A. Recent climate change in the Arabian Peninsula: Seasonal rainfall and temperature climatology of Saudi Arabia for 1979–2009. *Atmos. Res.* **2012**, *111*, 29–45. [CrossRef]

37. Tekeli, A.E.; Fouli, H. Evaluation of TRMM satellite-based precipitation indexes for flood forecasting over Riyadh City, Saudi Arabia. *J. Hydrol.* **2016**, *541 Pt A*, 471–479. [CrossRef]

38. El-Baz, F. Remote Sensing: Generating Knowledge about Groundwater. Arab Environment: Water. 2010 Report of the Arab Forum for Environment and Development (AFED). 2010. Available online: http://www.afedonline.org/Report2010/pdf/AR/prefacear.pdf. (accessed on 12 January 2017).

39. Edgell, H.S. Aquifers of Saudi Arabia and their geological framework. *Arab. J. Sci. Eng. Water Resour. Arab. Penins. Part I* **1997**, *22*, 3–31.

40. McClure, H.A. Ar Rub' Al Khali. In *Quaternary Period in Saudi Arabia*; Springer: Vienna, Austria, 1978; Volume 1, pp. 252–263.

41. El Hassan, I.M.; Algarni, D.A.; Dalbouh, F.M. Flood risk prediction using DEM and GIS as applied to Wijj Valley, Taif, Saudi Arabia. *J. Geomat.* **2014**, *8*, 86–89.

42. EMRL (Environmental Modeling Research Laboratory). *Watershed Modeling System (WMS) Version 10.1 Tutorial*; Brigham Young University: Provo, UT, USA, 2016.

43. FAO (Food and Agriculture Organization of the United Nations). *World Reference Base for Soil Resources. World Soil Resources Report, #84*; FAO: Rome, Italy, 1998; p. 88.

44. Ogden, F.L.; Saghafian, B. Green and Ampt infiltration with redistribution. *J. Irrig. Drain. Eng.* **1997**, *123*, 386–393. [CrossRef]

45. GSSHA Primer, 2012: GSSHA Model Description and Formulation. Available online: http://gsshawiki.com/gssha/GSSHA_Primer#Description (accessed on 18 November 2016).

46. Gsshawiki, 2016: Information of GSSHA Model. Available online: http://gsshawiki.com/gssha/Main_Page (accessed on 18 November 2016).

47. Chintalapudi, S.; Sharif, H.O.; Furl, C. High-resolution, Fully-distributed Event-based Hydrologic Simulations over a Large Watershed in Texas. *Arab. J. Sci. Eng.* **2016**, in press. [CrossRef]

48. Chow, V.T. A general formula for hydrologic frequency analysis. *Trans. Am. Geophys. Union* **1951**, *32*, 231–237. [CrossRef]

49. World Meteorological Organization (WMO). *Manual on Estimation of Probable Maximum Precipitation*; Oper. Hydrol. Rep., 1, WMO Pap. 332; WMO: Geneva, Switzerland, 1986.

50. McKay, G.A. *Statistical Estimates of Precipitation Extremes for the Prairie Provinces*; Canada Department of Agriculture, Prairie Farm Rehabilitation Administration (PFRA) Engineering Branch: Regina, SK, Canada, 1965.

51. SCS (Soil Conservation Service, U.S. Department of Agriculture). *A Method for Estimating Volume and Runoff in Small Watersheds, TP-149*; US Department of Agriculture, Soil Conservation Service: Washington, DC, USA, 1973.

*water*

MDPI

Article

# Improvements to Runoff Predictions from a Land Surface Model with a Lateral Flow Scheme Using Remote Sensing and In Situ Observations

Jong Seok Lee and Hyun Il Choi *

Department of Civil Engineering, Yeungnam University, 280 Daehak-Ro, Gyeongsan, Gyeongbuk 38541, Korea; ljs5219@gmail.com
* Correspondence: hichoi@ynu.ac.kr; Tel.: +82-53-810-2413

Academic Editors: Xianwei Wang and Hongjie Xie
Received: 30 December 2016; Accepted: 20 February 2017; Published: 22 February 2017

**Abstract:** Like most land surface models (LSMs) coupled to regional climate models (RCMs), the original Common Land Model (CoLM) predicts runoff from net water at each computational grid without explicit lateral flow (LF) schemes. This study has therefore proposed a CoLM+LF model incorporating a set of lateral surface and subsurface runoff computations controlled by topography into the existing terrestrial hydrologic processes in the CoLM to improve runoff predictions in land surface parameterizations. This study has assessed the new CoLM+LF using Earth observations at the 30-km resolution targeted for mesoscale climate applications, especially for surface and subsurface runoff predictions in the Nakdong River Watershed of Korea under study. Both the baseline CoLM and the new CoLM+LF are implemented in a standalone mode using the realistic surface boundary conditions (SBCs) and meteorological forcings constructed from remote sensing products and in situ observations, mainly by geoprocessing tools in a Geographic Information System (GIS) for the study domain. The performance of the CoLM and the CoLM+LF simulations are evaluated by the comparison of daily runoff results from both models with observations during 2009 at the Jindong stream gauge station in the study watershed. The proposed CoLM+LF, which can simulate the effect of runoff travel time over a watershed by an explicit lateral flow scheme, more effectively captures seasonal variations in daily streamflow than the baseline CoLM.

**Keywords:** lateral flow; surface runoff; subsurface runoff; topography; surface boundary condition; meteorological forcing; remote sensing; Geographic Information System; land surface model; Common Land Model

## 1. Introduction

Regional climate models (RCMs) that have been used to reproduce past and recent climate features are expected to provide credible predictions of future climate changes and impacts at regional or local scales. For the predictability of RCMs, the coupled models need more realistic and accurate calculations for interactions between the land surface and the atmosphere [1]. The Intergovernmental Panel on Climate Change (IPCC) also addressed the need for improvements to terrestrial land surface parameterizations in land surface models (LSMs) coupled to RCMs [2]. As the climate and hydrology modeling studies have developed toward physical sophistication and high resolution, LSMs coupled to RCMs have improved land surface parameterization schemes by incorporating sophisticated process interactions in the terrestrial hydrologic cycle [3–10]. However, some unrealistic parameterizations for terrestrial hydrologic schemes in LSMs have a direct or indirect influence on the complex and dynamic responses of land surface processes, which may cause serious errors in both terrestrial water and energy predictions.

The fact that most land surface parameterizations are currently limited to vertical fluxes in each single grid column may negatively affect the model's predictions for terrestrial water and energy processes, especially because the existing terrestrial hydrologic schemes in most LSMs are incapable of representing the lateral water movement induced by topographic features. As incoming precipitation is divided into evapotranspiration, soil moisture, surface and subsurface runoff by land surface processes in LSMs, runoff plays an important part in the terrestrial water budget. However, runoff is estimated just by the soil water budget without topographically controlled surface and subsurface flow schemes in most LSMs. Moreover, surface runoff used for neither soil moisture nor subsurface runoff calculation as the boundary condition results in a mass balance error in the terrestrial water cycle. The infiltration rate calculations do not take into consideration the role of surface flow depth, which can cause errors in both infiltration and surface flow calculations [11–13]. Such unrealistic and simplified parameterizations in the terrestrial hydrologic scheme may affect other key components in land surface water and energy cycles, limiting the predictability of LSMs.

The Common Land Model (CoLM), a Soil–Vegetation–Atmosphere Transfer (SVAT) model [14], already coupled to the mesoscale Climate–Weather Research and Forecasting model (CWRF), has been developed and updated for sophisticated land surface processes. The CWRF has built-in modules for realistic Surface Boundary Conditions (SBCs) and is the most comprehensive of the weather and climate models [15–17]. The CWRF's responses to various subgrid topographic representations and parameter selections were examined [18,19]. A scalable soil moisture transport scheme was developed for representing subgrid topographic control in land–atmosphere interactions of the CWRF [20,21]. The CWRF simulations were evaluated based on a continuous integration for the period 1979–2009 using a 30-km grid spacing over the North American domain [22]. The model improvement and performance of the CWRF were evaluated especially for possible impacts of precipitation, temperature, radiation, and extreme events' occurrence and magnitude to help with future climate projection [23]. Also, many studies [6–9,14,24–28] have evaluated the performance of the CoLM simulations in a standalone mode by the observational forcing data. Nonetheless, the mesoscale simulations of the CoLM at the 30-km resolution are found to still have problems in the terrestrial hydrologic scheme, especially streamflow runoff predictions. Choi et al. [10] have therefore developed a conjunctive surface-subsurface flow (CSSF) model based on a one-dimensional (1D) diffusion wave model for the surface flow routing scheme coupled with the 3D volume averaged soil-moisture transport (VAST) model for water flux in unsaturated soils [21]. However, the CSSF model cannot be widely implemented yet for watersheds all around the world because the VAST model requires closure parameterizations by regional estimates of subgrid and lateral soil moisture fluxes for each local watershed.

Based on an assumption that the lateral soil moisture movement may not be a major water flux in large spatial scale simulations, this study has focused on incorporating a routed surface flow scheme and an unrouted subsurface lateral runoff scheme into the existing 1D soil moisture transport formulation to evaluate the influence of a lateral flow scheme on runoff predictions in the CoLM. A new version of the CoLM, named CoLM+LF, is proposed in this study as it can explicitly route surface runoff from excess water of both infiltration and saturation, and estimate subsurface runoff induced by topographic controls. This study has implemented both the baseline CoLM and the new CoLM+LF at the CWRF 30-km grid resolution, focusing on surface and subsurface runoff predictions because the terrestrial hydrologic parameterizations including the runoff scheme must be tested and evaluated at the same resolution as the coupled climate models.

The principal aim of this study is to assess the improvement to runoff predictions from the new CoLM+LF with a lateral flow scheme using the best Earth observation data possible for model predictability. For evaluating the performance of runoff results simulated from both the baseline CoLM and the new CoLM+LF, this study has selected a watershed under study, the Nakdong River Watershed in Korea. All the CoLM and the CoLM+LF schemes were implemented at 30-km grid points in this study watershed without any downscaling and upscaling schemes for exchanges between the atmosphere and the land surface. For such direct applications of both the CoLM and the CoLM+LF, this study has constructed a set of realistic SBCs and meteorological forcing data based on high-quality Earth observations at the finest possible resolution from multifarious sources such as remote sensing products (see Section 3.2 for details), photography images, in situ observations, scanned and digitized maps, etc. for the study domain at 30-km (0.25°) resolution on the geographic coordinate system following Choi [29,30]. The raw data at finer resolutions were transformed into SBCs on the 30-km spacing grids in the study domain mainly by geoprocessing tools in a Geographic Information System (GIS), ArcGIS (ESRI, Redlands, CA, USA). Most meteorological forcing data to drive both the CoLM and the CoLM+LF simulations in a standalone mode were spatially interpolated onto the 30-km computational grids by the Inverse Distance Weighting (IDW) method in ArcGIS from observations at 19 gauge stations managed by the Korea Meteorological Administration (KMA) around the study watershed. For the performance assessment of the terrestrial hydrologic schemes in predicting runoff by the goodness-of-fit test, the results from both the CoLM and the CoLM+LF simulations through the sensitivity analysis of key parameters for runoff variables were compared with daily streamflow discharges during a year of 2009 observed at the Jindong stream gauge station managed by the Ministry of Land, Infrastructure, and Transport (MOLIT) near the drainage outlet of the Nakdong River Watershed, Korea. It is expected that the proposed CoLM+LF with a lateral flow scheme using realistic Earth observations can more effectively generate daily streamflow variations compared with the baseline CoLM runoff results from the local net excess water flux in each grid when ignoring the role of surface flow depth over the watershed, as most current LSMs do.

## 2. Model Description

Although the CSSF incorporates the 3D VAST for improvements to terrestrial hydrologic schemes, the implementation of the 3D VAST model requires five key parameters for the dependence of soil moisture variability or terrain features on the mean moisture flux, which should be regionally estimated for each watershed. Owing to a lack of applicable regional parameters for the use of the 3D VAST model, this study has proposed a new version of the CoLM, named the CoLM+LF, to improve runoff predictability. Hence, a set of topographically controlled surface and subsurface flow schemes have been combined with the existing 1D soil water transport scheme in the baseline CoLM. A sensitivity analysis of key parameters for runoff variables in both the baseline CoLM and the new CoLM+LF is also presented in this study.

### 2.1. Baseline Runoff Scheme in the CoLM

Surface runoff $R_s$ is generated by Hortonian runoff [31] due to infiltration excess and Dunnian runoff [32] induced by saturation excess as:

$$R_s = \underbrace{(1 - F_{imp})\max[\, 0, \ Q_w - I_{\max} \,]}_{Hortonian} + \underbrace{F_{imp}Q_w}_{Dunnian} , \tag{1}$$

where $F_{imp}$ is the impermeable area fraction of the saturated area and the frozen area. $Q_w$ is the available water supply rate such as rainfall, dewfall, and snowmelt rate on the surface. $I_{\max}$ is the maximum potential infiltration rate.

In the CoLM, subsurface runoff $R_{sb}$ comprises three components as:

$$R_{sb} = R_{sb,bas} + R_{sb,dra} + R_{sb,sat} , \tag{2}$$

where $R_{sb,bas}$, $R_{sb,dra}$, and $R_{sb,sat}$ denote subsurface runoff components from baseflow, bottom drainage, and saturation excess, respectively. The bottom drainage contribution to subsurface runoff is negligible when the actual bedrock is located within the model soil layers under the exponentially decay profile of the hydraulic conductivity. The saturation excess runoff rarely occurs by incorporating a maximum surface infiltration limit condition and the effective hydraulic conductivity function at the interface of unfrozen areas from Choi and Liang [9]. In this study watershed under such conditions, the baseflow is almost comparable to the whole subsurface runoff.

After Sivapalan et al. [33], the baseflow is calculated by the subsurface saturated lateral runoff equation based on the TOPMODEL [34] as:

$$R_{sb,bas} = \frac{\zeta K_s(0)}{f} e^{-\overline{\lambda}} e^{-fz_\nabla} , \tag{3}$$

where $\zeta$ is an anisotropic ratio of the lateral to the vertical hydraulic conductivities, $K_s(0)$ is the saturated hydraulic conductivity on the surface of the top soil layer, and $f$ is the decay factor of the saturated hydraulic conductivity. $\overline{\lambda}$ is the grid-averaged topographic index defined as $\lambda = \ln(a/\tan \beta)$ where $a$ is the drainage area per unit contour length and $\tan \beta$ is the local surface slope. $z_\nabla$ is the water table depth. Because the topographic index $\lambda$ has uncertainties in the regional and continental studies where the digital elevation model (DEM) data are generally available at coarse resolutions [35], a simplified parameterization of Equation (3) was used in some models [6,7,9,25] as:

$$R_{sb,bas} = R_{sb,max} \, e^{-fz_\nabla} , \tag{4}$$

where a single calibration parameter $R_{sb,max}$ is the maximum baseflow coefficient representing $\zeta K_s(0) e^{-\overline{\lambda}}/f$.

## 2.2. Lateral Flow Runoff Scheme in the CoLM+LF

For the interaction between surface and subsurface runoff components, the new scheme considers the influence of overland flow depth on both infiltration rate and surface runoff. Total available water supply rate $Q_t$ on the surface is computed by incorporating the surface flow depth $h$ during a computational time $\Delta t$ into the available water supply rate $Q_w$ as:

$$Q_t = Q_w + h/\Delta t , \tag{5}$$

Hence, the net surface runoff $R_n$ as a result of water exchange between surface and subsurface is

$$R_n = (1 - F_{imp})\max[\, 0, \, Q_t - I_{max}\,] + F_{imp}Q_t - h/\Delta t , \tag{6}$$

In addition, the CoLM+LF model utilizes the 1D non-inertia diffusion wave equation for a surface flow routing scheme as:

$$\frac{\partial h}{\partial t} + c_d \frac{\partial h}{\partial x_c} = D_h \frac{\partial^2 h}{\partial x_c^2} + R_n , \tag{7}$$

where $t$ is time and $x_c$ is a flow direction coordinate. $c_d$ is the diffusion wave celerity and $D_h$ is the hydraulic diffusivity. Note that the spatial and temporal variation of the surface water simulated by the 1D diffusion wave model in Equation (7) depends on the net water exchange flux $R_n$ between surface and subsurface in Equation (6).

The surface flow rate $Q_s$ using the Darcy–Weisbach formula can be written as:

$$Q_s = Bh^{3/2} \sqrt{\frac{8g}{f_d} \left( S_o - \frac{\partial h}{\partial x_c} \right)} , \tag{8}$$

where $B$ is the flow cross-section width and $g$ is the gravitational acceleration. $f_d$ is the Darcy–Weisbach friction resistance coefficient calculated for flow regimes such as laminar, transition, or turbulent, respectively, based on the Reynolds number of the surface flow. $S_o$ is the bottom slope in the flow direction.

Baseflow Equations (3) and (4) in the baseline CoLM are incapable of representing the frozen soil area and surface macropore effects as well as the hydraulic conductivity variation with soil textures for different layers. Moreover, both equations may either fail to capture observed recession curves or produce unrealistic remaining soil moisture content, as demonstrated in Choi and Liang [9]. Starting with the basic assumptions in TOPMODEL, therefore, the saturated lateral flow $q_b$ at a depth $z$ beneath a water table can be written as:

$$q_b(z) = F_{liq}(z)\zeta K_{sz}(z) \tan \beta , \qquad (9)$$

where $F_{liq}$ is the unfrozen part of soil water and $K_{sz}$ is the vertical saturated hydraulic conductivity.

The baseflow runoff $R_{sb,bas}$ for a grid cell area $A$ is calculated by integrating Equation (9) through the entire saturated soil layers and along the channel length $L$ connected to the grid outlet as:

$$R_{sb,bas} = \frac{Q_b}{A} = \frac{\int_L \int_{z_\nabla}^{z_N} q_b \, dz \, dL}{A} = \frac{\int_L \int_{z_\nabla}^{z_N} F_{liq}(z)\zeta K_{sz}(z) \tan \beta \, dz \, dL}{A} = \frac{\sum_{k=j}^{N} T(k) \tan \beta \cdot L}{A} , \qquad (10)$$

where $Q_b$ is the total baseflow from a grid cell. $T(k) = \int_{z_{k-1}}^{z_k} F_{liq}(z)\zeta K_{sz}(z) dz$ is a transmissivity varying nonlinearly with depth between vertical coordinates $z_k$ and $z_{k-1}$ of the layer $k$ where $z_{k-1}$ is replaced with $z_\nabla$ for the interface layer $j$ with the water table. $N$ is the total number of soil layers. To prevent Equation (10) from producing an unrealistic (negative or less than the residual) value for soil moisture content, a layer baseflow $q_b(k)$ for each discrete layer $k$ below the water table $z_\nabla$ is finally determined as:

$$q_b(k) = \min \left[ \frac{T(k) \tan \beta \cdot L}{A} , \frac{\left[\theta_{liq}(k) - \theta_r(k)\right] \Delta z_k'}{\Delta t} \right] , \qquad (11)$$

where $\Delta z_k' = \begin{cases} z_k - z_\nabla & \text{for} \quad k = j \\ z_k - z_{k-1} & \text{for} \quad k = j+1 \text{ to } N \end{cases}$ . $\theta_{liq}$ is the liquid soil moisture content and $\theta_r$ is the minimum soil moisture content for residual water in a soil.

The liquid water $\theta_{liq}(k)$ for each soil layer $k$ should be updated by the layer baseflow $q_b(k)$ for mass conservation as:

$$\theta_{liq}(k) = \theta_{liq}(k) - \frac{q_b(k)\Delta t}{\Delta z_k} , \qquad (12)$$

where $\Delta z_k = z_k - z_{k-1}$ is a layer thickness for the layer $k$.

## 3. Model Implementation

The new CoLM+LF model has incorporated a set of topographically controlled surface and subsurface flow schemes into the existing terrestrial hydrologic representation in the baseline CoLM. The performance of both the baseline CoLM and the new CoLM+LF in predicting runoff was evaluated over the second-largest watershed on complex terrain in Korea to study the impact of lateral flow induced by topography using the 30-km computational grid mesh targeted for regional applications. The model experiments were performed in a standalone mode for which both the CoLM and the CoLM+LF were driven by the realistic SBCs and meteorological forcing data at the 30-km grid scale, constructed by applications of Earth observation data and GIS in this study.

### 3.1. Study Watershed

The Nakdong River Watershed with high topographic heterogeneity in Korea was selected under study to evaluate the performance of the baseline runoff simulation in the CoLM and the lateral flow adopted runoff simulation in the CoLM+LF at the 30-km resolution. Figure 1 shows the study domain comprising 72 (8 × 9) computational grid cells at 30-km (0.25°) horizontal spacing overlaid with the main streamline and the watershed boundary of the Nakdong River on the geographic coordinate system. The Nakdong River Watershed is located between 127°29′~129°18′ E and 35°03′~37°13′ N with the size of 23,384.21 km$^2$ and the main channel length of 510.36 km. The Nakdong River flows from the Taebaek Mountains over the north region to the South Sea. The terrain elevation ranges from

1 to 1885 EL.m over the watershed, and the width of the river ranges from only a few meters in its upper reaches to several hundred meters towards its estuary. Major land cover types are comprised of irrigated cropland/pasture, cropland/woodland mosaic, savanna, and mixed forest in the USGS land cover classification. A stream gauge station is located at the Jindong Bridge where daily streamflow discharges were measured in 2009 by the MOLIT. There are 19 meteorological gauge stations managed by the KMA around the study watershed for hourly or daily observations. See Section 3.3 for details on observations and meteorological forcing data in the study domain.

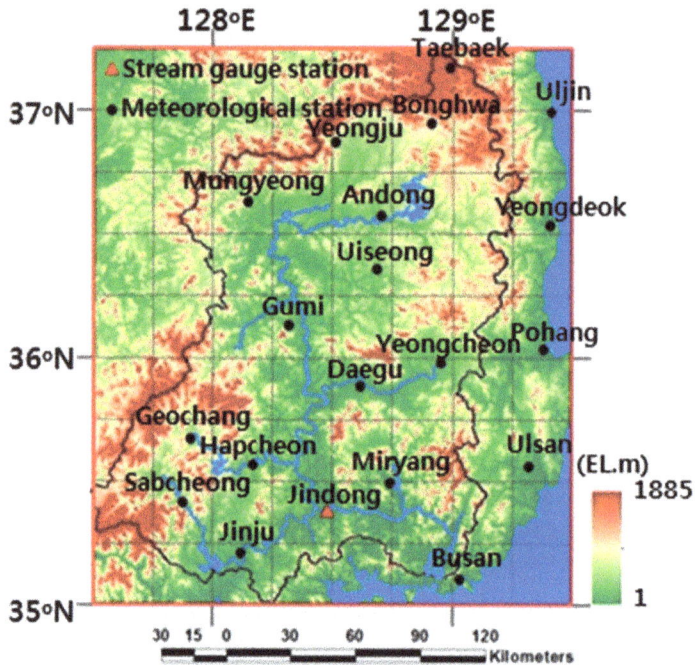

**Figure 1.** Location map for meteorological and stream gauge stations in the study watershed overlaid with the main streamline and the watershed boundary of the Nakdong River in Korea on the 30-km computational grid mesh.

### 3.2. Surface Boundary Conditions

This study has constructed a set of primary SBCs from high-quality and -resolution observational data over the Nakdong River Watershed. The primary SBCs required for the CoLM and the CoLM+LF implementations comprise three groups of SBC datasets related to vegetation, terrain, and flow direction features on the 30-km spacing grids transformed from various Earth observations at the finest resolution possible for the study domain.

The vegetative SBCs include land cover category, albedo, fractional vegetation cover, and monthly leaf area index in 2009. First of all, the raw data at much finer than the 30-km resolution on various data format and map projections were converted into ArcGIS raster data, and then the representative (majority or average) values were computed for each 30-km computational grid. The 30-km land cover category data was made from the 1-km USGS land cover classification with 24 categories [36], developed from the Advanced Very High Resolution Radiometer (AVHRR) satellite-derived Normalized Difference Vegetation Index (NDVI) composites. The one of the USGS 24 land cover types constituting the largest fraction is selected as the representative value for each 30-km grid. Provided that category 16 (water bodies) with the largest fraction is not the absolute majority in a grid, this grid's land cover type is not the water body but belongs in the category that constitutes the second largest fraction. As shown in Figure 2a, the study watershed consists of the USGS land cover categories 3 (irrigated cropland and pasture), 6 (cropland/woodland mosaic), 10 (savanna), and 15 (mixed forest) at the 30-km scale. The three water body (category 16) grids in the study domain were used as the standard identification for consistency of water bodies in all datasets of SBCs. Following Yucel [37], the 30-km albedo values were assigned with respect to the 30-km USGS land cover categories, distributed from 0.13 to 0.20 for the study watershed and given to 0.08 for water bodies as shown in Figure 2b. The high-resolution fractional vegetation cover was computed following Zeng et al. [38,39] from the 1-km NDVI data in the Système Pour l'Observation de la Terre-VEGETATION (SPOT-VGT) satellite products [40]. Figure 2c denotes spatial distributions of the 30-km fractional vegetation cover values ranging from 83.2% to 100%, calculated by averaging all 1-km values within each 30-km grid using a zonal statistic function, ZONALMEAN in ArcGIS geoprocessing tools. The raw leaf area index data is provided from the 1-km MOD 15 LAI data [41] by Moderate Resolution Imaging Spectroradiometer (MODIS) from the Terra (EOS AM) and Aqua (EOS PM) satellites. After abnormal values were removed by a smoothing filter method by Liang et al. [16] and then missing values were filled using an interpolation method by Choi [29], the monthly leaf area index data were calculated following Zeng et al. [39] for each 30-km grid by the ZONALMEAN function in ArcGIS. Figure 2d denotes leaf area index values in a range of 2.3 to 4.6 for the study domain in July 2009.

The terrain SBCs comprise surface elevation, bedrock depth, and soil sand/clay fraction profiles over the 11 soil layers at the 30-km resolution. After higher resolution raw data on various data formats and map projections were converted into ArcGIS raster data, the average of all raster values within each 30-km grid was calculated by the ZONALMEAN function in ArcGIS. The surface elevation data was constructed from the 90-m (3 arc-second) DEM provided by the National Aeronautics and Space Administration (NASA) Shuttle Radar Topographic Mission (SRTM) DEM dataset [42]. The 30-km surface elevation data ranges from 22.8 to 910.0 EL.m for the study domain as shown in Figure 3a. The bedrock depth and soil sand/clay fraction profiles were constructed by the Harmonized World Soil Database (HWSD) [43], developed by the Land Use Change (LUC) project of International Institute for Applied Systems Analysis (IIASA) and the Food and Agriculture Organization of the United Nations (FAO). The spatial distribution of the bedrock depth is in the range from 0.8 to 135.2 cm below the surface ground as shown in Figure 3b. Figure 3c,d denote the 30-km soil composition fraction results for the first soil model layer, ranging between 24.9% and 36.3% for sand, and between 34.4% and 58.9% for clay.

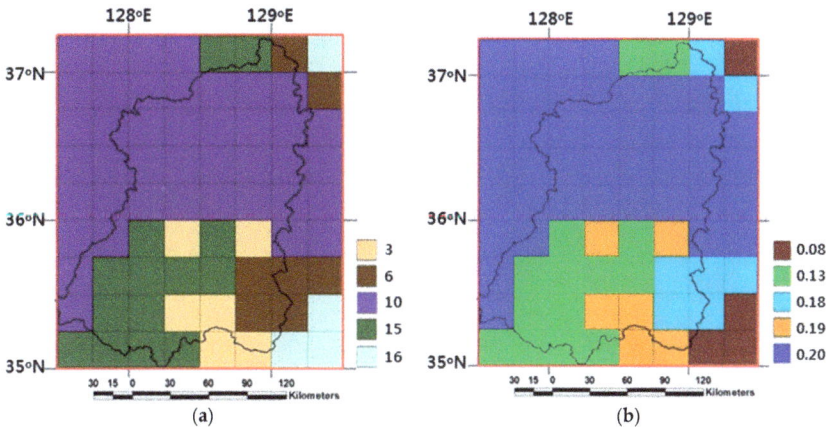

Figure 2. *Cont.*

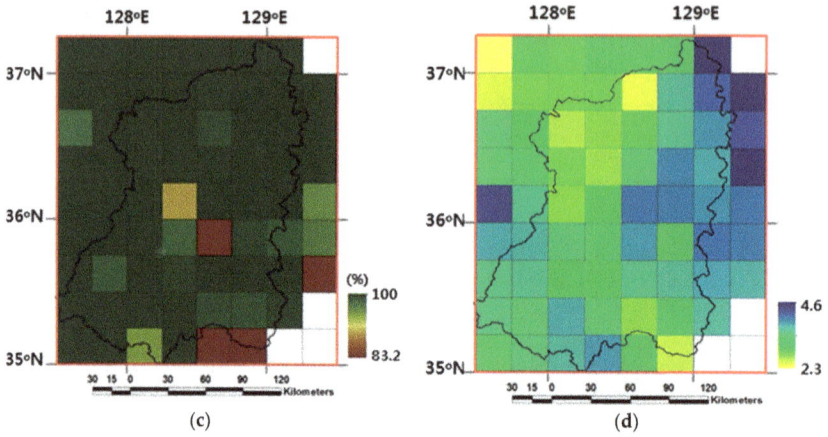

**Figure 2.** Spatial distributions of vegetative surface boundary conditions for (**a**) land cover category (3: irrigated cropland and pasture, 6: cropland/woodland mosaic, 10: savanna, 15: mixed forest, and 16: water bodies); (**b**) albedo; (**c**) fractional vegetation cover; and (**d**) July's leaf area index on 30-km computational grids over the study domain.

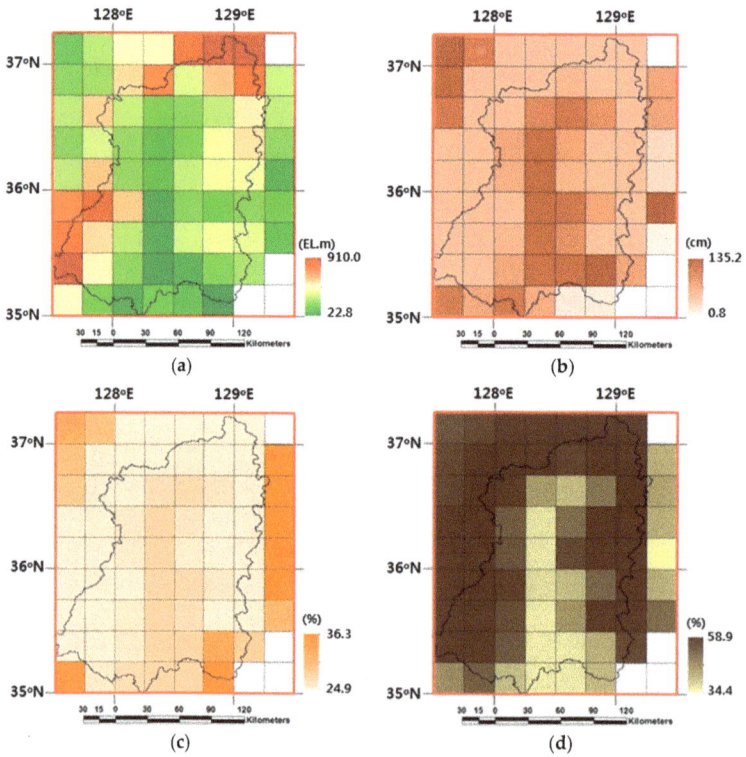

**Figure 3.** Spatial distributions of terrain surface boundary conditions for (**a**) surface elevation; (**b**) bedrock depth; (**c**) the first soil layer's sand fraction; and (**d**) the first soil layer's clay fraction on 30-km computational grids over the study domain.

The new CoLM+LF requires additional SBCs on the lateral flow information such as flow direction and accumulation data for each 30-km grid to perform the topographically controlled surface and subsurface flow computations. These fields were constructed by our own upscaling method using the reference flow information data at finer resolution possible, the 90-m (3 arc-second) terrain and drainage data provided by the Hydrological data and maps based on SHuttle Elevation Derivatives at multiple Scales (HydroSHEDS) [44], derived from the NASA SRTM DEM. The HydroSHEDS geospatial dataset is known to provide the realistic stream flow direction data through a sequence of algorithms such as void-filling, filtering, hydrologic conditioning, stream burning, manual corrections, and upscaling techniques on the original SRTM DEM for improvements to data information accuracy suitable for hydrologic study applications. Figure 4 compares the difference between the 30-km flow information datasets derived from our own upscaling method using the 90-m HydroSHEDS data and from the eight direction flow model using the 30-km SBC of surface elevation in ArcGIS. The 30-km flow information result upscaled by our own method properly represents and is much closer to the main streamline of the Nakdong River in the study domain. The 30-km lateral flow SBCs have a significant influence on the lateral surface and subsurface flow computation as well as the drainage area delineation.

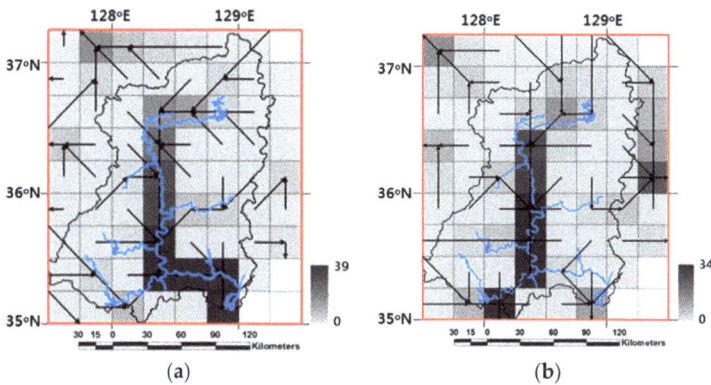

**Figure 4.** Comparison of lateral flow surface boundary conditions for the CoLM+LF flow direction (arrows) and flow accumulation (grey-black pixels) (**a**) from our own upscaling method; and (**b**) from ArcGIS eight direction method on 30-km computational grids with overlays of the main stream networks (blue lines) and the watershed boundary (black polygon) over the study domain.

To prevent any inconsistency over all the 30-km SBCs due to different sources of the raw data, all constructed SBCs were adjusted by the 30-km land cover category as the standard identification for land and water body grids.

### 3.3. Meteorological Forcing Data

The meteorological forcing variables are required for both the CoLM and the CoLM+LF simulations in the offline mode. Most meteorological forcing data to drive both model simulations during the year of 2009 were also constructed onto the 30-km computational grids by the ZONALMEAN function in ArcGIS after in situ observations at 19 KMA meteorological gauge stations around the study watershed were spatially interpolated onto the study domain with average values weighted by the inverse of the distance from the gauge point. The 30-km daily meteorological forcing data constructed in this study are precipitation (mm), snow (cm), air pressure (hPa), vapor pressure (hPa), air temperature ($^\circ$C), specific humidity (%), zonal and meridional wind speeds (m/s), and downward short wave radiation (MJ/m$^2$). The daily meteorological forcing data were linearly interpolated for the computational time step of 600 seconds in the both models.

Figure 5 denote spatial distributions of selective meteorological forcing data for the study domain on 5 July, one day of precipitation events in 2009.

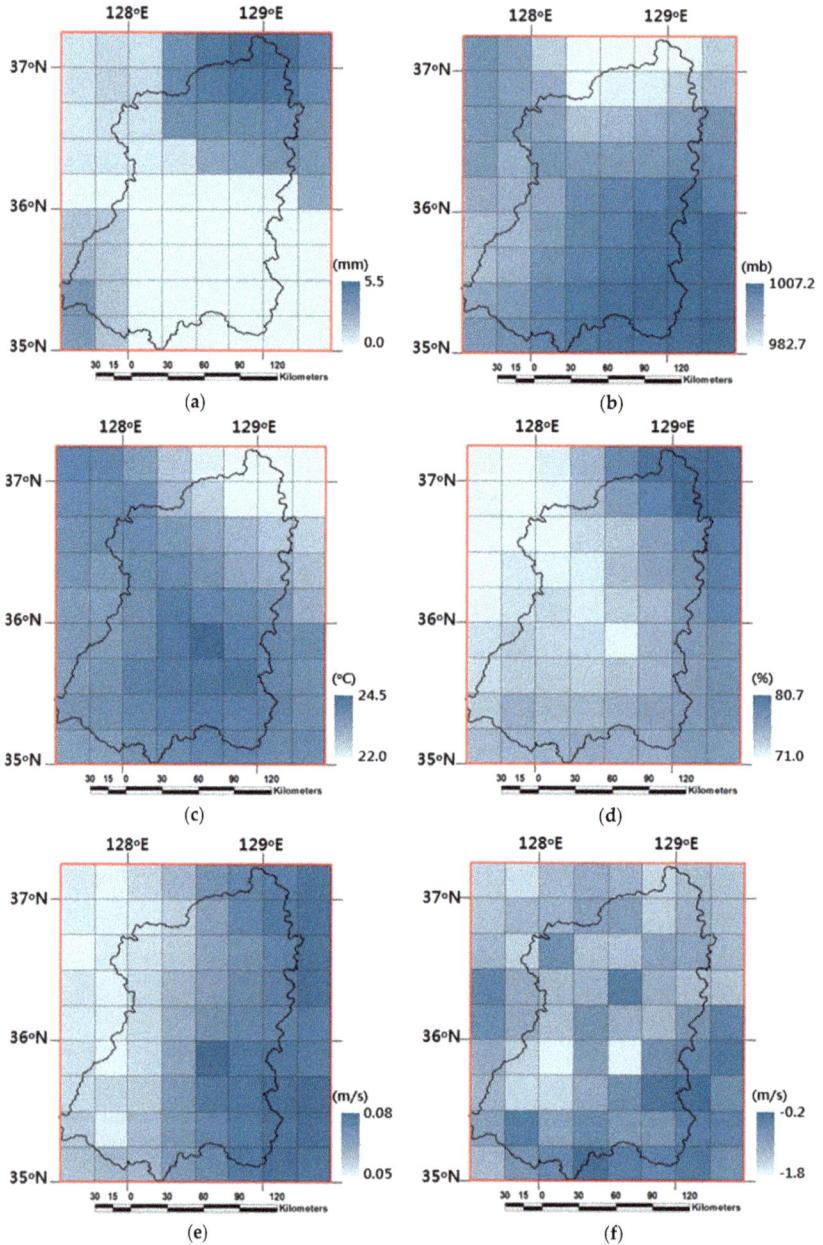

**Figure 5.** Spatial distributions of selective meteorological forcing variables for (**a**) precipitation; (**b**) air pressure; (**c**) air temperature; (**d**) specific humidity; (**e**) zonal wind speed; and (**f**) meridional wind speed on 30-km computational grids over the study domain on 5 July, one of the precipitation event days in 2009.

*3.4. Initial Conditions*

To resolve uncertainty in the initial conditions, both CoLM and CoLM+LF under the assumed initial conditions on 1 January 2009 were run six times repeatedly without interruption for the whole of 2009. The sixth set of results from CoLM and CoLM+LF was saved for analysis and interpretation.

## 4. Results

The runoff results simulated from both CoLM and CoLM+LF at the 30-km resolution were compared with streamflow discharges observed at the Jindong gauge station by the MOLIT around the outlet of the study watershed. For the performance assessment of both model simulations, the relative agreement of the model results with the observations was evaluated by the Nash–Sutcliffe efficiency (*NSE*) [45] and the absolute value of the relative bias (*ARB*), a set for assessing the goodness of fit as suggested by McCuen et al. [46] as:

$$NSE = 1 - \frac{\sum\limits_{i=1}^{n} (O_i - S_i)^2}{\sum\limits_{i=1}^{n} (O_i - \overline{O})^2},\tag{13}$$

$$ARB = \frac{\left| \sum\limits_{i=1}^{n} (O_i - S_i) \right|}{\overline{O}},\tag{14}$$

where *n* is total number of data, $O_i$ is the observed data and $S_i$ is the simulated data at day *i*, and $\overline{O}$ is the average of $O_i$. *NSE* can assess the model predictive ability and *ARB* is used to compute total volume error in model predictions.

*4.1. Runoff Results from the Baseline CoLM*

Total runoff is comprised of surface and subsurface runoff parts. Since total runoff for a watershed was calculated by the sum of watershed-wide averages of surface runoff and subsurface runoff in the baseline CoLM simulations, total runoff was calculated from average values over all upstream grid cells of the target grid with the Jindong gauge station. Total runoff simulation results were compared with daily specific discharges (discharge per unit drainage area) observed at the Jindong stream gauge station in the study watershed during the year of 2009. Following Choi and Liang [9], the sensitivity of the baseline CoLM was examined to two calibration parameters, the hydraulic conductivity decay factor *f* and the maximum baseflow coefficient $R_{sb,max}$ for runoff simulations in the Nakdong River Watershed.

Figure 6 denotes both *NSE* and *ARB* scores for runoff results under a set of two calibration parameters in a wide range from 2 to 9 m$^{-1}$ for the hydraulic conductivity decay factor *f* and from $1 \times 10^{-5}$ to $9 \times 10^{-1}$ mm/s for the maximum baseflow coefficient $R_{sb,max}$. *NSE* scores are low and negative on the whole analysis, and the maximum score of $-0.140$ occurs with a tolerable *ARB* of 0.367 when the hydraulic conductivity decay factor *f* is 7 m$^{-1}$ and the maximum baseflow coefficient $R_{sb,max}$ is $7 \times 10^{-1}$ mm/s. These two calibration parameters play a significant role in the baseflow generation for the baseline CoLM without the topographically controlled lateral flow scheme.

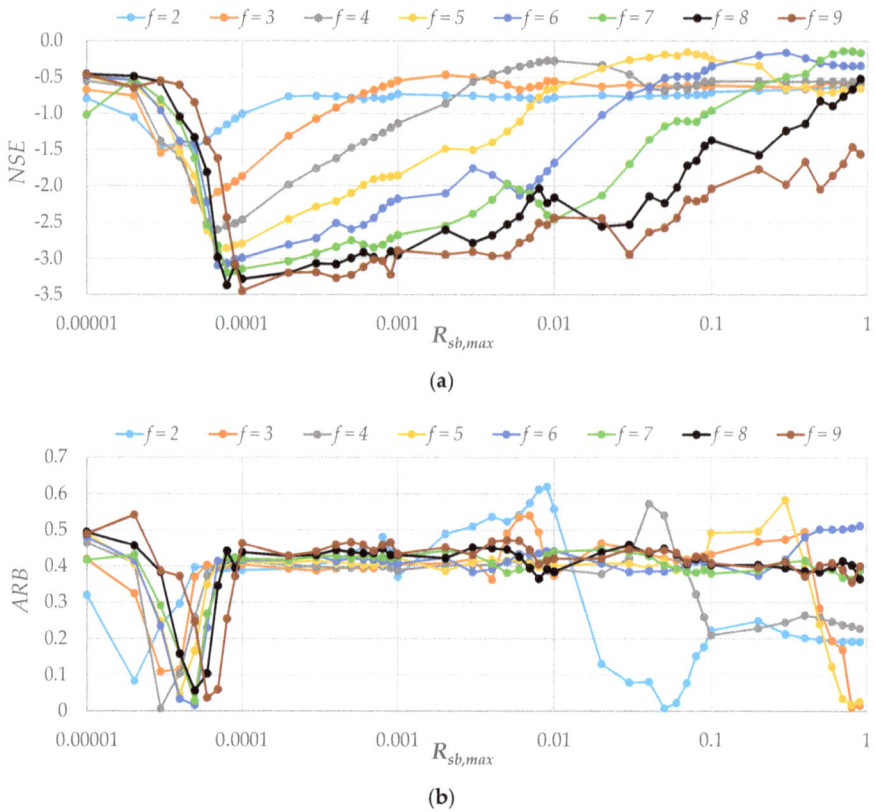

**Figure 6.** Comparison of the model performance evaluation results for assessing the goodness of fit by (**a**) the Nash–Sutcliffe efficiency (*NSE*); and (**b**) the absolute value of the relative bias (*ARB*) with respect to the hydraulic conductivity decay factor $f$ and the maximum baseflow coefficient $R_{sb,max}$ in the baseline CoLM for total runoff during 2009 at the Jindong stream gauge station in the study watershed.

### 4.2. Runoff Results from the New CoLM+LF

In the CoLM+LF that can simulate surface outflow at each grid, total runoff was calculated by the sum of the specific discharge of surface flow and the upstream grid-averaged subsurface runoff for a target grid point as:

$$R_{tot} = \frac{Q_s}{(n_{fa} + 1)A} + \overline{R}_{sb} \tag{15}$$

where $R_{tot}$ is total runoff, $n_{fa}$ is the flow accumulation number at the target grid point, and $\overline{R}_{sb}$ is the averaged subsurface runoff for the total grid cells located upstream of the target grid point.

The CoLM+LF runoff simulations were examined to two calibration parameters, the hydraulic conductivity decay factor $f$ values from 2 to 9 m$^{-1}$ and the hydraulic conductivity anisotropic ratio $\zeta$ values from 10 to $10^6$. Figure 7 represents both *NSE* and *ARB* scores between total runoff results from the CoLM+LF and specific discharges observed at the Jindong stream gauge station for the study watershed in 2009. Overall, both *NSE* and *ARB* scores are much better than those for the baseline CoLM results, and the hydraulic conductivity decay factor $f$ of 3 m$^{-1}$ and the hydraulic conductivity anisotropic ratio $\zeta$ of $10^4$ are calibrated in considering synchronization performance results for *NSE* of 0.774 and *ARB* of 0.012.

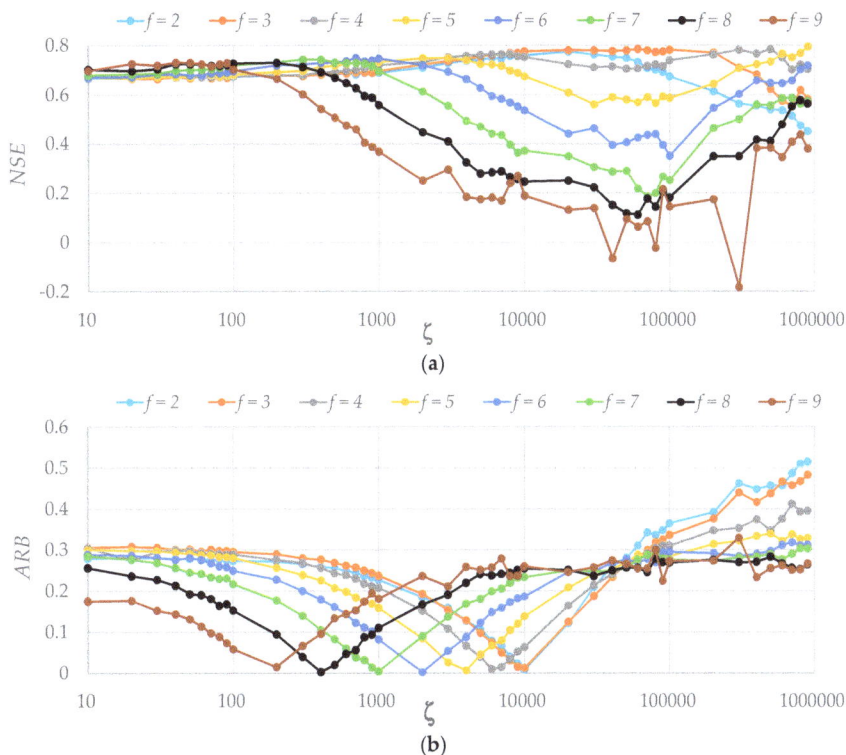

**Figure 7.** Comparison of the model performance evaluation results for assessing the goodness of fit by (**a**) the Nash–Sutcliffe efficiency (*NSE*); and (**b**) the absolute value of the relative bias (*ARB*) with respect to the hydraulic conductivity decay factor $f$ and the hydraulic conductivity anisotropic ratio $\zeta$ in the new CoLM+LF for total runoff during 2009 at the Jindong stream gauge station in the study watershed.

*4.3. Comparison of Runoff Results*

Figure 8 compares the time series of daily specific discharges (total runoff) during 2009 simulated from the baseline CoLM and the new CoLM+LF under each calibrated parameter set, along with observations at the Jindong gauge station in the study watershed. Table 1 summarizes model performance results measured by $NSE = -0.140$ and $ARB = 0.367$ for the baseline CoLM under calibrated parameters of $f = 7 \text{ m}^{-1}$ and $R_{sb,\max} = 7 \times 10^{-1} \text{ mm/s}$, and $NSE = 0.774$ and $ARB = 0.012$ for the new CoLM+LF under calibrated parameters of $f = 3 \text{ m}^{-1}$ and $\zeta = 10^{4}$, respectively. As shown in Figure 8 and Table 1, the seasonal variability of observed streamflow is realistically captured by the runoff result from the CoLM+LF incorporating the lateral surface and subsurface flow schemes, whereas the baseline CoLM generates high peak discharges due to immediate response to precipitation events without surface flow routing or runoff travel time effect over the watershed, leading to the overestimated runoff result in total volume. Moreover, the runoff result from the baseline CoLM without considering impacts of surface flow depth on the infiltration rate may make errors in both infiltration and surface flow calculations. The new CoLM+LF scheme significantly improves the performance of the baseline CoLM simulation in representing the watershed runoff prediction.

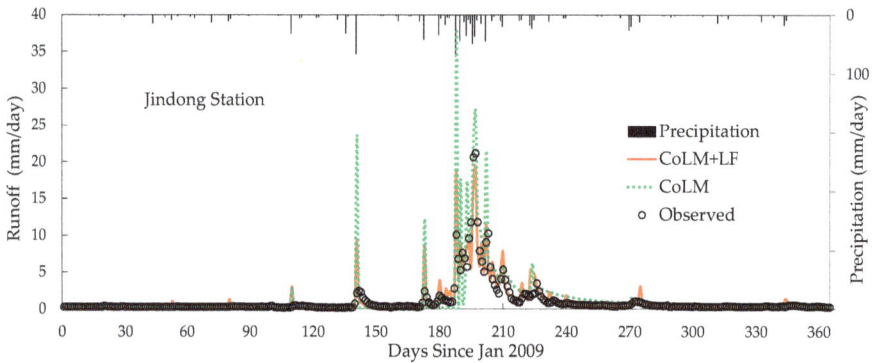

**Figure 8.** Comparison of daily time series of model simulated specific discharges of total runoff from the baseline CoLM and the new CoLM+LF under each calibrated parameter set, along with the daily observations from the Jindong stream gauge station in the study watershed during 2009. The observed hyetographs of total precipitation averaged for the study watershed are presented along the secondary vertical axis.

**Table 1.** Comparison of the model performance measured by the Nash–Sutcliffe coefficient, *NSE* and the absolute relative bias, *ARB* for daily runoff results in the study watershed during 2009 simulated from the baseline CoLM under the calibrated values of the hydraulic conductivity decay factor $f$ and the maximum baseflow coefficient $R_{sb,max}$ and from the new CoLM+LF under the calibrated values of the hydraulic conductivity decay factor $f$ and the hydraulic conductivity anisotropic ratio $\zeta$.

| Models | Model Performance | | Calibration Parameters | |
|---|---|---|---|---|
| | *NSE* | *ARB* | $f$ | $R_{sb,max}$ or $\zeta$ |
| CoLM | −0.140 | 0.367 | $7 \text{ m}^{-1}$ | $7 \times 10^{-1} \text{ mm/s}$ |
| CoLM+LF | 0.774 | 0.012 | $3 \text{ m}^{-1}$ | $10^4$ |

Figure 9 compares surface runoff and subsurface runoff components separately from both the baseline CoLM and the new CoLM+LF simulations under each calibrated parameter set. The baseline CoLM simulates surface runoff with much higher and shaper peaks than the new CoLM+LF result. Although the baseline CoLM simulation generates much enhanced baseflow by larger values of the two calibration parameters to capture the recession parts in the observed hydrograph, even the selected set of two calibration parameters still generates unrealistic high peaks due to quick response of surface runoff to precipitation events in the baseline CoLM. On the contrary, the lateral surface and subsurface flow computations play an important part in capturing the seasonal streamflow patterns in the new CoLM+LF result. The CoLM+LF incorporating the lateral surface flow scheme generates lower peaks and declining recession curves in surface runoff by the surface flow routing effect, and the baseflow is effectually generated by the surface flow depth contribution to infiltration with a lower $f$ value than the baseline CoLM simulation. It is found that the watershed runoff is much more realistically predicted in the new CoLM+LF parameterizations by interactions between the lateral surface flow and the subsurface flow controlled by topography.

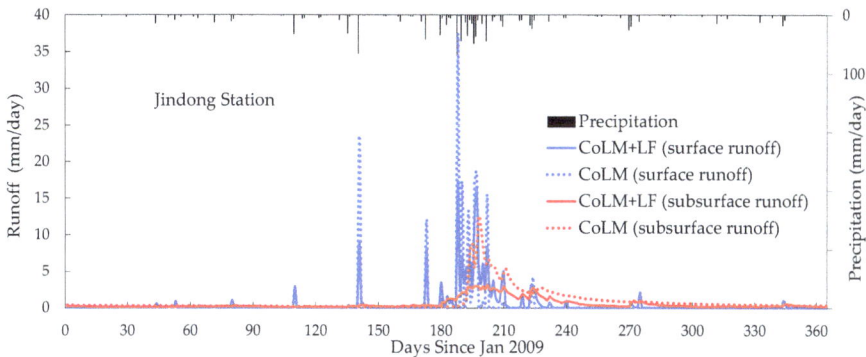

**Figure 9.** Comparison of daily time series of surface runoff and subsurface runoff simulated from the baseline CoLM and the new CoLM+LF under each calibrated parameter set at the Jindong stream gauge station in 2009. The observed hyetographs of total precipitation averaged for the study watershed are presented along the secondary vertical axis.

## 5. Discussion

Like most existing LSMs, the baseline CoLM predicts runoff from local net excess water flux after excluding surface evapotranspiration and soil moisture flux from precipitation. A disregard for the role of lateral surface and subsurface runoff in the terrestrial hydrologic cycle may make significant errors in the baseline CoLM simulations. This study has therefore introduced a new model, the CoLM+LF that incorporates the 1D diffusion wave surface flow model and the 1D topographically controlled baseflow scheme into the existing formulations for the terrestrial hydrologic cycle in the baseline CoLM. The baseline CoLM and the new CoLM+LF were implemented for the Nakdong River Watershed in Korea in a standalone mode by the realistic SBCs and meteorological forcings on the 30-km grid mesh for mesoscale applications by direct interactions between hydrological and atmospheric components. This study has collected high quality observational data mainly by multispectral remote sensing products and in situ observations from finer possible spatial data. The various raw data were spatially transformed into the proper representative values on the 30-km spacing grids in the study domain by our own program codes and geoprocessing tools in ArcGIS. The new CoLM+LF simulations need the additional SBCs such as the 30-km flow direction and accumulation data to implement an explicit lateral flow scheme. Since it is problematic to directly use the eight direction flow model provided by ArcGIS on the 30-km resolution grids, our own upscaling method properly generated the 30-km flow direction result, which coincides better with the Nakdong River stream network, compared with the result from ArcGIS. This study has successfully provided a primary set of SBCs and daily meteorological forcing data during the year 2009 for the study domain, including the entire Nakdong River Watershed from the comprehensive Earth observations.

This study has evaluated the performance of the CoLM and the CoLM+LF in offline simulations of daily runoff through the sensitivity analysis of key parameters for runoff variables at a watershed scale by comparison of the simulated results with observations at the Jindong gauge station in the study watershed. Surface runoff generated simply from the local net excess water in each grid disappears at the next computational time step in the baseline CoLM. Since surface runoff calculated at every time step irrespective of the remaining surface runoff at the previous time step makes no contribution to infiltration, larger values of the two calibration parameters (decay factor $f$ and the maximum baseflow coefficient $R_{sb,max}$) were estimated for the baseline CoLM to enhance both infiltration and baseflow generation. As the baseflow equation in the baseline CoLM is apt to underestimate baseflow, as demonstrated in Choi and Liang [9], runoff recession curves observed at the Jindong gauge station cannot be captured by runoff results generated with $R_{sb,max}$ values in the order of $10^{-4}$ mm/s used in previous studies [6,7,9,25]. Accordingly, the baseline CoLM simulations that predict sharp and steep peaks in surface runoff were tuned by larger values for $f$ of 7 m$^{-1}$ and $R_{sb,max}$ of $7 \times 10^{-1}$ mm/s to enhance baseflow generation. Such simplistic assumptions and crude parameterizations for the lateral surface and subsurface runoff processes in the baseline CoLM may cause significant model errors and consequently unrealistic model parameters by calibration. On the other hand, the new CoLM+LF incorporating an explicit surface flow routing scheme can facilitate infiltration by the surface flow depth contribution, leading to the enhanced baseflow generation as well. Moreover, baseflow can be effectually generated in the new CoLM+LF by a new formulation for baseflow controlled by topography which can also depict the effects of surface macropores and vertical hydraulic conductivity changes. In the new CoLM+LF with a set of the lateral surface and subsurface runoff schemes, lower peaks and smoother recession curves of surface runoff were generated under relatively smaller value for $f$ of 3 m$^{-1}$ due to the surface flow routing effect, and the baseflow were successfully generated by the surface flow depth contribution to infiltration and topographically controlled baseflow scheme with the hydraulic conductivity anisotropic ratio $\zeta$ of $10^4$.

This study has demonstrated that the CoLM+LF incorporating the topographically controlled surface and subsurface flow computations realistically can predict the temporal variation of the spatial distribution of streamflow runoff at a watershed scale, while the baseline CoLM may generate unrealistic surface runoff and infiltration results, which are important components for terrestrial water distribution and movement. The new CoLM+LF provides improved runoff modeling capability to the baseline CoLM for better streamflow predictions affecting the terrestrial hydrologic cycle crucial to climate variability and change studies. The new CoLM+LF is expected to be a helpful and essential tool for water resource management and hydrological impact assessment, particularly in regions with complex topography. This proposed CoLM+LF targeted for mesoscale climate application and watershed scale hydrologic analysis at relatively large spatial scales needs to be implemented for comprehensive terrestrial hydrologic simulations with long-term observations after climatological data are constructed over various watersheds including this study domain. The next study is planning to perform an analysis on uncertainties in the SBCs constructed from various remote sensing products and examine the sensitivity of model predictability to the spatial and temporal resolutions of input data. The next model also needs to include aquifer recharge, deep aquifer groundwater, channel flow routing, and regulation storage for further investigation.

**Acknowledgments:** This research was supported by a Yeungnam University research grant (215A380201) in 2015 and the BK21 plus program through the National Research Foundation (NRF) funded by the Ministry of Education of Korea.

**Author Contributions:** Jong Seok Lee and Hyun Il Choi conceived and designed the new model experiments. Jong Seok Lee constructed the surface boundary conditions and meteorological forcing data for the study domain; Hyun Il Choi analyzed the sensitivity of model calibration parameters and wrote the manuscript draft.

**Conflicts of Interest:** The authors declare no conflict of interest.

## References

1. Sellers, P.J.; Los, S.O.; Tucker, C.J.; Justice, C.O.; Dazlich, D.A.; Collatz, G.J.; Randall, D.A. A revised land surface parameterization (SiB2) for atmospheric GCMs. Part II: The generation of global fields of terrestrial biophysical parameters from satellite data. *J. Clim.* **1996**, *9*, 706–737. [CrossRef]
2. Bates, B.C.; Kundzewicz, Z.W.; Wu, S.; Palutikof, J.P. *Climate Change and Water*; Technical Paper of Intergovernmental Panel on Climate Change; IPCC Secretariat: Geneva, Switzerland, 2008; p. 210.
3. Stieglitz, M.; Rind, D.; Famiglietti, J.; Rosenzweig, C. An efficient approach to modeling the topographic control of surface hydrology for regional modeling. *J. Clim.* **1997**, *10*, 118–137. [CrossRef]
4. Chen, J.; Kumar, P. Topographic influence of the seasonal and interannual variation of water and energy balance of basins in North America. *J. Clim.* **2001**, *14*, 1989–2014. [CrossRef]
5. Warrach, K.; Stieglitz, M.; Mengelkamp, H.T.; Raschke, E. Advantages of a topographically controlled runoff simulation in a soil-vegetation-atmosphere transfer model. *J. Hydrometeorol.* **2002**, *3*, 131–148. [CrossRef]

6. Niu, G.Y.; Yang, Z.L. The versatile integrator of surface and atmosphere processes (VISA) Part II: Evaluation of three topography based runoff schemes. *Glob. Planet. Chang.* **2003**, *38*, 191–208. [CrossRef]

7. Niu, G.Y.; Yang, Z.L.; Dickinson, R.E.; Gulden, L.E. A simple TOPMODEL-based runoff parameterization (SIMTOP) for use in GCMs. *J. Geophys. Res.* **2005**, *110*. [CrossRef]

8. Oleson, K.W.; Niu, G.Y.; Yang, Z.L.; Lawrence, D.M.; Thornton, P.E.; Lawrence, P.J.; Stockli, R.; Dickinson, R.E.; Bonan, G.B.; Levis, S.; et al. Improvements to the community land model and their impact on the hydrological cycle. *J. Geophys. Res.* **2008**, *113*. [CrossRef]

9. Choi, H.I.; Liang, X.Z. Improved terrestrial hydrologic representation in mesoscale land surface models. *J. Hydrometeorol.* **2010**, *11*, 797–809. [CrossRef]

10. Choi, H.I.; Liang, X.Z.; Kumar, P. A conjunctive surface–Subsurface flow representation for mesoscale land surface models. *J. Hydrometeorol.* **2013**, *14*, 1421–1442. [CrossRef]

11. Schmid, B.H. On overland flow modelling: Can rainfall excess be treated as independent of flow depth? *J. Hydrol.* **1989**, *107*, 1–8. [CrossRef]

12. Wallach, R.; Grigorin, G.; Rivlin, J. The errors in surface runoff prediction by neglecting the relationship between infiltration rate and overland flow depth. *J. Hydrol.* **1997**, *200*, 243–259. [CrossRef]

13. Singh, V.; Bhallamudi, S.M. A complete hydrodynamic border-strip irrigation model. *J. Irrig. Drain. Eng.* **1997**, *122*, 189–197. [CrossRef]

14. Dai, Y.; Zeng, X.; Dickinson, R.E.; Baker, I.; Bonan, G.B.; Bosilovich, M.G.; Denning, A.S.; Dirmeyer, P.A.; Houser, P.R.; Niu, G.; et al. The common land model. *Bull. Am. Meteor. Soc.* **2003**, *84*, 1013–1023. [CrossRef]

15. Liang, X.Z.; Choi, H.I.; Kunkel, K.E.; Dai, Y.; Joseph, E.; Wang, J.X.L.; Kumar, P. *Development of the Regional Climate-Weather Research and Forecasting Model (CWRF): Surface Boundary Conditions*; ISWS SR 2005-01; Illinois State Water Survey: Champaign, IL, USA, 2005; p. 32.

16. Liang, X.Z.; Choi, H.I.; Kunkel, K.E.; Dai, Y.; Joseph, E.; Wang, J.X.L.; Kumar, P. Surface boundary conditions for mesoscale regional climate models. *Earth Interact.* **2005**, *9*, 1–28. [CrossRef]

17. Liang, X.Z.; Xu, M.; Gao, W.; Kunkel, K.E.; Slusser, J.; Dai, Y.; Min, Q.; Houser, P.R.; Rodell, M.; Schaaf, C.B.; et al. Development of land surface albedo parameterization bases on Moderate Resolution Imaging Spectroradiometer (MODIS) data. *J. Geophys. Res.* **2005**, *110*. [CrossRef]

18. Liang, X.Z.; Xu, M.; Zhu, J.; Kunkel, K.E.; Wang, J.X.L. Development of the regional climate-weather research and forecasting model (CWRF): Treatment of topography. In Proceedings of the 2005 WRF/MM5 User's Workshop, Boulder, CO, USA, 27–30 June 2005; p. 5.

19. Liang, X.Z.; Xu, M.; Choi, H.I.; Kunkel, K.E.; Rontu, L.; Geleyn, J.F.; Müller, M.D.; Joseph, E.; Wang, J.X.L. Development of the regional Climate-Weather Research and Forecasting model (CWRF): Treatment of subgrid topography effects. In Proceedings of the 7th Annual WRF User's Workshop, Boulder, CO, USA, 19–22 June 2006; p. 5.

20. Choi, H.I. 3-D Volume Averaged Soil-Moisture Transport Model: A Scalable Scheme for Representing Subgrid Topographic Control in Land-Atmosphere Interactions. Ph.D. Dissertation, University of Illinois at Urbana-Champaign, Champaign, IL, USA, 2006; p. 189.

21. Choi, H.I.; Kumar, P.; Liang, X.Z. Three-dimensional volume-averaged soil moisture transport model with a scalable parameterization of subgrid topographic variability. *Water Resour. Res.* **2007**, *43*, W04414. [CrossRef]

22. Liang, X.Z.; Xu, M.; Yuan, X.; Ling, T.; Choi, H.I.; Zhang, F.; Chen, L.; Liu, S.; Su, S.; Qiao, F.; et al. Regional Climate-Weather Research and Forecasting Model (CWRF). *Bull. Am. Meteorol. Soc.* **2012**, *93*, 1363–1387. [CrossRef]

23. Liang, X.Z.; Chen, L.; He, Y.; Gan, Y. Evaluation of CWRF performance in simulating US climate variations. In Proceedings of the 2013 Fall Meeting, AGU, San Francisco, CA, USA, 9–13 December 2013.

24. Qian, T.; Dai, A.; Trenberth, K.E.; Oleson, K.W. Simulation of global land surface conditions from 1948 to 2004: Part I: Forcing data and evaluations. *J. Hydrometeorol.* **2006**, *7*, 953–975. [CrossRef]

25. Niu, G.Y.; Yang, Z.L. Effects of frozen soil on snowmelt runoff and soil water storage at a continental scale. *J. Hydrometeor.* **2006**, *7*, 937–952. [CrossRef]

26. Niu, G.Y.; Yang, Z.L.; Dickinson, R.E.; Gulden, L.E.; Su, H. Development of a simple groundwater model for use in climate models and evaluation with gravity recovery and climate experiment data. *J. Geophys. Res.* **2007**, *112*, D07103. [CrossRef]

27. Lawrence, P.J.; Chase, T.N. Representing a new MODIS consistent land surface in the Community Land Model (CLM3.0). *J. Geophys. Res.* **2007**, *112*, G01023. [CrossRef]

28. Lawrence, D.M.; Thornton, P.E.; Oleson, K.W.; Bonan, G.B. The partitioning of evapotranspiration into transpiration, soil evaporation, and canopy evaporation in a GCM: Impacts on land-atmosphere interaction. *J. Hydrometeorol.* **2007**, *8*, 862–880. [CrossRef]

29. Choi, H.I. Parameterization of high resolution vegetation characteristics using remote sensing products for the Nakdong River Watershed, Korea. *Remote Sens.* **2013**, *5*, 473–490. [CrossRef]

30. Choi, H.I. Application of a Land Surface Model Using Remote Sensing Data for High Resolution Simulations of Terrestrial Processes. *Remote Sens.* **2013**, *5*, 6838–6856. [CrossRef]

31. Horton, R.E. The Role of Infiltration in the Hydrologic Cycle. *Eos Trans. Am. Geophys. Union* **1933**, *14*, 446–460. [CrossRef]

32. Dunne, T.; Black, R.G. An Experimental Investigation of Runoff Production in Permeable Soils. *Water Resour. Res.* **1970**, *6*, 478–490. [CrossRef]

33. Sivapalan, M.; Beven, K.J.; Wood, E.F. On hydrologic similarity: 2. A scaled model of storm runoff production. *Water Resour. Res.* **1987**, *23*, 2266–2278. [CrossRef]

34. Beven, K.J.; Kirkby, M.J. A physically based, variable contributing area model of basin hydrology. *Hydrol. Sci. Bull.* **1979**, *24*, 43–69. [CrossRef]

35. Kumar, P.; Verdin, K.L.; Greenlee, S.K. Basin level statistical properties of topographic index for North America. *Adv. Water Resour.* **2000**, *23*, 571–578. [CrossRef]

36. The USGS Land Cover Institute (LCI). Available online: http://landcover.usgs.gov/landcoverdata.php (accessed on 30 December 2016).

37. Yucel, I. Effects of implementing MODIS land cover and albedo in MM5 at two contrasting US regions. *J. Hydrometeorol.* **2006**, *7*, 1043–1060. [CrossRef]

38. Zeng, X.; Dickinson, R.E.; Walker, A.; Shaikh, M.; DeFries, R.S.; Qi, J. Derivation and evaluation of global 1–km fractional vegetation cover data for land modeling. *J. Appl. Meteorol.* **2000**, *39*, 826–839. [CrossRef]

39. Zeng, X.; Shaikh, M.; Dai, Y.; Dickinson, R.E.; Myneni, R. Coupling of the common land model to the NCAR community climate model. *J. Clim.* **2002**, *15*, 1832–1854. [CrossRef]

40. SPOT-VEGETATION PROGRAMME. Available online: http://www.vgt.vito.be/ (accessed on 30 December 2016).

41. MOD 15—Leaf Are Index (LAI) and Fractional Photosynthetically Active Radiation (FPAR). Available online: http://cybele.bu.edu/download/manuscripts/mod15.pdf (accessed on 30 December 2016).

42. SRTM 90m Digital Elevation Database v4.1. Available online: http://www.cgiar-csi.org/data/srtm-90m-digital-elevation-database-v4-1 (accessed on 30 December 2016).

43. Harmonized World Soil Database v 1.2. Available online: http://webarchive.iiasa.ac.at/Research/LUC/External-World-soil-database/HTML/ (accessed on 30 December 2016).

44. (Hydrological data and maps based on SHuttle Elevation Derivatives at multiple Scales) HydroSHEDS. Available online: http://hydrosheds.cr.usgs.gov/index.php (accessed on 30 December 2016).

45. Nash, J.E.; Sutcliffe, J.V. River flow forecasting through conceptual models part I—A discussion of principles. *J. Hydrol.* **1970**, *10*, 282–290. [CrossRef]

46. McCuen, R.H.; Knight, Z.; Cutter, A.G. Evaluation of the Nash-Sutcliffe efficiency index. *J. Hydrol. Eng.* **2006**, *11*, 597–602. [CrossRef]

![water logo] *water*

MDPI

*Article*

# Optimal Combinations of Non-Sequential Regressors for ARX-Based Typhoon Inundation Forecast Models Considering Multiple Objectives

**Huei-Tau Ouyang** [1,*], **Shang-Shu Shih** [2] and **Ching-Sen Wu** [1]

[1] Department of Civil Engineering, National Ilan University, Yilan 26047, Taiwan; olivercswu@niu.edu.tw
[2] Department of Civil Engineering, National Taiwan University, Taipei 10617, Taiwan; uptreeshih@ntu.edu.tw
* Correspondence: htouyang@niu.edu.tw; Tel.: +886-3-935-7400

Received: 19 April 2017; Accepted: 11 July 2017; Published: 14 July 2017

**Abstract:** Inundation forecast models with non-sequential regressors are advantageous in efficiency due to their rather fewer input variables required to be processed. This type of model is nevertheless rare mainly because of the difficulty in finding the proper combination of regressors for the model to perform accurate prediction. A novel methodology is proposed in this study to tackle the problem. The approach involves integrating a Multi-Objective Genetic Algorithm (MOGA) with forecast models based on ARX (Auto-Regressive model with eXogenous inputs) to transfer the search for the optimal combination of non-sequential regressors into an optimization problem. An innovative approach to codifying any combinations of model regressors into binary strings is developed and employed in MOGA. The Pareto optimal sets of three types of models including linear ARX (LARX), nonlinear ARX with Wavelet function (NLARX-W), and nonlinear ARX with Sigmoid function (NLARX-S) are searched for by the proposed methodology. The results show that the optimal models acquired through this approach have good inundation forecasting capabilities in every aspect in terms of accuracy, time shift error, and error distribution.

**Keywords:** typhoon; inundation; ARX model; non-sequential regressors; multi-objective genetic algorithm

---

## 1. Introduction

Typhoon is a common weather phenomenon in subtropical areas and usually occurs between July and October of each year. Heavy rainfall brought by the typhoon during the event usually results in serious inundation problems in low-lying areas, which not only causes property loss to the local population but also threatens their safety. Due to restrictions on engineering funding, structural protective measures are constrained by the designed limits. Once the typhoon scale exceeds the designed protective limit, people must rely on nonstructural means for disaster relief during the event, such as evacuating people from areas in potentially high flooding risk. Among nonstructural measures the accurate forecast of the inundation level in the areas within the next several hours is a critical factor in the decision-making and planning of disaster relief actions.

Relevant studies on inundation forecast technology are quite ample and can generally be divided into either the numerical simulation or the black-box modeling. The numerical simulation is based on theoretical deduction through the understanding of the mechanism from rainfall to inundation. The advantage of this type of method is the completed support from the theoretical basis for the physical mechanism of inundation. The simulation result often has a high degree of accuracy, which renders the method a powerful tool of inundation forecast. However, the disadvantage of this method is its high demand for computing resources and CPU time, which makes it unsuitable to provide the real-time forecast required in the quick disaster prevention and rescue response during the typhoon attack. On the other hand, the black-box modeling relies on a different approach by deeming the process from rainfall to inundation as a black box. It does not delve into the internal physical mechanism but instead analyzes the input and output data of the system to simulate the relationship between them. These types of models cannot explain the physical mechanism involved in the system, but they can correctly and effectively simulate the response of the system, and the computing speed is faster than numerical models [1]. These practical benefits render black-box modeling a suitable forecasting tool for decision making and rescue planning during the typhoon period.

Abundant studies with regard to black-box modeling for inundation forecasting can be found in literature, for example, Liong et al. [2] developed a river stage forecasting model based on an artificial neural network (ANN) and yielded a very high degree of prediction accuracy even for up to seven lead days. Campolo et al. [3] developed a flood forecasting model based on ANN that exploits real-time information available for the basin of the River Arno to predict the basin's water level evolution. Keskin et al. [4] proposed a flow prediction method based on an adaptive neural-based fuzzy inference system (ANFIS) coupled with stochastic hydrological models. Shu and Ouarda [5] proposed a methodology using ANFIS for flood quantile estimation at ungauged sites and demonstrated that the ANFIS approach has a much better generalization capability than other alternatives. Kia et al. [6] develop a flood model based on ANN using various flood causative factors in conjunction with geographic information system (GIS) to model flood-prone areas in southern Malaysia. Lin et al. [7] proposed a real-time regional forecasting model to yield 1- to 3-h lead time inundation maps based on K-means cluster analysis incorporated with support vector machine (SVM). Tehrany et al. [8] proposed a methodology for flood susceptibility mapping by combining SVM and weights-of-evidence (WoE) models and demonstrated that the ensemble method outperforms the individual methods. Del Giudice et al. [9] developed a methodology that formulated models with increasing detail and flexibility, describing their systematic deviations using an autoregressive bias process. Chang and Tsai [10] proposed a spatial–temporal lumping of radar rainfall for modeling inflow forecasts to mitigate time-lag problems and improve forecasting accuracy.

In view of the above literature, most of the approaches employ sequential data as model inputs. Less has been explored for forecasting models with non-sequential data inputs. This type of model is efficient because there are rather fewer inputs required to be processed. The challenge for these models, however, lies in selecting the appropriate combination of non-sequential variables to be used in the inputs. This study aims to propose a methodology to tackle this difficulty. The approach proceeds by integrating a Multi-Objective Genetic Algorithm (MOGA) with models based on ARX (Auto-Regressive models with eXogenous inputs) to search for the optimal combination of non-sequential regressors for model inputs. Three types of ARX-based models are tested by the proposed methodology, including linear ARX (LARX), nonlinear ARX with Wavelet function (NLARX-W), and nonlinear ARX with Sigmoid function (NLARX-S). The models are assessed by a number of indexes to examine their performance on various aspects, and the characteristics of the models selected from the Pareto optimal sets located by MOGA with the best performance in each index are compared and discussed. The remainder of this paper is arranged as follows: Section 2 illustrates the environmental background of the study area, the ARX models, and the optimal models obtained by MOGA; Section 3 discusses the characteristics of the acquired optimal models, and finally the conclusions are drawn in Section 4 based on the findings.

## 2. Materials and Methods

### 2.1. Study Area

Yilan County (Figure 1), located in northeastern Taiwan, is known for its rainy weather. With about 200 rainy days per year, the annual average precipitation is around 2000–2500 mm. The terrain is surrounded by mountains in the west and coastline in the east. It is frequently hit by typhoons in the summer and autumn of each year. On average, two to three typhoons hit Taiwan every year and among which 45% land in Yilan County [11]. The heavy rainfall during the typhoon attacking period often causes severe inundation in the low-lying areas. Among these inundation-prone areas, the situation in Donsan area (Figure 1) is most extreme. The drainage area of Donsan is about 34 km². Ground level in the area is very low and flat. The ground elevation in more than one-third of the area is below 2 m, as seen in Figure 1. During typhoon invasion, these low grounds are often flooded, causing heavy damages and tremendous losses to the properties.

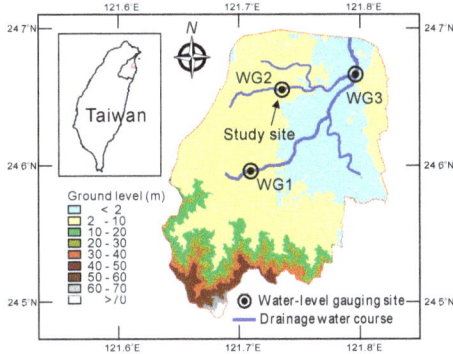

**Figure 1.** Donsan area in Yilan County, Taiwan.

The high tendency in severe flooding within the area has raised an urgent need for effective disaster preventive measures. In order to monitor the local inundation condition during the typhoon period, Taiwan Water Resources Agency established a surveillance network for the area in 2011. It includes three water-level gauging stations, as well as a data transmission system that can receive the precipitation observations of the Quantitative Precipitation Estimation and Segregation Using Multiple Sensor (QPESUMS, [12]). QPESUMS consists of eight Doppler radar stations that each scans an area with a radius of approximately 230 km. The system provides quantitative precipitation estimation by integrating observations from weather radars and rainfall readings from 406 automatic gauges and 45 ground stations in Taiwan. QPESUMS also provides rainfall forecasts by tracking and extrapolating the movement paths of storm cells based on radar readings. During the typhoon period, QPESUMS delivers the rainfall data in a frequency of every 10 min through a network connection to the surveillance system. The water level gauging stations return the local inundation water-level data with the same frequency through radio transmission.

Since the monitoring system of Donsan area was established, it has collected data of 10 typhoons over the years. Figure 2 shows the recorded water level of each station and QPESUMS rainfall data during Typhoon Trami in 2013. Detailed information of each typhoon event is listed in Table 1. These data not only provide the local rainfall and water level information during the typhoon period but can also be used for the establishment of water-level forecast models. In the present study, the WG2 gauging site is selected as the study object to establish the water level forecast model.

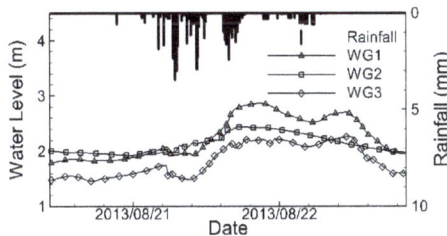

**Figure 2.** Water level and rainfall records during Typhoon Trami in Donsan area, Taiwan.

**Table 1.** Historical typhoon records in Donsan area.

| Typhoon | Year | Time of Official Typhoon Sea Warning Issued h/Day/Month (UTC) | Affecting Period (h) | Maximum Rainfall Intensity (mm/h) | Cumulative Rainfall (mm) |
|---------|------|----------------------------------------------|------------|---------------------|---------------------|
| Songda | 2011 | 0230/27/May | 36 | 28.5 | 213 |
| Nanmadol | 2011 | 0530/27/August | 99 | 26.5 | 172 |
| Saola | 2012 | 2030/30/July | 90 | 37.5 | 537 |
| Soulik | 2013 | 0830/11/July | 63 | 30.0 | 133 |
| Trami | 2013 | 1130/20/August | 45 | 22.0 | 461 |
| Usagi | 2013 | 2330/20/September | 93 | 23.0 | 165 |
| Matmo | 2014 | 1730/21/July | 54 | 35.5 | 98 |
| Fung-wong | 2014 | 0830/19/September | 72 | 41.0 | 72 |
| Soudelor | 2015 | 1130/6/August | 69 | 87.5 | 421 |
| Dujuan | 2015 | 0830/27/September | 57 | 42.0 | 218 |

*2.2. Model Construction*

The present study adopts ARX to build the water-level forecast model for the quick calculation feature. ARX is a kind of black-box model which can be divided into the linear model and nonlinear model according to the relationship between the input and output. The nonlinear model can be further subdivided into various types according to the different nonlinear functions applied. The present study respectively aims at linear ARX, as well as ARX with Wavelet function and ARX with Sigmoid function in the nonlinear ARX, to discuss the inundation forecast performance of these three types of models.

With regard to emergency response to the typhoon, the water level is a greater concern than runoff. Therefore, the purpose of the forecast model is to establish the relationship between rainfall and water level, rather than the more commonly seen rainfall and runoff relationship. Although few studies have focused on the relationship between rainfall and water level, relevant works can be found in the literature [13–15].

2.2.1. Linear ARX

Linear ARX (LARX) is an extension of the AR model [16] in time series analysis by adding the influence of other exogenous inputs. The equation is as follows:

$$H(t+1) + a_1 H(t) + \cdots + a_{n_a} H(t-n_a) = b_1 R(t-n_k) + \cdots + b_{n_b} R(t-n_b-n_k+1) + e(t) \quad (1)$$

in which $H$ represents the system output; $R$ is the exogenous input; $n_a$ and $n_b$ are the number of terms of $H$ and $R$, respectively; $n_k$ is the time lag of $R$; $e$ is white noise; $a_1$ through $a_{n_a}$ and $b_1$ through $b_{n_b}$ are the coefficient of each term, respectively. $H(t+1)$ through $H(t-n_a)$ and $R(t-n_k)$ through $R(t-n_b-n_k+1)$ are called the regressors which take in the known data to forecast $H(t+1)$. In LARX, the relationship between $H(t)$ and the regressors is linear, as shown in Equation (1). In this study, $H$ represents the water level at the site of WG2 in Donsan area, while $R$ is the local rainfall data provided by QPESUMS.

2.2.2. Nonlinear ARX

Nonlinear ARX extends linear ARX by using a nonlinear relationship between the forecast and the regressors, as shown in the equation below:

$$H(t+1) = \psi(H(t), \ldots, H(t-n_a), R(t-n_k), \ldots, R(t-n_b-n_k+1)) \quad (2)$$

where $\psi$ is the nonlinearity estimator which has the following form:

$$\psi(x) = \sum_{k=1}^{n} \alpha_k \kappa(\beta_k(x-\gamma_k)) \quad (3)$$

in which $x$ is the row vector consists of the regressors, $n$ is the number of terms of nonlinearity estimator, $\alpha_k$ and $\beta_k$ are the coefficients of each term, and $\gamma_k$ is the mean value of the regressor vector. $\kappa$ is the nonlinear unit, which can adopt different nonlinear functions.

Two different types of nonlinear ARX models are employed in the present study, viz. the nonlinear ARX with Wavelet function (NLARX-W) and the nonlinear ARX with Sigmoid function (NLARX-S). In NLARX-W, the nonlinear unit $\kappa$ adopts the Wavelet function [17], which has the following formula:

$$\kappa(s) = \left(\dim(s) - s^T s\right) e^{-0.5 s^T s} \tag{4}$$

where dim represents the dimension of the vector $s$; $e$ is the exponential function.

On the other hand, in NLARX-S, $\kappa$ adopts the Sigmoid function as shown in the equation below:

$$\kappa(s) = (e^s + 1)^{-1} \tag{5}$$

The model structure of LARX, NLARX-W, and NLARX-S is determined by the selection of regressors which are to be determined later on through the optimization process.

### 2.2.3. Rainfall Data Analysis

To analyze the relationship between water level and rainfall, correlation analysis on typhoon data records was carried out. The definition of the correlation coefficient (CC) is shown below [18]:

$$CC(x,y) = \frac{cov(x,y)}{\sigma_x \sigma_y} = \frac{\sum_{i=1}^{n}(x_i - \overline{x})(y_i - \overline{y})}{\sqrt{\sum_{i=1}^{n}(x_i - \overline{x})^2}\sqrt{\sum_{i=1}^{n}(y_i - \overline{y})^2}} \tag{6}$$

in which $cov$ is the covariance between variables $x$ and $y$; $\sigma_x$ and $\sigma_y$ are the standard deviations of $x$ and $y$, respectively; $n$ is the quantity of the data points. CC ranges between $-1$ and $1$, where $-1$ represents the perfect negative correlation between $x$ and $y$, $+1$ means the perfect positive correlation, and 0 means that there is no correlation between $x$ and $y$.

The rainfall data provided by QPESUMS are 10-min rainfall increments. However, it has been reported by Ouyang [19] that the variation in water level is slower than the variation in rainfall, and often a certain amount of rainfall accumulation is required for the water to build up. To search for the most correlated cumulative rainfall to the water level at the study site, the procedure of Ouyang [19] was adopted here. The moving cumulative rainfall data of the typhoon events with accumulation duration ranging from 1 h to 30 h were first constructed. The correlations of each cumulative rainfall data set and the water level data were then calculated. The results are as shown in Figure 3. The circle dots in the figure represent the average CC of the typhoon events, and the error bars denote the CC distribution of the events. As shown in the figure, the average CC is first gradually increasing along with an increase in the accumulation duration, and then followed by a gradual recession with a further increase in the accumulation duration. As shown in the figure, the maximum average CC appears when the accumulation duration is about 18 h. This indicates that the water level at the study site is most correlated with the 18 h cumulative rainfall which shall be adopted as the input data of the ARX models.

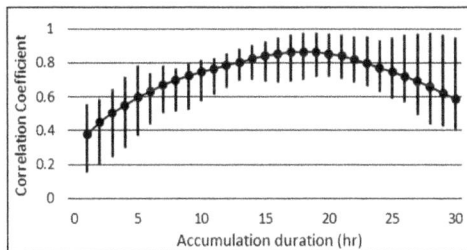

**Figure 3.** Correlations between water level and cumulative rainfall with various accumulation durations.

### 2.2.4. Regressors

The ARX simulation depends on the selection of regressors. With different combinations of regressors, the performance of the model varies. In the selection of model regressors, two methods are commonly utilized in the literature. The first one is Sequential Time Series (see for example, [20–23]) where a sequence of time series data from the current time to a certain time before is used as model regressors, for example, $R(t), R(t-1), \ldots, R(t-m)$, in which $R$ represents the cumulative rainfall. The second common method is Pruned Sequential Time Series (see for example [24,25]), which retains only the most relevant period of data sequence as the model input to prevent excessive variables from influencing the model simulations. For example, $R(t-a), \ldots, R(t-b)$, where $t-a$ through $t-b$ represent the time period when $R$ is most correlated to the water level.

In addition to the above two methods, a third method namely Non-Sequential Time Series has been proposed by Talei et al. [26]. In their study, two antecedent rainfall data $R(t-T_1)$ and $R(t-T_2)$ are selected as the model input, in which $T_1$ and $T_2$ are two non-sequential time points determined through a search test. Their results show that the models acquired in this method perform better than those acquired in the two aforementioned methods. In the study of Talei et al. [26], $T_1$ and $T_2$ were determined by a search process using the method of exhaustion. However, this approach of exhaustion search requires tremendous computing resources and CPU time, which makes it only suitable for the combination of a few regressors. The present study improves this aspect by adopting genetic algorithm (GA) for the optimal selection of model regressors, such that the number of model regressors is no longer limited and a vast amount of models with various combinations of regressors can be explored.

In the present study, the input variables of the models are selected non-sequentially from the combination of $R(t), R(t-1), \ldots, R(t-m)$, where $R$ represents the 18-h cumulative rainfall which, according to the previous data analysis, has been shown most correlated to the water-level. $m$ defines the range of regressors to be selected. The greater the value of $m$, the more possible combinations of regressors and thus candidate models can be explored. Considering the computing time required for the search process, $m$ is set to be 10 in the present study. The water level forecast is also related to water level data $H(t), \ldots, H(t-n)$. For simplicity, $n$ is also set to be 10 herein. The relation between the regressors and output of each model is summarized below:

$$H(t+1) = \psi[\text{combination of } R(t), R(t-1), \cdots, R(t-10), \text{ and } H(t), H(t-1), \cdots, H(t-10)] \quad (7)$$

in which $\psi$ represents the function of ARX.

### 2.2.5. Assessing Indexes

In order to search for the optimal models, the following indexes are employed to evaluate the water level forecast capacity of each candidate model.

#### Coefficient of Efficiency (CE)

CE is an index designed to evaluate the performance of a hydrological model [27]. CE is defined as follows

$$CE = 1 - \frac{\sum_{t=1}^{n_t}[H_{obs}(t) - H_{est}(t)]^2}{\sum_{t=1}^{n_t}[H_{obs}(t) - \overline{H}_{obs}]^2} \quad (8)$$

where $H_{obs}$ and $H_{est}$ are the observed and estimated water levels, respectively; $\overline{H}_{obs}$ is the average of observed water level, and $n_t$ is the number of data points. A CE value closer to 1 indicates that the predicted water levels fit more for the observations.

#### Relative Time Shift Error (RTS)

It has been shown by Talei and Chua [28] that a certain time shift error might emerge when using past data to forecast into the future. The forecasted water-level hydrograph displays a shifted delay in time from the observations. In order to assess the time shift error of the models, the process of Talei and Chua [28] is adopted in this study. The water-level hydrograph predicted by the model is first shifted back in time from 0 to $k$ data points, and the CE of each shifted hydrograph is then calculated by comparing to the observations. The shift step $\delta$ associated to the maximum CE is considered as the time shift error of the model. The definition of Relative Time Shift is thus as follows

$$\text{RTS} = \frac{\bar{\delta}}{k} \tag{9}$$

where $\bar{\delta}$ denotes the average $\delta$ of the typhoon events, and $k$ is the forecast horizon (i.e., prediction time steps; each time step is 10 min in the present study). RTS ranges between 0 and 1. A smaller RTS represents less time shift in the predictions.

Threshold Statistic for a Level of $x\%$ ($\text{TS}_x$)

$\text{TS}_x$ ([29,30]) was employed to assess the error distribution in the forecasted water-level hydrographs. $\text{TS}_x$ is defined as follows

$$\text{TS}_x = \frac{y_x}{n} \times 100 \tag{10}$$

where $n$ is the total amount of data points; $y_x$ is the amount of forecasted data points with absolute relative error $|RE_t|$ less than a specified criteria of $x\%$. $RE_t$ is defined as

$$RE_t = \frac{H_t^o - H_t^c}{d_t^o} \times 100 \tag{11}$$

where $H_t^o$ and $H_t^c$ are the observed and predicted water level at time $t$, respectively; and $d_t^o$ is the observed water depth. Using water depth instead of water level as the denominator in Equation (11) has several benefits. First, the value of $|RE_t|$ thus defined ranges between 0 and 1, where $|RE_t| = 0$ indicates perfect prediction and 1 denotes a bad prediction deviated from the observation by the entire water depth. Second, a better accuracy in error measurement is achieved since the scale of water level might be in tens or hundreds of meters (depends on the location where the data were recorded) whereas in water depth it is often in meters only. A greater $\text{TS}_x$ means that the forecasted water-level hydrograph contains more predicted data points with the absolute relative error less than $x\%$, and thus the model performance is better. The present study adopts 15% as the threshold value (i.e., $\text{TS}_{15}$).

### 2.2.6. Data Processing

The water level and cumulative rainfall data are standardized using the following equation [31]

$$y_n = 0.1 + 0.8 \left( \frac{y_i - y_{min}}{y_{max} - y_{min}} \right) \tag{12}$$

where $y_n$ is the data after standardization; $y_i$ is the raw data; $y_{max}$ and $y_{min}$ are the maximum and minimum of raw data, respectively. The process of data standardization concentrates the dispersive data in a defined interval, which has shown great improvement on the predictions [31].

### 2.3. Model Optimization

The three indexes of CE, RTS, and $\text{TS}_{15}$ each evaluates the model's water level forecast accuracy, the time shift error, and the error distribution, respectively. All the three indexes provide valuable information for disaster prevention action during typhoon attack and shall be considered at the same time. However, it is difficult to weigh the importance of the three indexes. In order to find the models with good performance in all aspects, the multi-objective genetic algorithm (MOGA) was adopted in the present study as a tool for the search.

### 2.3.1. Multi-Objective Genetic Algorithm

The theoretical basis of GA was developed based on the Darwinian natural selection theory. After Holland [32] developed a firm mathematical foundation for the algorithm, GA has been widely applied to various fields to solve optimization problems that could not be tackled by traditional methods. In GA, each individual in a population is deemed as a possible solution to the optimization problem at hand. Based on the specified objective functions, the performance of each individual is evaluated and compared, and the individuals with better performance shall have a greater chance to pass its gene to the next generation. Through this procedure, the overall performance of the entire population gradually evolves and improves. After several generations of evolution, the individuals with optimal performance shall emerge, and these optimal individuals are considered as the optimal solutions of the problem. The algorithm has been shown capable of locating the global optima [33] and is particularly suitable for solving multi-objective optimization problems [34].

### 2.3.2. Objective Functions

Based on the definition of CE, RTS, and TS$_{15}$, three design goals of the optimal model can be identified, which are the maximum CE, minimum RTS, and maximum TS$_{15}$, respectively. Also, according to Equations (8)–(10), the upper limit of CE and TS$_{15}$ is 1, and RTS is always a positive number; therefore, the objective functions shall be defined as follows:

$$\text{Objective 1:} \quad \text{minimize} \left(1 - \overline{CE}\right) \tag{13}$$

$$\text{Objective 2:} \quad \text{minimize} \ \overline{RTS} \tag{14}$$

$$\text{Objective 3:} \quad \text{minimize} \left(1 - \overline{TS}_{15}\right) \tag{15}$$

in which $\overline{CE}$, $\overline{RTS}$, $\overline{TS}_{15}$ represent the averaged value of the three indexes for the typhoon events, respectively. The design goal of Objective 1 and Objective 3 is to get $\overline{CE}$ and $\overline{TS}_{15}$ closest to 1, while the design goal of Objective 2 is to get $\overline{RTS}$ closest to 0.

### 2.3.3. Codification of Regressor Combinations

The design variables of the optimization problem are the various possible combinations of model regressors consisting of $R(t)$ through $R(t-10)$ and $H(t)$ through $H(t-10)$, as shown in Equation (7). To utilize MOGA to search for the optimal models, the combination of the regressors has to be codified into a chromosome. In the present study, this procedure is accomplished by using binary bit string codifications, as illustrated in Figure 4. The combination of regressors is represented by a binary bit string chromosome consisting of 22 genes each associated to a specific regressor. If the value of a gene is 1, the regressor associated to this gene shall be selected as an input variable of the model. Contrarily, for a gene with a value of 0, the associated regressor shall not be selected as input. Each chromosome represents a specific combination of regressors and thus a candidate model. With 22 genes in one chromosome, the total candidate models thus formed in the search space are $2^{22} \approx 4.19 \times 10^6$.

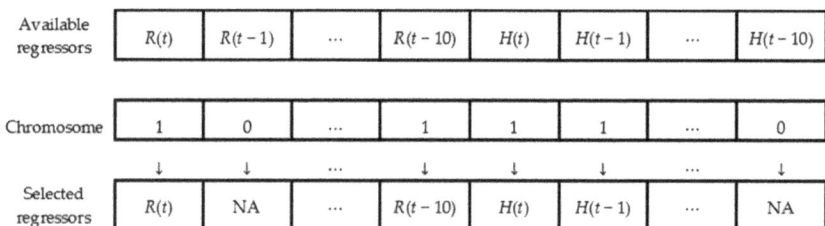

| Available regressors | $R(t)$ | $R(t-1)$ | $\cdots$ | $R(t-10)$ | $H(t)$ | $H(t-1)$ | $\cdots$ | $H(t-10)$ |
|---|---|---|---|---|---|---|---|---|
| Chromosome | 1 | 0 | $\cdots$ | 1 | 1 | 1 | $\cdots$ | 0 |
| | ↓ | ↓ | $\cdots$ | ↓ | ↓ | ↓ | $\cdots$ | ↓ |
| Selected regressors | $R(t)$ | NA | $\cdots$ | $R(t-10)$ | $H(t)$ | $H(t-1)$ | $\cdots$ | NA |

**Figure 4.** Binary bit string codification of a combination of regressors.

For model calibration, 6 out of the 10 events were selected, and the rest events were employed for validation. The selection of the calibration events is based on total cumulative rainfall to represent large, middle, and small amounts of rainfalls. The events thus selected are Songda, Nanmadol, Soulik, Fung-wong, Soudelor, and Dujuan. The three performance indexes of CE, RTS and $TS_{15}$ are calculated and averaged over the validation events. The optimal models for the three types of LARX, NLARX-W, and NLARX-S are searched for using MOGA. With regard to MOGA setting, the population size is set to 200 with Pareto fraction of 0.35. The maximum generation of evolution is 500, and the stopping criterion of the evolution is set to 50 generations of stalls.

## 3. Results and Discussion

The result acquired by MOGA is the Pareto optimal model set, in which every model is un-dominated, that is, at least one of the three indexes of the model is not exceeded by another model. For each model type of LARX, NLARX-W, and NLARX-S, the three models with the best performance in the three indexes, respectively, are selected among the Pareto optimal set. The results are shown in Table 2. The selected models are named according to their model type and the featuring index, L1 represents the model with the maximum $\overline{CE}$ in LARX, L2 represents the one with the minimum $\overline{RTS}$ in LARX, and L3 represents the maximum $\overline{TS}_{15}$ model in LARX. Similarly, W1, W2 and W3, as well as S1, S2, and S3, respectively represent the models with the maximum $\overline{CE}$, minimum $\overline{RTS}$, and maximum $\overline{TS}_{15}$ in NLARX-W and NLARX-S. The scores of the nine optimal models on the three indexes and the corresponding chromosomes (i.e., selected regressors) are listed in Table 2. As shown, the selected regressors associated to each of the nine optimal models appear to be in a non-sequential pattern. This supports the result of Talei et al. [26] that a model with non-sequential inputs has better performance than the ones with sequential or pruned sequential inputs.

**Table 2.** Optimal models and the corresponding chromosomes (L1–L3: optimal Linear Auto-Regressive model with eXogenous inputs (LARX) models; W1–W3: optimal nonlinear ARX with Wavelet function (NLARX-W) models; S1–S3: optimal nonlinear ARX with Sigmoid function (NLARX-S) models; $R(t-1)$ through $R(t-10)$: antecedent cumulative rainfall data; $H(t-1)$ through $H(t-10)$: antecedent water level data).

| Index | Model | | | | | | | | |
|---|---|---|---|---|---|---|---|---|---|
| | L1 | L2 | L3 | W1 | W2 | W3 | S1 | S2 | S3 |
| CE | 0.859 | 0.245 | 0.852 | 0.933 | 0.269 | 0.911 | 0.946 | 0.734 | 0.929 |
| RTS | 0.403 | 0.251 | 0.500 | 0.125 | 0.108 | 0.278 | 0.111 | 0.098 | 0.167 |
| $TS_{15}$ | 0.590 | 0.472 | 0.618 | 0.728 | 0.356 | 0.743 | 0.725 | 0.426 | 0.768 |
| Available regressor | Chromosome (Selected regressor) | | | | | | | | |
| $R(t-1)$ | 1 | 0 | 0 | 0 | 0 | 0 | 1 | 1 | 1 |
| $R(t-2)$ | 0 | 0 | 1 | 0 | 0 | 0 | 1 | 0 | 0 |
| $R(t-3)$ | 1 | 1 | 1 | 1 | 1 | 1 | 1 | 1 | 1 |
| $R(t-4)$ | 1 | 0 | 1 | 1 | 0 | 1 | 1 | 0 | 1 |
| $R(t-5)$ | 0 | 0 | 0 | 0 | 0 | 0 | 1 | 1 | 1 |
| $R(t-6)$ | 1 | 1 | 0 | 0 | 0 | 0 | 0 | 1 | 1 |
| $R(t-7)$ | 0 | 0 | 0 | 0 | 0 | 0 | 1 | 0 | 0 |
| $R(t-8)$ | 0 | 0 | 0 | 1 | 0 | 1 | 1 | 1 | 1 |
| $R(t-9)$ | 0 | 0 | 0 | 0 | 0 | 0 | 0 | 1 | 1 |
| $R(t-10)$ | 0 | 0 | 0 | 0 | 0 | 0 | 1 | 1 | 1 |
| $H(t-1)$ | 0 | 0 | 1 | 0 | 0 | 0 | 0 | 0 | 0 |
| $H(t-2)$ | 0 | 1 | 0 | 0 | 1 | 1 | 0 | 0 | 1 |
| $H(t-3)$ | 0 | 0 | 0 | 0 | 1 | 0 | 1 | 0 | 0 |
| $H(t-4)$ | 0 | 1 | 0 | 0 | 1 | 0 | 0 | 0 | 0 |
| $H(t-5)$ | 1 | 0 | 0 | 0 | 0 | 0 | 0 | 0 | 0 |
| $H(t-6)$ | 0 | 0 | 1 | 0 | 0 | 0 | 0 | 1 | 0 |
| $H(t-7)$ | 0 | 1 | 0 | 0 | 0 | 0 | 0 | 0 | 0 |
| $H(t-8)$ | 0 | 1 | 0 | 1 | 0 | 0 | 0 | 1 | 0 |
| $H(t-9)$ | 0 | 0 | 0 | 0 | 1 | 0 | 0 | 0 | 1 |
| $H(t-10)$ | 0 | 0 | 0 | 0 | 0 | 0 | 0 | 0 | 0 |
| Prediction lead: $k = 18$ (3 h) | | | | | | | | | |

The prediction lead time is crucial for disaster relief action during typhoon attack. An appropriate choice of the prediction lead time has to be linked to the properties of the hydrological behavior of the watershed. It has been observed in the study area that the variation of water level often lags behind the rainfall (for example, as observed in Figure 2). This time lag behavior between the rainfall and the water level presents a characteristic of the watershed, which can be used as an index for the selection of the appropriate prediction lead time. For that, the corrections between the cumulative rainfall data and the water level data shifted back in time by various time lags are analyzed. The results are as shown in Figure 5. The circle dots represent the average CC of the typhoon events and the error bars denote the maximum and the minimum CCs of the events. As seen, the average CC reaches the peak at the time lag of $t - 3$. This indicates that the water level is most correlated to the cumulative rainfall with 3 h of lead in the study area. The prediction lead time in the present study is thus selected to be 3 h in accordance with this hydrological characteristic of the watershed.

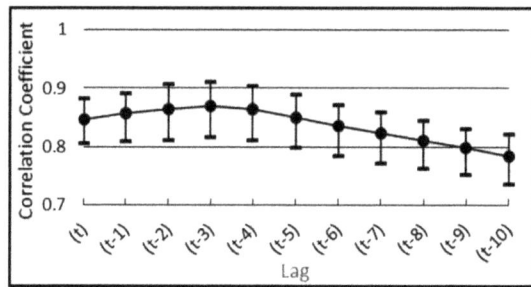

**Figure 5.** Correlations between cumulative rainfall and water level with various time lags.

Figure 6a–c compare the performance of the nine models in CE, RTS, and $TS_{15}$. In the comparison of CE, as shown in Figure 6a, the performance of Model Type 1 (L1, W1, S1) and Model Type 3 (L3, W3, S3) is quite good, in which CE all reaches above 0.8. The CEs of Type 1 Models are seen a little higher than Type 3 Models, but the differences are not significant. The CE performance of Model Type 2 is relatively poorer, especially for L2 and W2. This is because Model Type 2 features on reducing the time shift error. Nevertheless, it is noted that the CE of Model S2 still reaches above 0.7. As shown in Figure 6a, the CEs of the nonlinear models (W- and S-series of models) are somewhat higher than the linear L-series of models, thus indicating that the relationship between rainfall and water level at the site of WG2 is nonlinear. In the comparison of nonlinear models, the CEs of NLARX-S models (S1, S2, S3) appear to be higher than NLARX-W models (W1, W2, W3), and that difference is particularly evident with S2 and W2. In the comparison of time shift errors, as shown in Figure 6b, the RTS of Model Type 2 (L2, W2, S2) is seen lower than Type 1 (L1, W1, S1) and Type 3 (L3, W3, S3). The RTS of S2 in all models is the lowest, showing the minimum time shift error in the forecast, which is followed by W2 with a little increase in RTS; both of which are nonlinear models. The RTS of L2 is the worst in Type 2 models, and as shown in Figure 6b, even W1 and S1 which feature on CE have shown somewhat better performance on RTS than L2. Also, the RTS of the linear L1 and L3 models are much higher than the nonlinear W- and S-series of models. This yet again implies the nonlinearity between the rainfall and the water-level at the study area. Figure 6c is the comparison of the nine models in $TS_{15}$ performance. As shown in the figure, both Type 1 models ((L1, W1, S1) and Type 3 models (L3, W3, S3) display high performance on $TS_{15}$. Among the nine models, the $TS_{15}$ of S3 is the highest, which is closely followed by W3. The linear L3 has the worst $TS_{15}$ in Type 3 models, which as shown in Figure 6c, is even lower than the nonlinear W1 and S1 that features on CE. In summary, the comparisons in Figure 6 show that the overall performance of nonlinear models is better than linear models on every aspect. Comparisons in the nonlinear models show that the NLARX-S models perform a little better than the NLARX-W models, but the differences are not very significant.

The results of the model predictions might be related to the hydrological behavior of the watershed. As has been shown in Figure 5, the correction between the water level and the rainfall in the study area is most prominent with a time lag of 3 h. It is noted in Table 2 that the model regressors optimized by MOGA all include $R(t - 3)$ in the inputs. This result might reflect the hydrological behavior of the study area. It is also noted that Type 1 and Type 3 models which exhibit higher CE and $TS_{15}$ values include not only $R(t - 3)$ but also $R(t - 4)$ or $R(t - 2)$ as their regressors. As seen in Figure 2, the correction between the water level and the rainfall in the study area is also high with time lags of $t - 2$ and $t - 4$. This might explain the rather good results of CE and $TS_{15}$ scores achieved by Type 1 and Type 3 models.

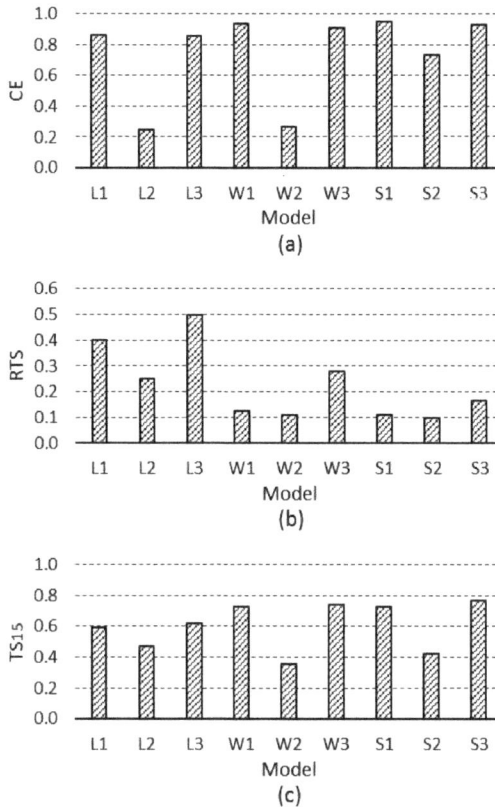

**Figure 6.** Performance comparison of the optimal models (**a**) Coefficient of efficiency (CE), (**b**) Relative time shift error (RTS) and (**c**) Threshold statistic for a level of $x\%$ $TS_{15}$.

Figure 7 is the comparison of validated water level hydrograph and measured data of Type 1 models, with a prediction lead time of 3 h. Figure 7a–d show the validation results of Typhoon Saola, Matmo, Trami and Usagi, respectively. As seen in the figures, the models generally give reasonable forecasts comparing to the data. It is noted that all the three models exhibit certain degrees of time shift errors on the rising limb of the hydrographs. Since the rising limb of the flood wave is the most important phase for flood forecasting activities, the delayed forecasts pose a limitation of the models.

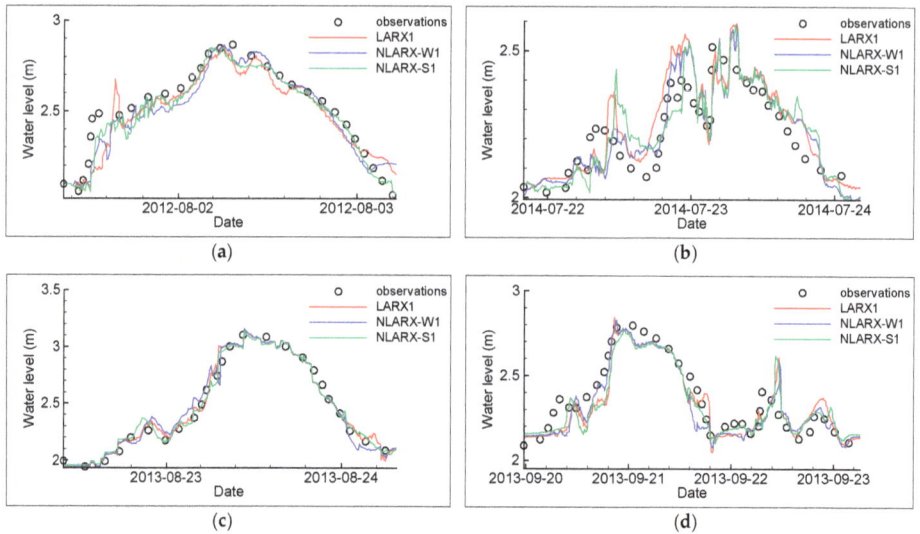

**Figure 7.** Comparison of model validation and measured data for (**a**) Typhoon Saola, (**b**) Typhoon Matmo, (**c**) Typhoon Trami, and (**d**) Typhoon Usagi (3-h lead time).

The above model performance comparison is mainly based on 3 h of prediction lead time. In order to investigate the performance under other lead times, the models are applied to the forecasts with prediction lead times varying from 0.5 h to 3 h, and the three indexes of CE, RTS, and $TS_{15}$ associated to each prediction lead time are calculated. The results are shown in Figure 8a–c. Figure 8a shows the CE variations of Type 1 models (L1, W1, and S1) along with the prediction lead. As shown in the figure, it appears that the CEs of all the three models increase with smaller prediction leads. For 0.5 h of prediction lead, the CEs of all models reach above 0.95, among which S1 is the highest. As the prediction lead time increases, the CEs of the three models gradually decrease. CE of L1 drops below 0.9 after the lead time passes 2 h, while W1 and S1 maintain above 0.9 up to 3 h of prediction lead. As the figure shows, for prediction leads from 0.5 to 3 h, CE performance of the three models shows S1 as optimal, closely followed by W1, and L1 is the worst. Figure 8b shows the RTS of Type 2 models (L2, W2, S2) varying along with prediction lead time from 0.5 to 3 h. As shown in the figure, the RTS of the three models gradually increases as the prediction lead time increases. All three models appear to have higher RTS with longer prediction leads, indicating greater relative time shift errors there. Comparing the three models, the RTS performance of S2 appears to be the best, and L2 is the worst. Figure 8c shows the $TS_{15}$ of Type 3 models (L3, W3, S3) varying along with prediction lead time. As shown in the figure, the $TS_{15}$ of all three models increase with smaller prediction leads. With 0.5 h of lead time, $TS_{15}$ of the three models all reach above 0.9, showing very good forecasts. As the prediction lead time increases, the $TS_{15}$ of the three models gradually decreases. The $TS_{15}$ performance of the three models shows S3 as the best, closely followed by W3, and L3 as the worst.

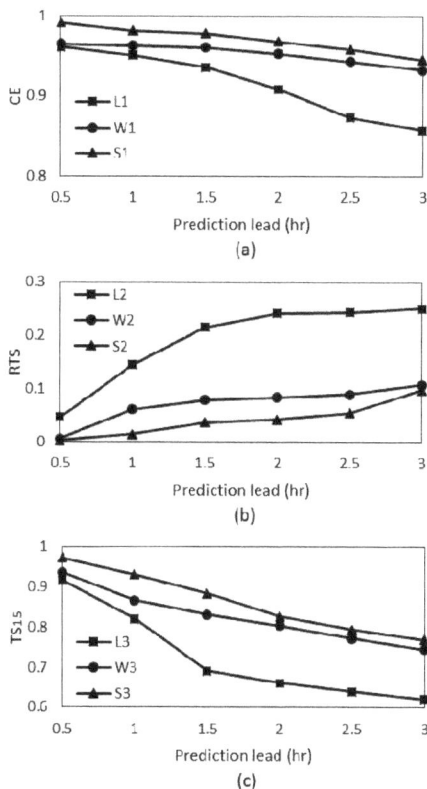

**Figure 8.** Variation of assessing indexes of the optimal models with respect to different prediction lead times (a) CE, (b) RTS and (c) $TS_{15}$.

## 4. Conclusions

A methodology for the determination of the optimal combination of non-sequential regressors for typhoon inundation forecasting models has been developed. By integrating a MOGA with ARX-based models, the proposed methodology is capable of locating the optimal models that conform to multiple objectives in terms of high prediction accuracy, low time shift error, and low threshold statistics. Testing results show that the nine optimal models obtained by the MOGA all display non-sequential patterns in the resultant combinations of regressors, which signifies the superiority of this type of model as well as the capacity of the proposed methodology in locating the optimal non-sequential regressors. In comparing the resultant models obtained by the proposed methodology, the results show that the overall performance of nonlinear models (NLARX-W and NLARX-S) is significantly better than linear models (LARX), revealing a nonlinear relationship between the rainfall and water level at the study area. On all the three assessing indexes of CE, RTS, and $TS_{15}$, the evaluation results of the three types of models present, in general, NLARX-S as the best, NLARX-W falling slightly behind, and LARX as the worst. These results provide a new approach to constructing better models for typhoon inundation level forecasts.

**Acknowledgments:** This research was supported by the Ministry of Science and Technology in Taiwan under grant No. MOST 105-2625-M-197-001. Support from the Water Resources Agency in Taiwan is also acknowledged.

**Author Contributions:** Huei-Tau Ouyang designed the framework of the study; Shang-Shu Shih and Ching-Sen Wu participated in data collection and final approval of the writing.

**Conflicts of Interest:** The authors declare no conflict of interest.

## References

1.  Karlsson, M.; Yakowitz, S. Rainfall-runoff forecasting methods, old and new. *Stoch. Hydrol. Hydraul.* **1987**, *1*, 303–318. [CrossRef]
2.  Liong, S.Y.; Lim, W.H.; Paudyal, G.N. River stage forecasting in Bangladesh: Neural network approach. *J. Comput. Civ. Eng.* **2000**, *14*, 1–8. [CrossRef]
3.  Campolo, M.; Soldati, A.; Andreussi, P. Artificial neural network approach to flood forecasting in the River Arno. *Hydrol. Sci. J.* **2003**, *48*, 381–398. [CrossRef]
4.  Keskin, M.E.; Taylan, D.; Terzi, O. Adaptive neural-based fuzzy inference system (ANFIS) approach for modeling hydrological time series. *Hydrol. Sci. J.* **2006**, *51*, 588–598. [CrossRef]
5.  Shu, C.; Ouarda, T.B.M.J. Regional flood frequency analysis at ungauged sites using the adaptive neuro-fuzzy inference system. *J. Hydrol.* **2008**, *349*, 31–43. [CrossRef]
6.  Kia, M.B.; Pirasteh, S.; Pradhan, B.; Mahmud, A.R.; Sulaiman, W.N.A.; Moradi, A. An artificial neural network model for flood simulation using GIS: Johor River Basin, Malaysia. *Environ. Earth Sci.* **2012**, *67*, 251–264. [CrossRef]
7.  Lin, G.F.; Lin, H.Y.; Chou, Y.C. Development of a real-time regional-inundation forecasting model for the inundation warning system. *J. Hydroinf.* **2013**, *15*, 1391–1407. [CrossRef]
8.  Tehrany, M.S.; Pradhan, B.; Jebur, M.N. Flood susceptibility mapping using a novel ensemble weights-of-evidence and support vector machine models in GIS. *J. Hydrol.* **2014**, *512*, 332–343. [CrossRef]
9.  Del Giudice, D.; Reichert, P.; Bareš, V.; Albert, C.; Rieckermann, J. Model bias and complexity–Understanding the effects of structural deficits and input errors on runoff predictions. *Environ. Model. Softw.* **2015**, *64*, 205–214. [CrossRef]
10. Chang, F.J.; Tsai, M.J. A nonlinear spatio-temporal lumping of radar rainfall for modeling multi-step-ahead inflow forecasts by data-driven techniques. *J. Hydrol.* **2016**, *535*, 256–269. [CrossRef]
11. Pan, T.Y.; Chang, L.Y.; Lai, J.S.; Chang, H.K.; Lee, C.S.; Tan, Y.C. Coupling typhoon rainfall forecasting with overland-flow modeling for early warning of inundation. *Nat. Hazards* **2014**, *70*, 1763–1793. [CrossRef]
12. Gourley, J.J.; Maddox, R.A.; Howard, K.W.; Burgess, D.W. An exploratory multisensor technique for quantitative estimation of stratiform rainfall. *J. Hydrometeorol.* **2002**, *3*, 166–180. [CrossRef]
13. Thirumalaiah, K.; Deo, M.C. Real-time flood forecasting using neural networks. *Comput. Aided Civ. Infrastruct. Eng.* **1998**, *13*, 101–111. [CrossRef]
14. Bazartseren, B.; Hildebrandt, G.; Holz, K.P. Short-term water level prediction using neural networks and neuro-fuzzy approach. *Neurocomputing* **2003**, *55*, 439–450. [CrossRef]
15. Yu, P.S.; Chen, S.T.; Chang, I.F. Support vector regression for real-time flood stage forecasting. *J. Hydrol.* **2006**, *328*, 704–716. [CrossRef]
16. Yule, G.U. On a method of investigating periodicities in disturbed series, with special reference to Wolfer's sunspot numbers. *Philos. Trans. R. Soc. Lond. Ser. A Contain. Pap. Math. Phys. Charact.* **1927**, *226*, 267–298. [CrossRef]
17. Grossmann, A.; Morlet, J. Decomposition of Hardy functions into square integrable wavelets of constant shape. *Siam J. Math. Anal.* **1984**, *15*, 723–736. [CrossRef]
18. Gayen, A.K. The frequency distribution of the product-moment correlation coefficient in random samples of any size drawn from non-normal universes. *Biometrika* **1951**, *38*, 219–247. [CrossRef] [PubMed]
19. Ouyang, H.T. Multi-objective optimization of typhoon inundation forecast models with cross-site structures for a water-level gauging network by integrating ARMAX with a genetic algorithm. *Nat. Hazards Earth Syst. Sci.* **2016**, *16*, 1897–1909. [CrossRef]
20. Furundzic, D. Application example of neural networks for time series analysis: Rainfall–runoff modeling. *Signal Process.* **1998**, *64*, 383–396. [CrossRef]
21. Tokar, A.S.; Markus, M. Precipitation-runoff modeling using artificial neural networks and conceptual models. *J. Hydrol. Eng.* **2000**, *5*, 156–161. [CrossRef]
22. Riad, S.; Mania, J.; Bouchaou, L.; Najjar, Y. Predicting catchment flow in a semi-arid region via an artificial neural network technique. *Hydrol. Process.* **2004**, *18*, 2387–2393. [CrossRef]

23. Chua, L.H.; Wong, T.S.; Sriramula, L.K. Comparison between kinematic wave and artificial neural network models in event-based runoff simulation for an overland plane. *J. Hydrol.* **2008**, *357*, 337–348. [CrossRef]

24. Mitra, S.; Hayashi, Y. Neuro-fuzzy rule generation: Survey in soft computing framework. *IEEE Trans. Neural Netw.* **2000**, *11*, 748–768. [CrossRef] [PubMed]

25. Nayak, P.C.; Sudheer, K.P.; Jain, S.K. Rainfall-runoff modeling through hybrid intelligent system. *Water Resour. Res.* **2007**, *43*. [CrossRef]

26. Talei, A.; Chua, L.H.C.; Wong, T.S. Evaluation of rainfall and discharge inputs used by Adaptive Network-based Fuzzy Inference Systems (ANFIS) in rainfall–runoff modeling. *J. Hydrol.* **2010**, *391*, 248–262. [CrossRef]

27. Nash, J.E.; Sutcliffe, J.V. River flow forecasting through conceptual models part I—A discussion of principles. *J. Hydrol.* **1970**, *10*, 282–290. [CrossRef]

28. Talei, A.; Chua, L.H. Influence of lag time on event-based rainfall–runoff modeling using the data driven approach. *J. Hydrol.* **2012**, *438*, 223–233. [CrossRef]

29. Jain, A.; Ormsbee, L.E. Short-term water demand forecast modeling techniques—Conventional methods versus AI. *J. Am. Water Works Assoc.* **2012**, *94*, 64–72.

30. Jain, A.; Varshney, A.K.; Joshi, U.C. Short-term water demand forecast modeling at IIT Kanpur using artificial neural networks. *Water Resour. Manag.* **2001**, *15*, 299–321. [CrossRef]

31. Rajurkar, M.P.; Kothyari, U.C.; Chaube, U.C. Artificial neural networks for daily rainfall—Runoff modeling. *Hydrol. Sci. J.* **2002**, *47*, 865–877. [CrossRef]

32. Holland, J.H. Genetic algorithms and the optimal allocation of trials. *SIAM J. Comput.* **1973**, *2*, 88–105. [CrossRef]

33. Goldberg, D.E.; Holland, J.H. Genetic algorithms and machine learning. *Mach. Learn.* **1988**, *3*, 95–99. [CrossRef]

34. Bagchi, T.P. *Multi-Objective Scheduling by Genetic Algorithms*; Springer Science & Business Media: New York, NY, USA, 1999; pp. 143–145.

*water*  MDPI

*Article*

# Exploring Jeddah Floods by Tropical Rainfall Measuring Mission Analysis

**Ahmet Emre Tekeli**

Civil Engineering Department, Çankırı Karatekin University, Çankırı 18100, Turkey;
ahmetemretekeli@karatekin.edu.tr; Tel.: +90-376-212-9582

Received: 13 June 2017; Accepted: 10 August 2017; Published: 16 August 2017

**Abstract:** Estimating flash floods in arid regions is a challenge arising from the limited time preventing mitigation measures from being taken, which results in fatalities and property losses. Here, Tropical Rainfall Measuring Mission (TRMM) Multi Satellite Precipitation Analysis (TMPA) Real Time (RT) 3B2RT data are utilized in estimating floods that occurred over the city of Jeddah located in the western Kingdom of Saudi Arabia. During the 2000–2014 period, six floods that were effective on 19 days occurred in Jeddah. Three indices, constant threshold (CT), cumulative distribution functions (CDFs) and Jeddah flood index (JFI), were developed using 15-year 3-hourly 3B42RT. The CT calculated, as 10.37 mm/h, predicted flooding on 14 days, 6 of which coincided with actual flood-affected days (FADs). CDF thresholds varied between 87 and 93.74%, and JFI estimated 28 and 20 FADs where 8 and 7 matched with actual FADs, respectively. While CDF and JFI did not miss any flood event, CT missed the floods that occurred in the heavy rain months of January and December. The results are promising despite that only rainfall rates, i.e., one parameter out of various flood triggering mechanisms, i.e., soil moisture, topography and land use, are used. The simplicity of the method favors its use in TRMM follow-on missions such as the Global Precipitation Measurement Mission (GPM).

**Keywords:** Jeddah; floods; TRMM; 3B42RT; Saudi Arabia; GPM

---

## 1. Introduction

Disasters, which can be categorized as either man-made or natural, are defined as events or dangerous cases that may lead to injury or loss of human life with/without property loss [1] in addition to the interruptions they cause on human activities [2]. Out of 31 natural disasters, 28 are the result of meteorological events [3]. Among these 28 meteorology-based disasters, floods are the most common [4]. Despite the limited areas in which they occur [5], flash floods are the most commonly faced, the most deadly and the most challenging [6,7]. It is the limited response time that makes flash floods challenging. Hapuarachi et al. [8] identified excess rainfall as the main driving mechanism for flash floods.

The vulnerability of arid and semi-arid regions to flash floods has been indicated to be equal to that of regions with heavy rain. Moreover, Zipser et al. [9] mentioned the occurrence of the strongest convective storms, and Haggag and El-Badry [10] indicated the rapid formation of flash floods in arid and semi-arid regions. The Kingdom of Saudi Arabia (KSA) is well known for its dry climatic conditions [11] and is classified as a semi-arid region [12,13]. However, the floods that occurred over Jeddah, Makkah and Riyadh in KSA indicate flash floods risks in the semi-arid kingdom (Figure 1a). The risks as well as the impact of flash floods increase due to the rainfall intensity and lack of mitigation implementations [11].

(a)

(b)

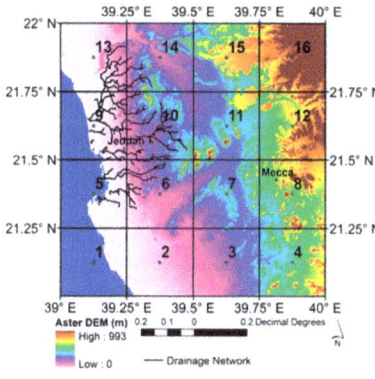

(c)

**Figure 1.** Flooding event in Jeddah from 24–26 November 2009 (**a**) Source: https://ontheredsea. wordpress.com/tag/jeddah/, location of Jeddah in KSA (**b**) Source: http://www.weather-forecast. com/system/images/249/original/Jeddah.jpg?1299372620, Digital Elevation Model of Jeddah city and surroundings obtained from ASTER DEM (30 m) (**c**).

Negri et al. [14] identified the establishment of early warning systems as the most effective way to reduce life and property damage in flash flood cases. Despite improvements in numerical weather predictions, it is not easy to detect flash floods. This situation increases the importance of rainfall observations in flash flood estimations.

Ground-based rain gauges have been the main source of data for rainfall observations. However, as mentioned in Negri et al. [14], in these ground-based stations, data transfer problems can be seen during flooding events, and maintenance of the stations are needed after flooding [15]. In addition Borga et al. [5] indicated the inadequacy of ground-based rain gauges in showing the spatial variability of rainfall. Moreover, insufficiency in spatial and temporal coverage, particularly for rainfall observations over Jeddah, is mentioned in Deng et al. [16].

Ground radar and satellite-based remote sensing are free from such disadvantages and are being used more frequently in both research and operational applications [17]. In particular, satellite-based remote sensing provides new techniques to monitor extreme rainfall events in an uninterrupted manner and enables implementation of new flood warning systems [15,18]. Rainfall intensities at high spatial (1 km to 10 km) and temporal (30 min–3 h) resolutions can be obtained in near real time [19]. Borga et al. [5] mentioned that as remotely sensed precipitation became a major component in flood warning systems, mortality decreased due to timely warnings provided by satellite-based rainfall intensities [18,20].

In this study, three different flood indexes, namely, the constant threshold (CT), cumulative distribution functions (CDFs) and Jeddah flood index (JFI) that are based on the Tropical Rainfall Measuring Mission (TRMM) satellite 3B42RT rainfall rates are compared in forecasting of flooding events in Jeddah. Thus, the main objective is assessing the forecasting capabilities of TRMM 3B42RT-based indices in identifying the Jeddah floods.

## 2. Study Area and Data Sets

### 2.1. Study Area

With a population of 4.2 million over an area of 1600 km$^2$, Jeddah, the second largest city in KSA after the capital Riyadh is the largest sea port on the Red Sea coast (Figure 1b) and has been a commercial hub in KSA [21]. Jeddah extends in a northern to southern direction bordered by the Red Sea in the west and by mountains in the east (Figure 1c). Drainage extending from those mountains in the east crosses the city, transporting surface flow to the city and ultimately merges with Red Sea (Please Refer to Figure 8 in Youssef et al. [12]).

With air temperatures dropping to 15 °C in winter and reaching as high as 52 °C in summer, Jeddah can be classified as having a hot, arid climate [21]. Despite Jeddah being located in the rainiest region of KSA, precipitation is still low, below the potential evaporation point, and indicates high temporal and spatial variability [12]. The monthly maximum, average and minimum air temperatures (°C) and monthly cumulative precipitation (mm) observed in Jeddah are provided in Figure 2 [21]. As seen in Figure 2, the summer period can be very hot (max 52 °C) with no rainfall. Based on precipitation values in Figure 2, Jeddah exhibits three main seasons. The high rain season covers November, December and January, the low rain season covers February, March, April and October and the dry season covers June, July, August and September. Most of the rain is received in brief thunderstorms during the high rain season [12]. Table 1 presents the Jeddah floods gathered from various sources. Figure 2 indicates that, out of six floods observed over Jeddah, 3 occurred in the high season (January, November, December), 2 in the low season (April, October) and 1 in the dry (July) season.

**Table 1.** Major flood events observed over Jeddah city in KSA.

| Start | End | Flood Affected | |
|---|---|---|---|
| Date | Date | Days | Data Source |
| 28 April 2005 | 28 April 2005 | 1 | [22] |
| 29 October 2006 | 31 October 2006 | 3 | [23] |
| 24 November 2009 | 26 November 2009 | 3 | [10,16,22,24–28] |
| 13 July 2010 | 16 July 2010 | 4 | [29] |
| 30 December 2010 | * | 1 | [10,25] |
| 25 January 2011 | 31 January 2011 | 7 | [10,24–26,28] |

Note: * Related information could not be found.

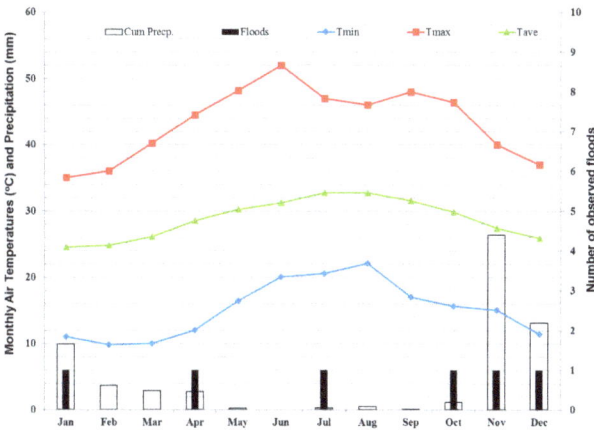

**Figure 2.** Monthly maximum, average and minimum air temperatures (°C), monthly cumulative precipitation (mm) and number of observed floods for Jeddah (Data source: https://en.wikipedia.org/wiki/Jeddah#Climate).

### 2.2. Satellite Data

The Tropical Rainfall Measuring Mission (TRMM) is a collaboration between the Japan Aerospace Exploration Agency (JAXA) and National Aeronautics and Space Administration (NASA). It was the first satellite to carry a precipitation radar. Besides the radar, microwave imaging and lightning sensors are used in rainfall detection [30]. Using multi-channel microwave and infrared observations from satellites [31], TRMM Multi-satellite Precipitation Analysis (TMPA) data produce the "best" precipitation estimate between 50° N–50° S. Precipitation products have high temporal (3 h) and high spatial resolution (0.25° × 0.25°). Real-time (RT) data (3B42RT) are provided to users 6–9 h following the data reception and research products (3B42) are available 15 days following the end of month [32].

Usability of TRMM 3B42 data for water resource applications and high performance over KSA was demonstrated by Almazroi [33] and Kheimi and Gutub [34]. Tekeli and Fouli [35] indicated the capability of 3B42RT data in forecasting floods in Riyadh, KSA. Similar to Tekeli and Fouli [35] version 7 (V7) of 3B42RT data posted online in May of 2012 [36] is used here. Figure 3 presents TRMM 3B42RT pixel coverage over Jeddah. Despite Haggag and El-Badry [10] stating the underestimation of rainfall fields by TRMM 3B42 data, Figure 4 shows the flooding event on 25 November 2009 as detected by 3B42RT data. Details of TRMM 3B42RT data production algorithms can be obtained from the TRMM website [37], and data can be downloaded from the web address; https://pmm.nasa.gov/data-access/downloads/trmm.

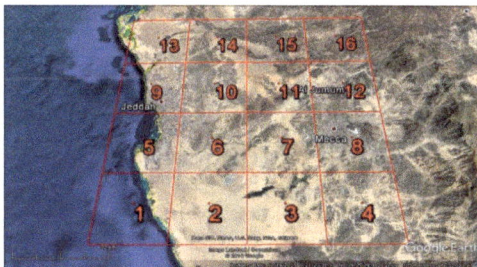

**Figure 3.** Coverage of TRMM 3B42RT pixels over Jeddah.

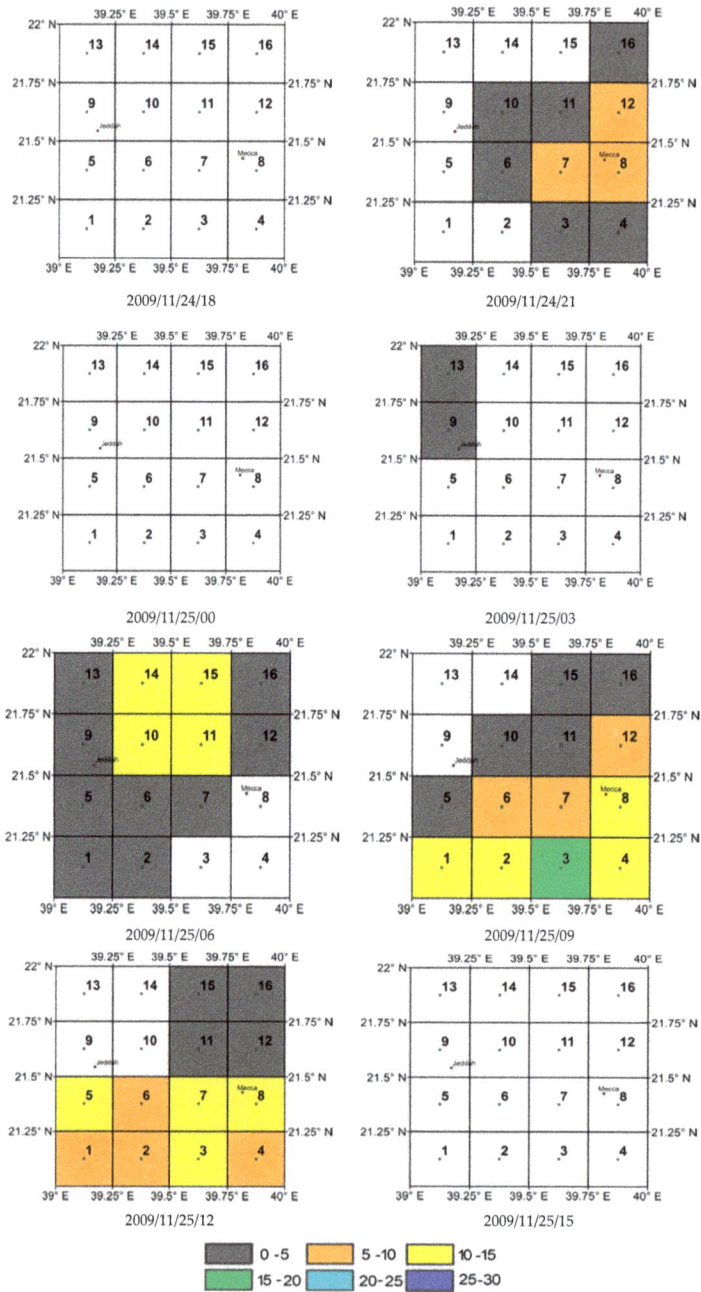

**Figure 4.** 3B42RT rainfall rates for 24–25 November 2009 (in mm/h) indicated in year/month/day/hour.

## 3. Methodology

Hapuarachi et al. [8] and Alfieri and Thielen [38] reviewed flash flood occurrences and found that rainfall comparison (RC) provides good estimates despite its simplicity, as it just requires Quantitative

Precipitation Estimates (QPE). Borga et al. [5] mentioned that event detection is the most important step in flash flood warnings. Thus, the comparison of rainfall amounts, RC, with thresholds has been used in detecting flash floods [39]. However, as Hamada et al. [40] indicated, regional differences should be considered in determining thresholds. In addition, a large database is needed for flash flood threshold determination, as flash floods are not frequent occurrences. For this study, 3-h interval 3B42RT TRMM data covering the years from 2000 to 2014 are used to obtain three indices, namely constant threshold (CT), cumulative distribution function (CDF) and Jeddah flood index (JFI).

*3.1. Constant Threshold*

The intensity duration frequency (IDF) curves indicate the relationship between intensity (i), duration (d) and frequency (f) of rainfall and provide rainfall intensity for a given (selected) rainfall duration and frequency [41]. Equation (1) represents the IDF curve for Jeddah that was developed by Ewea et al. [42] using storms recorded at Mudaylif station for 27 years covering the period 1975–2001.

$$i(mm/h) = (236.63\ln(Tr) + 388.48) * (D) (0.0107\ln(Tr) - 0.7869) \tag{1}$$

where Tr is the return period, D is rainfall duration and i is the rainfall intensity in mm/h. Table 2 shows the calculated rainfall intensities that would cause flooding over Jeddah for different return periods based on Equation (1).

Taking the return period as 2.33 for average annual flooding [43] and rainfall duration as 180 min (as 3B42RT data are 3 hourly), a flood causing the rainfall intensity threshold is determined as 10.37 mm/h.

**Table 2.** Calculated rainfall intensities using Equation (1) for Jeddah city in KSA.

| Tr (Years) | Duration (min) | I (mm/h) |
| --- | --- | --- |
| 2.33 | 180 | 10.37 |
| 5 | 180 | 14.13 |
| 25 | 180 | 23.11 |
| 50 | 180 | 27.44 |
| 100 | 180 | 32.08 |

*3.2. Cumulative Distribution Function (CDF)*

Three-hourly TRMM 3B42RT rainfall intensities over the pixels (pixels 5, 6, 9, 10, 13 and 14—See Figure 3) are aggregated monthly, pixel wise, and cumulative distribution functions (CDFs) are obtained for each pixel for each month. Figure 5 presents monthly CDFs covering the 2000–2014 period for the flood observed months.

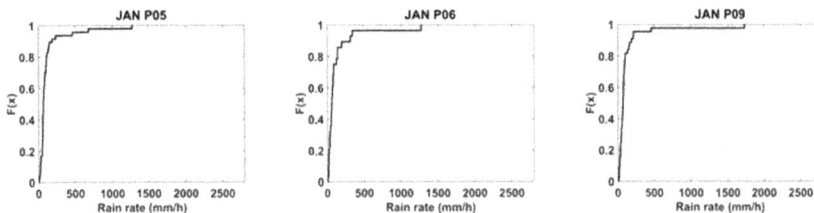

**Figure 5.** *Cont.*

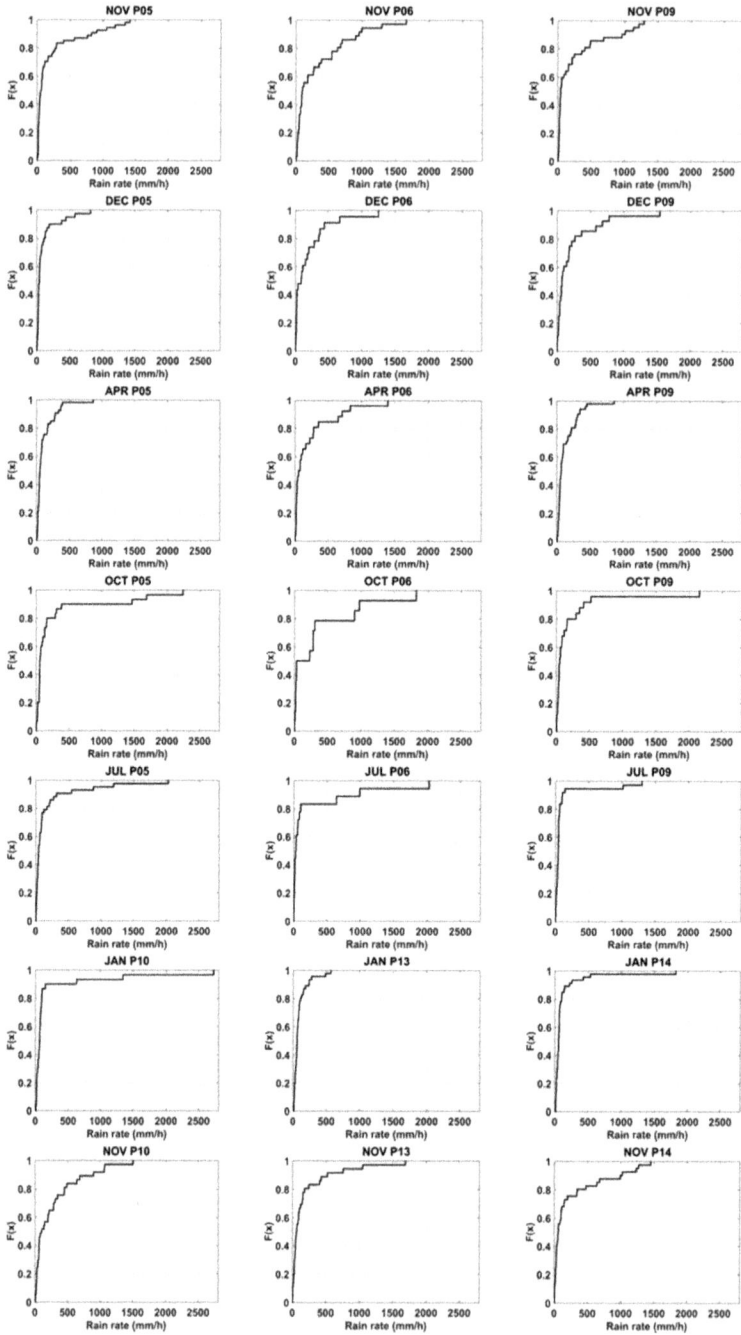

**Figure 5.** *Cont.*

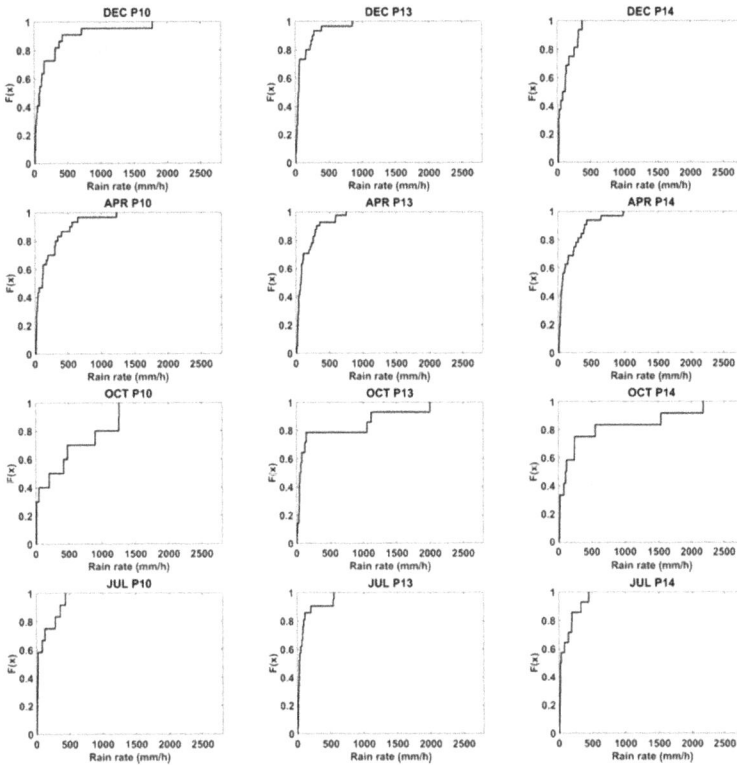

**Figure 5.** Cumulative distribution functions of 3B42RT pixels over Jeddah. $P$ indicates the pixel numbers given in Figure 3. (Rain rates should be multiplied by 0.01).

Different characteristics of CDFs can be seen in Figure 5. This is in agreement with [35,44], whereas it is opposite to [45].

*3.3. Jeddah Flood Index (JFI)*

Tekeli and Fouli [35] proposed the Riyadh Flood Precipitation Index (RFPI), which is a modified version of the European Precipitation Climatology Index (EPIC) proposed by Alfieri et al. [7] and by Alfieri and Thielen [38] for extreme rain storm and flash flood early warning. Modification to the EPIC implemented by [35] included the monthly calculation instead of yearly values in EPIC. In this study, the Jeddah Flood Index (JFI) was developed similar to the monthly case proposed by [35]. Thus, JFI can be given by the following equation.

$$\text{JFI} = \frac{P_i}{\frac{\sum_{j=1}^{N} Max(P_i)}{N}} \tag{2}$$

$P_i$ indicates the 3-h rain rate in 3B42RT data, and $N$ is the number of years with available 3B42RT data. The denominator is the monthly average of the maximum three-hour interval of rain rates found in 3B42RT. Monthly values of the denominator are based on 3-h intervals for 15-year (2000–2014) TRMM data for the pixels covering Jeddah, summarized in Table 3.

**Table 3.** Calculated monthly denominator values for JFI (Equation (2)) for Jeddah city in KSA (mm/h).

|           | **P5** | **P6** | **P9** | **P10** | **P13** | **P14** |
|-----------|--------|--------|--------|---------|---------|---------|
| January   | 2.48   | 2.32   | 2.42   | 3.33    | 1.78    | 2.70    |
| February  | 1.18   | 1.59   | 1.09   | 0.73    | 0.83    | 1.07    |
| March     | 0.86   | 0.87   | 1.45   | 0.76    | 1.00    | 1.26    |
| April     | 2.26   | 3.06   | 2.40   | 3.25    | 2.37    | 2.50    |
| May       | 1.82   | 3.07   | 1.98   | 1.90    | 1.63    | 1.42    |
| June      | 1.13   | 0.48   | 1.28   | 0.38    | 0.87    | 0.64    |
| July      | 2.87   | 3.66   | 1.57   | 1.07    | 0.99    | 1.51    |
| August    | 2.60   | 5.01   | 3.15   | 4.08    | 1.85    | 3.06    |
| September | 0.43   | 0.69   | 0.39   | 3.04    | 0.52    | 5.16    |
| October   | 5.29   | 5.26   | 3.70   | 9.04    | 3.35    | 5.91    |
| November  | 4.27   | 6.92   | 4.08   | 4.83    | 2.66    | 4.60    |
| December  | 1.78   | 3.34   | 3.53   | 5.34    | 2.10    | 1.75    |

High correlations of observed high flows with EPIC values (1, 1.5) are mentioned in [7]. Tekeli and Fouli [35] tested values of 1, 1.5 and 2 for RFPI. Since there are no discharge observations for the study period, for JFI threshold is selected as 1.

## 4. Results and Discussion

Flood events and minimum and maximum rainfall rates obtained from TRMM 3B42RT data are presented in Figure 6. High rainfall rates matched well with flood events except for August, during which no floods were reported. Moreover, both frequency and magnitude of the high rainfall rates are more common during the high rain season (November, December and January period). Thus, seasonality of rainfall patterns of Jeddah are also detected well by 3B42RT.

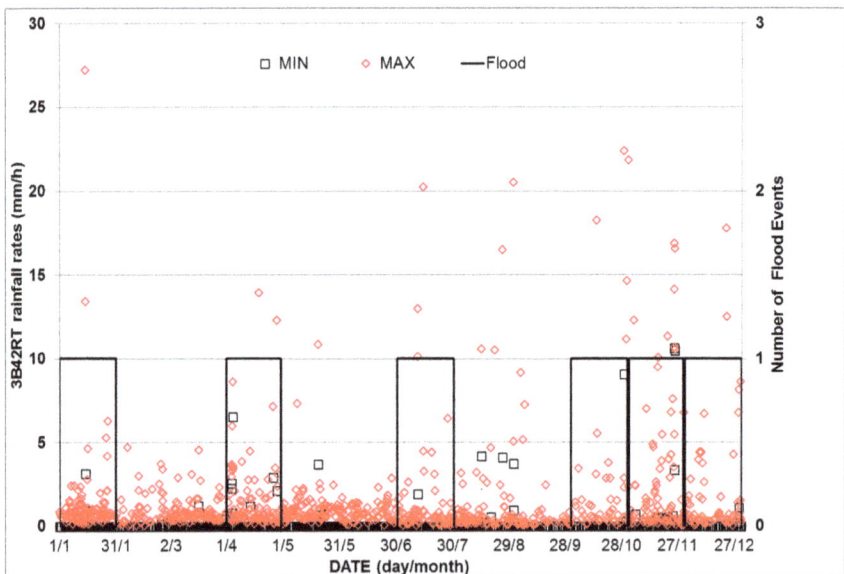

**Figure 6.** Temporal distribution of major flood events over Jeddah throughout the year and minima (**black squares**) and maxima (**red diamonds**) rainfall rates obtained from 3B42RT data for Jeddah between 2000 and 2014.

It is seen that 3B42RT intensities increase towards to the east of Jeddah (Figure 4). This is in line with the topography where high mountains on the east cause large amounts of rainfall that flow down quickly to Jeddah city [16,24]. Moreover, Haggag and El-Badry [10] also mentioned that eastern parts of the catchment receive 220mm/year more rainfall than other parts.

For the constant threshold (CT), rainfall rates equal or greater than 10.37 mm/h obtained from Equation (1) were searched for in all three-hour interval 3B42RT TRMM data covering the 2000–2014 period. Tekeli and Fouli [35] and Hamada et al. [40] used 90% and 99.9% as threshold values in cumulative distribution functions (CDFs). As an initial value, CDF 90% is used as a threshold value in this study. For each 3-h interval covering the 2000–2014 period, the Jeddah Flood Index (JFI) of the 3B42RT pixels are calculated, and values greater than the threshold value 1, are searched.

Based on the above-mentioned thresholds for each method, estimated flood events are summarized in Table 4. Estimated events that match with the real flood observations are framed by a thick dark border. CT yielded 20 estimated flood events on 14 different days. This led to the identification of 8 events on 6 days. For CDFs, a 90% thresholds resulted in too many false alarms, especially for January, April and December—all of which are in the high rain season. JFI, with a threshold value of 1.00, yielded the highest number flood estimates. It is easily seen that other than December and January, the observed floods are seen when all CT, CDF and JFI indicated flooding.

**Table 4.** Estimated flood dates and time (UTC) according to the constant threshold (10.37 mm/h), cumulative distribution function (90%) and Jeddah flood index (1.00).

| Year | Month | Day | Hour | CT | CDF | JFI | Year | Month | Day | Hour | CT | CDF | JFI |
|------|-------|-----|------|----|-----|-----|------|-------|-----|------|----|-----|-----|
| 2003 | 1 | 15 | 12 |   | X | X | 2000 | 10 | 14 | 21 | X | X | X |
| 2003 | 1 | 16 | 0 |   | X | X | 2006 | 10 | 29 | 3 | X | X | X |
| 2004 | 1 | 8 | 18 |   | X | X | 2006 | 10 | 30 | 12 | X | X | X |
| 2004 | 1 | 12 | 0 |   | X |   | 2006 | 10 | 30 | 15 | X | X | X |
| 2008 | 1 | 10 | 21 |   | X |   | 2006 | 10 | 31 | 15 | X | X | X |
| 2009 | 1 | 12 | 15 |   | X |   |      |    |    |    |   |   |   |
| 2010 | 1 | 7 | 0 |   | X | X | 2000 | 11 | 13 | 18 |   |   | X |
| 2011 | 1 | 14 | 18 | X | X | X | 2000 | 11 | 14 | 0 |   |   | X |
| 2011 | 1 | 14 | 21 | X | X | X | 2000 | 11 | 16 | 18 |   | X | X |
| 2011 | 1 | 25 | 18 |   | X | X | 2003 | 11 | 23 | 15 |   |   | X |
| 2011 | 1 | 26 | 0 |   | X |   | 2003 | 11 | 24 | 9 |   |   | X |
| 2011 | 1 | 26 | 3 |   | X |   | 2003 | 11 | 25 | 6 | X | X | X |
| 2011 | 1 | 26 | 6 |   | X | X | 2003 | 11 | 25 | 9 | X | X | X |
| 2011 | 1 | 26 | 9 |   | X | X | 2003 | 11 | 25 | 12 | X | X | X |
|      |   |    |   |   |   |   | 2004 | 11 | 3 | 15 | X | X | X |
| 2004 | 4 | 3 | 21 |   |   | X | 2009 | 11 | 25 | 6 | X | X | X |
| 2000 | 4 | 4 | 0 |   | X | X | 2009 | 11 | 25 | 12 | X | X | X |
| 2000 | 4 | 4 | 3 |   | X | X | 2012 | 11 | 18 | 21 |   |   | X |
| 2000 | 4 | 4 | 6 |   | X | X | 2012 | 11 | 23 | 15 |   |   | X |
| 2000 | 4 | 4 | 12 |   | X | X | 2013 | 11 | 10 | 9 |   |   | X |
| 2004 | 4 | 23 | 6 |   |   | X | 2014 | 11 | 16 | 12 |   | X | X |
| 2005 | 4 | 18 | 15 | X | X | X | 2014 | 11 | 21 | 21 | X | X | X |
| 2005 | 4 | 25 | 21 |   | X | X |      |    |    |    |   |   |   |
| 2005 | 4 | 28 | 3 | X | X | X | 2001 | 12 | 31 | 12 |   | X | X |
| 2005 | 4 | 28 | 9 |   |   | X | 2004 | 12 | 3 | 18 |   | X | X |
| 2006 | 4 | 4 | 21 | X |   | X | 2004 | 12 | 8 | 0 |   | X | X |
| 2006 | 4 | 9 | 9 |   |   | X | 2009 | 12 | 22 | 18 | X | X | X |
| 2006 | 4 | 13 | 21 |   | X | X | 2009 | 12 | 22 | 21 | X | X | X |
| 2006 | 4 | 27 | 0 |   | X | X | 2010 | 12 | 9 | 12 |   | X | X |
| 2008 | 4 | 12 | 21 |   |   | X | 2010 | 12 | 10 | 0 |   | X | X |
| 2010 | 4 | 3 | 21 |   | X | X | 2010 | 12 | 11 | 0 |   | X | X |

Table 4. *Cont.*

| Year | Month | Day | Hour | CT | CDF | JFI | Year | Month | Day | Hour | CT | CDF | JFI |
|------|-------|-----|------|----|-----|-----|------|-------|-----|------|----|-----|-----|
| 2012 | 4 | 4 | 18 | | | X | 2010 | 12 | 29 | 6 | | X | X |
| | | | | | | | 2010 | 12 | 29 | 9 | | X | X |
| 2001 | 7 | 14 | 21 | | | X | 2010 | 12 | 30 | 6 | | | X |
| 2005 | 7 | 29 | 0 | | X | | 2014 | 12 | 8 | 12 | | | X |
| 2008 | 7 | 20 | 15 | | | X | | | | | | | |
| 2008 | 7 | 24 | 21 | | | X | | | | | | | |
| 2010 | 7 | 11 | 18 | X | X | X | | | | | | | |
| 2010 | 7 | 11 | 21 | | X | X | | | | | | | |
| 2010 | 7 | 14 | 15 | X | X | X | | | | | | | |
| 2010 | 7 | 14 | 18 | | X | X | | | | | | | |
| 2010 | 7 | 19 | 0 | | X | X | | | | | | | |
| 2011 | 7 | 10 | 3 | | | X | | | | | | | |
| 2012 | 7 | 27 | 15 | | X | X | | | | | | | |
| 2014 | 7 | 6 | 18 | | X | | | | | | | | |

Note: Estimated events that match with the real flood observations are framed by a thick dark border.

Dönmez and Tekeli [44] reduced false flood alarms by updating the CDF thresholds. Since they knew the places where flooding occurred, they used the CDFs of the respective pixels and performed updates accordingly. Unfortunately, the case is not the same in this study. In the approach proposed by [35], they used the constant threshold value of (3 mm/h) derived from intensity duration curves to determine the respective CDFs. Similarly, in this study, the value obtained from the intensity duration curve, 10.37 mm/h, is used to determine thresholds. The months October and November indicated values higher than the constant value. Thus, these months were used to derive the new thresholds. New values were determined as 91.68%, 91.62%, 93.74%, 87.30%, 88.87% and 87.46% for respective pixels 5, 6, 9, 10, 13 and 14. The first three values (91.68%, 91.62%, 93.74%) are within the 90–99.9% range mentioned in [35,40]. However, the last three (87.30%, 88.87% and 87.46%) are not within the range. Using these as thresholds and using the respective months' CDFs, flood estimations of the CDF-based method are updated. In addition, JFI values are also updated based on the actual flood occurrences. Estimated flood events based on updated thresholds are presented in Table 5. Both updates drastically reduced false flood alarms.

Table 5. Estimated flood dates and time (UTC) according to the constant threshold (10.37 mm/h), and updated cumulative distribution function and Jeddah flood index.

| Year | Month | Day | Hour | CT | CDF | JFI | Year | Month | Day | Hour | CT | CDF | JFI |
|------|-------|-----|------|----|-----|-----|------|-------|-----|------|----|-----|-----|
| 2003 | 1 | 15 | 12 | | | | 2000 | 10 | 14 | 21 | X | X | |
| 2003 | 1 | 16 | 0 | | | | 2006 | 10 | 29 | 3 | X | X | X |
| 2004 | 1 | 8 | 18 | | | | 2006 | 10 | 30 | 12 | X | X | |
| 2004 | 1 | 12 | 0 | | | | 2006 | 10 | 30 | 15 | X | X | |
| 2008 | 1 | 10 | 21 | | | | 2006 | 10 | 31 | 15 | X | X | X |
| 2009 | 1 | 12 | 15 | | | | | | | | | | |
| 2010 | 1 | 7 | 0 | | | | 2000 | 11 | 13 | 18 | | | |
| 2011 | 1 | 14 | 18 | X | X | X | 2000 | 11 | 14 | 0 | | | |
| 2011 | 1 | 14 | 21 | X | X | X | 2000 | 11 | 16 | 18 | | | |
| 2011 | 1 | 25 | 18 | | X | X | 2003 | 11 | 23 | 15 | | | |
| 2011 | 1 | 26 | 0 | | | | 2003 | 11 | 24 | 9 | | | |
| 2011 | 1 | 26 | 3 | | | | 2003 | 11 | 25 | 6 | X | X | X |
| 2011 | 1 | 26 | 6 | | | | 2003 | 11 | 25 | 9 | X | X | X |
| 2011 | 1 | 26 | 9 | | | | 2003 | 11 | 25 | 12 | X | X | X |

Table 5. *Cont.*

| Year | Month | Day | Hour | CT | CDF | JFI | Year | Month | Day | Hour | CT | CDF | JFI |
|------|-------|-----|------|----|-----|-----|------|-------|-----|------|----|-----|-----|
|      |       |     |      |    |     |     | 2004 | 11 | 3  | 15 | X | X | X |
| 2004 | 4 | 3  | 21 |   |   |   | 2009 | 11 | 25 | 6  | X | X | X |
| 2000 | 4 | 4  | 0  |   | X |   | 2009 | 11 | 25 | 12 | X | X | X |
| 2000 | 4 | 4  | 3  |   | X | X | 2012 | 11 | 18 | 21 |   |   |   |
| 2000 | 4 | 4  | 6  |   |   |   | 2012 | 11 | 23 | 15 |   |   |   |
| 2000 | 4 | 4  | 12 |   | X |   | 2013 | 11 | 10 | 9  |   |   |   |
| 2004 | 4 | 23 | 6  |   |   |   | 2014 | 11 | 16 | 12 |   |   |   |
| 2005 | 4 | 18 | 15 | X | X | X | 2014 | 11 | 21 | 21 |   |   |   |
| 2005 | 4 | 25 | 21 |   | X | X |      |    |    |    |   |   |   |
| 2005 | 4 | 28 | 3  | X | X | X | 2001 | 12 | 31 | 12 |   | X | X |
| 2005 | 4 | 28 | 9  |   |   |   | 2004 | 12 | 3  | 18 |   | X |   |
| 2006 | 4 | 4  | 21 |   |   | X | 2004 | 12 | 8  | 0  |   | X |   |
| 2006 | 4 | 9  | 9  |   |   |   | 2009 | 12 | 22 | 18 | X | X | X |
| 2006 | 4 | 13 | 21 | X | X | X | 2009 | 12 | 22 | 21 | X | X |   |
| 2006 | 4 | 27 | 0  |   |   | X | 2010 | 12 | 9  | 12 |   | X |   |
| 2008 | 4 | 12 | 21 |   |   |   | 2010 | 12 | 10 | 0  |   |   |   |
| 2010 | 4 | 3  | 21 |   | X |   | 2010 | 12 | 11 | 0  |   | X |   |
| 2012 | 4 | 4  | 18 |   |   |   | 2010 | 12 | 29 | 6  |   | X | X |
|      |   |    |    |   |   |   | 2010 | 12 | 29 | 9  |   | X | X |
| 2001 | 7 | 14 | 21 |   |   |   | 2010 | 12 | 30 | 6  |   | X | X |
| 2005 | 7 | 29 | 0  |   |   |   | 2014 | 12 | 8  | 12 |   |   |   |
| 2008 | 7 | 20 | 15 |   |   |   |      |    |    |    |   |   |   |
| 2008 | 7 | 24 | 21 |   |   |   |      |    |    |    |   |   |   |
| 2010 | 7 | 11 | 18 | X | X | X |      |    |    |    |   |   |   |
| 2010 | 7 | 11 | 21 |   | X | X |      |    |    |    |   |   |   |
| 2010 | 7 | 14 | 15 | X | X | X |      |    |    |    |   |   |   |
| 2010 | 7 | 14 | 18 |   | X |   |      |    |    |    |   |   |   |
| 2010 | 7 | 19 | 0  |   | X |   |      |    |    |    |   |   |   |
| 2011 | 7 | 10 | 3  |   |   |   |      |    |    |    |   |   |   |
| 2012 | 7 | 27 | 15 |   | X |   |      |    |    |    |   |   |   |
| 2014 | 7 | 6  | 18 |   |   |   |      |    |    |    |   |   |   |

Note: Estimated events that match with the real flood observations are framed by a thick dark border.

Table 6 summarizes the estimated number of flood alarms before and after updating under "Estimated/Updated" columns for CDF and JFI and under "Detected" columns number of missed, false and true detection number is presented for CT, CDF and JFI. Table 6 indicates that CT, CDF and JFI estimated 14, 28 and 20 flood-affected days (FADs) where 6, 8 and 7 matched with actual FADs. CT, with a ratio of 6/14, seems to be superior to CDF (8/28) and JFI (7/20). However, as CT missed January and December floods, the second best, JFI, is selected as the main flood estimation index.

Table 6. Comparison of flood alarms after updating the thresholds.

| Month | Actual Flood Occurrence | CT | | CDF | | JFI | |
|-------|-------------------------|-----------|----------|---------------------|-------------------|---------------------|-------------------|
|       |                         | Estimated | Detected | Estimated/Updated | Detected | Estimated/Updated | Detected |
| JAN | 1 | 1 | 1M | 10/2 | 2T + 8F/1T + 1F | 7/2 | 2T + 5F/1T + 1F |
| APR | 1 | 2 | 1T + 1F | 8/6 | 1T + 7F/1T + 5F | 13/7 | 1T + 12F/1T + 6F |
| JUL | 1 | 2 | 1T + 1F | 6/4 | 1T + 5F/1T + 3F | 8/2 | 1T + 7F/1T + 1F |
| OCT | 1 | 4 | 3T + 1F | 4/4 | 3T + 1F/3T + 1F | 4/2 | 3T + 1F/2T + 0F |
| NOV | 1 | 4 | 1T + 3F | 6/4 | 1T + 5F/1T + 3F | 13/3 | 1T + 12F/1T + 2F |
| DEC | 1 | 1 | 0T + 1F | 8/8 | 0T + 8F/1T + 7F | 10/4 | 1T + 9F/1T + 3F |
| Overall detection | | | 6/14 | | 8/28 | | 7/20 |

Note: M, F and T denotes missed, false and true detection respectively.

For the 14 January 2011, all indexes (Table 5, CT, CDF, JFI) indicated flooding. However, nothing was mentioned in the published literature. However, as can be seen from the cumulative precipitation figure (Figure 7) for the World Meteorological Organization (WMO) station 41024, which is located in Jeddah, heavy rains were recorded. Also, video clips dated the 14 January 2011 are found on the internet (https://www.youtube.com/watch?v=TiA1AhWrZTs and https://www.youtube.com/watch?v=1mgDiQ4iDUg) showing the flooded streets. Thus, the event on the 14 January 2011 can be treated as a correct detection. Moreover, the TRMM 3B42RT rainfall rates for the 25–26 January 2011 were lower than those for the 14 January 2011.

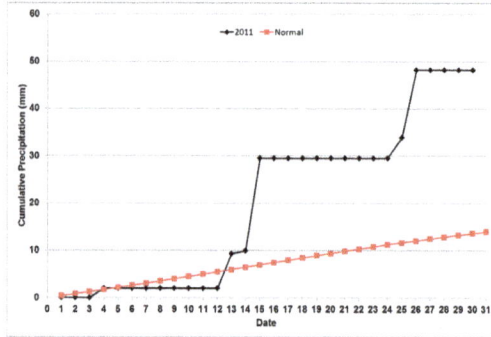

**Figure 7.** Cumulative precipitation plot for WMO station 41024 in Jeddah.

This is evident from the case that only CDF and JFI indicated flooding for the 25 January 2011. Despite the smaller rainfall intensities, the flood (25–31 January 2011) damage seemed more extensive than on the 14 January 2011. This might have occurred due to soil being saturated by rainfall on the 14 January 2011. Soil moisture is indicated as the second major flood triggering component, and the effect of soil moisture on flood estimations over Riyadh city is shown in [46].

Table 7 shows the flood estimations based on the updates (Table 5) and rain observations for WMO station 41024. In Table 7, flood observed dates are indicated in black boxes. Unfortunately, data for WMO 41024 are available back to 2008 on the website [47]. Table 7 indicates that for CT, 5 out of 6 days; for CDF, 9 out of 14 days; and for JFI, 7 out of 8 days of rain were observed at WMO station 41024. These high detection rates (0.83, 0.64 and 0.88) for CT, CDF and JFI, respectively, show the dependability of the methods. Higher detection rates of JFI with respect to CT support JFI being better in flood estimations.

**Table 7.** Updated flood estimations and rain observations at WMO station 41024.

| Year | Month | Day | Hour | CT | CDF | JFI | WMO41024 |
|------|-------|-----|------|----|-----|-----|----------|
| 2011 | 1 | 14 | 18 | X | X | X | Rain |
| 2011 | 1 | 14 | 21 | X | X | X | Rain |
| 2011 | 1 | 25 | 18 |  | X | X | Rain |
| 2010 | 4 | 3 | 21 |  | X |  | No Rain |
| 2010 | 7 | 11 | 18 | X | X | X | Rain |
| 2010 | 7 | 11 | 21 |  | X | X | Rain |
| 2010 | 7 | 14 | 15 | X | X | X | No Rain |
| 2010 | 7 | 14 | 18 |  | X |  | No Rain |
| 2010 | 7 | 19 | 0 |  | X |  | No Rain |
| 2012 | 7 | 27 | 15 |  | X |  | No Rain |
| 2009 | 11 | 25 | 6 | X | X | X | Rain |
| 2009 | 11 | 25 | 12 | X | X | X | Rain |
| 2014 | 11 | 21 | 21 | X | X |  | Rain |
| 2009 | 12 | 22 | 18 | X | X | X | Rain |
| 2009 | 12 | 22 | 21 | X | X |  | Rain |
| 2010 | 12 | 9 | 12 |  | X |  | Rain |
| 2010 | 12 | 11 | 0 |  | X |  | No Rain |
| 2010 | 12 | 29 | 6 |  | X | X | Rain |
| 2010 | 12 | 29 | 9 |  | X | X | Rain |
| 2010 | 12 | 30 | 6 |  | X | X | Rain |

Note: Estimated events that match with the real flood observations are framed by a thick dark border.

## 5. Conclusions

This is the first study to show that during floods in Jeddah, TRMM 3B42RT indicated high rainfall intensities. In addition, the seasonal variation of flood occurrences could be represented by 3B42RT data. Moreover, the rainfall timing and rates seemed to match the Weather Research Forecasting (WRF) Model simulations performed by [16]. The movements of storms from the northwest to southeast seen in 3B42RT images were in parallel with the model results of [16].

Using 3B42RT data, three different indices, constant threshold (CT), cumulative distribution functions (CDFs) and Jeddah Flood Index (JFI), were developed and compared for Jeddah flood detection capability. For the whole TRMM 3B42RT data period, i.e., 2000–1014, CT, CDF and JFI estimated 14, 28 and 20 flood affected days (FADs) where 6, 8 and 7 matched with actual FADs, leading to detection ratios of 6/14, 8/28 and 7/20, respectively. WMO has a station in Jeddah (with id: 41024); on the web, these data go back to 2008. After 2008, CT, CDF and JFI showed 6, 14 and 8 FADs where 5, 9 and 7 days of these, respectively, matched with 41024 rain records. Despite the higher 6/14 ratio of CT, since CT missed January and December floods and because of the higher rain match ratio of JFI with 41024, JFI is considered as the best index to indicate floods in Jeddah.

Accuracy assessments of all three methods were performed by using the flood information obtained from International Disasters Database and by combining information obtained from various papers. Thus, the accuracy of flood information (both location and time) is very important for valid assessments. The need for good documentation of flood events was also mentioned in [22].

Rainfall rate is one of the various flood triggering mechanisms considered in this study. Other parameters such as soil moisture (SM), land use and topography can be helpful in accurate flood predictions. Reductions in false flood alarms by using ancillary soil moisture information in flood estimations was presented by [46]. The occurrence of the 25–31 January 2011 flood despite its lower rainfall rates with respect to the 11–14 January 2011 3B42RT rainfall rates indicate the importance of SM. Thus, future flood estimation studies should consider incorporating SM values.

The task of TRMM has already been terminated. Nevertheless, the methodology presented can be implemented in follow-up missions such as the Global Precipitation Measuring Mission (GPM).

**Acknowledgments:** The data providers of TRMM 3B42RT Version 7 are acknowledged.

**Conflicts of Interest:** The author declares no conflicts of interest.

## References

1. EMA. *Hazards, Disasters and Survival*; Emergency Management Australia: Dickson, ACT, Australia, 2002.
2. Ergünay, O. Acil yardim planlamasi ve Afet yönetimi. *Uzman Der Derg.* **1999**, *10*, 6–7.
3. AFAD. Available online: https://www.afad.gov.tr/tr/IcerikListele.aspx?ID=153 (accessed on 13 July 2016).
4. International Federation of Red Cross and Red Crescent Societies. *World Disasters Report*; International Federation of Red Cross and Red Crescent Societies: New Delhi, Indian, 2003.
5. Borga, M.; Stoffel, M.; Marchi, L.; Marra, F.; Jakob, M. Hydrogeomorphic response to extreme rainfall in headwater systems: Flash floods and debris flows. *J. Hydrol.* **2014**, *518*, 194–205. [CrossRef]
6. Jonkman, S.N.; Kelman, I. An Analysis of the Causes and Circumstances of Flood Disaster Deaths. *Disasters* **2005**, *29*, 75–97. [CrossRef] [PubMed]
7. Alfieri, L.; Velasco, D.; Thielen, J. Flash flood detection through a multi-stage probabilistic warning system for heavy precipitation events. *Adv. Geosci.* **2011**, *29*, 69–75. [CrossRef]
8. Hapuarachchi, H.A.P.; Wang, Q.J.; Pagano, T.C. A review of advances in flash flood forecasting. *Hydrol. Process.* **2011**, *25*, 2771–2784. [CrossRef]
9. Zipser, E.J.; Cecil, D.J.; Liu, C.; Nesbitt, S.W.; Yorty, D.P. Where are the most intense thunderstorms on earth? *Bull. Am. Meteorol. Soc.* **2006**, *87*, 1057–1071. [CrossRef]
10. Haggag, M.; El-Badry, H. Mesoscale numerical study of quasi-stationary convective system over Jeddah in November 2009. *Atmos. Clim. Sci.* **2013**, *3*, 73–86. [CrossRef]

11. Al Saud, M. Assessment of flood hazard of Jeddah area 2009, Saudi Arabia. *J. Water Res. Prot.* **2010**, *2*, 839–847. [CrossRef]

12. Youssef, A.M.; Sefry, S.A.; Pradhan, B.; Alfadail, E.B. Analysis on causes of flash flood in Jeddah city (Kingdom of Saudi Arabia) of 2009 and 2011 using multi-sensor remote sensing data and GIS. *Geomat. Nat. Hazards Risk* **2016**, *7*, 1018–1042. [CrossRef]

13. Peel, M.C.; Finlayson, B.L.; McMahon, T.A. Updated world map of the Köppen-Geiger climate classification. *Hydrol. Earth Syst. Sci.* **2007**, *11*, 1633–1644. [CrossRef]

14. Negri, A.J.; Burkardt, N.; Golden, J.H.; Halverson, J.B.; Huffman, G.J.; Larsen, M.C.; McGinley, J.A.; Updike, R.G.; Verdin, J.P.; Wieczorek, G.F. The hurricane-flood-landslide continuum. *Bull. Am. Meteorol. Soc.* **2005**, *86*, 1241–1247. [CrossRef]

15. Asante, K.O.; Macuacuca, R.D.; Artan, G.A.; Lietzow, R.W.; Verdin, J.P. Developing a flood monitoring system from remotely sensed data for the Limpopo Basin. *IEEE Trans. Geosci. Remote Sens.* **2007**, *45*, 1709–1714. [CrossRef]

16. Deng, L.; McCabe, M.F.; Stenchikov, G.; Evans, J.P.; Kucera, P.A. Simulation of flash flood producing storm events in Saudi Arabia using the Weather Research and Forecasting Model. *J. Hydrom.* **2015**, *16*, 615–630. [CrossRef]

17. Wardah, T.; Abu Bakar, S.H.; Bardossy, A.; Maznorizan, M. Use of geostationary meteorological satelliteimages in convective rain estimation for flash-flood forecasting. *J. Hydrol.* **2008**, *356*, 283–298. [CrossRef]

18. Hong, Y.; Adler, R.F.; Negri, A.; Huffman, G.J. Flood and landslide applications of near real-time satellite rainfall products. *Nat. Hazards* **2007**, *43*, 285–294. [CrossRef]

19. Hong, Y.; Adler, R.F.; Huffman, G.J.; Pierce, H.; Gebremichael, M.; Hossain, F. Applications of TRMM-Based Multi-Satellite Precipitation Estimation for Global Runoff Prediction: Prototyping a Global Flood Modeling System. In *Satellite Rainfall Applications for Surface Hydrology*, 1st ed.; Gebremichael, M., Hossain, F., Eds.; Springer: New York, NY, USA, 2010.

20. Hatim, O.S.; Al-Zahrani, M.; El-Hassan, A. Physically, Fully-Distributed Hydrologic Simulations Driven by GPM Satellite Rainfall over an Urbanizing Arid Catchment in Saudi Arabia. *Water* **2017**, *9*, 163. [CrossRef]

21. Wikipedia. Available online: https://en.wikipedia.org/wiki/Jeddah#Climate (accessed on 30 April 2017).

22. Alamri, Y.A. Rains and Floods in Saudi Arabia. Crying of the Sky or of the People? *Saudi Med. J.* **2011**, *32*, 3.

23. The Saudi Blog. Available online: http://the-saudi-blog.blogspot.com.tr/2006/10/jeddah-suffers.html (accessed on 15 August 2017).

24. Almazroui, M.; Raju, P.V.S.; Yusef, A.; Hussein, M.A.A.; Omar, M. Simulation of extreme rainfall event of November 2009 over Jeddah, Saudi Arabia: The explicit role of topography and surface heating. *Theor. Appl. Climatol.* **2017**, 1–13. [CrossRef]

25. Ameur, F. Floods in Jeddah, Saudi Arabia: Unusual phenomenon and huge losses. In Proceedings of the 3rd European Conference on Flood Risk Management, FLOODrisk 2016, E3S Web of Conferences 7, 04019, Lyon, France, 17–21 September 2016. [CrossRef]

26. Subyani, A.M. Flood Hazards Analysis of Jeddah City, Western Saudi Arabia. *JAKU Earth Sci.* **2011**, *23*, 35–48. [CrossRef]

27. Al-Khalaf, A.K.; Basset, H.A. Diagnostic study of a severe thunderstorm over Jeddah. *Atmos. Clim. Sci.* **2013**, *3*, 150–164.

28. International Disasters Database. Available online: http://emdat.be/disaster_list/index.html (accessed on 30 April 2017).

29. Global Hazards. Available online: https://www.ncdc.nocc.gov/sotc/hazards/201007 (accessed on 15 August 2017).

30. TRMM. Available online: https://pmm.nasa.gov/TRMM/trmm-instruments (accessed on 31 January 2017).

31. Wanders, N.; Pan, M.; Wood, E.F. Correction of real time satellite precipitation with multi sensor satellite observations of land surface variables. *Remote Sens. Environ.* **2015**, *160*, 206–221. [CrossRef]

32. Huffman, G.J.; Adler, R.F.; Bolvin, D.T.; Gu, G.; Nelkin, E.J.; Bowman, K.P.; Hong, Y.; Stocker, E.F.; Wolff, D.B. The TRMM Multisatellite Precipitation Analysis (TMPA): Quasi-Global, Multilayer, Combined-Sensor Precipitation Estimates at Fine Scales. *J. Hydrom.* **2007**, *8*, 38–55. [CrossRef]

33. Almazroui, M. Calibration of TRMM rainfall climatology over Saudi Arabia during 1998–2009. *Atmos. Res.* **2011**, *99*, 400–414. [CrossRef]

34. Kheimi, M.M.; Gutub, S. Assessment of remotely sensed precipitation products across the Saudi Arabia region. In Proceedings of the 6th International Conference on Water Resources and Arid Environments, Riyadh, Saudi Arabia, 16–17 December 2014; pp. 315–327.

35. Tekeli, A.E.; Fouli, H. Evaluation of TRMM satellite-based precipitation indexes for flood forecasting over Riyadh City, Saudi Arabia. *J. Hydrol.* **2016**, *31*, 1243. [CrossRef]

36. Duan, Z.; Bastiaanssen, W.G.M. First results from Version 7 TRMM 3B43 precipitation product in combination with a new downscaling-calibration procedure. *Remote Sens. Environ.* **2013**, *131*, 1–13. [CrossRef]

37. TRMM. Available online: http://trmm.gsfc.nasa.gov/3b42.html (accessed on 31 January 2017).

38. Alfieri, L.; Thielen, J. A European precipitation index for extreme rain-storm and flash flood early warning. *Meteorol. Appl.* **2012**, *22*, 3–13. [CrossRef]

39. Zehe, E.; Sivapalan, M. Threshold behaviour in hydrological systems as (human) geo-ecosystems: Manifestations, controls, implications. *Hydrol. Earth Syst. Sci.* **2009**, *13*, 1273–1297. [CrossRef]

40. Hamada, A.; Murayama, Y.; Takayabu, Y.N. Regional Characteristics of Extreme Rainfall Extracted from TRMM PR Measurements. *J. Clim.* **2014**, *27*, 8151–8169. [CrossRef]

41. Usul, N. *Engineering Hydrology*, 3rd ed.; METU Press: Ankara, Turkey, 2005.

42. Ewea, H.A.; Elfeki, A.M.; Al-Amri, N.S. Development of intensity-duration-frequency curves for the Kingdom of Saudi Arabia. *Geomat. Nat. Hazards Risk* **2016**, 1–15. [CrossRef]

43. Viessman, W.; Lewis, G.L. *Introduction to Hydrology*, 5th ed.; Prentice Hall: Upper Saddle River, NJ, USA, 2002.

44. Dönmez, S.; Tekeli, A.E. Comparison of TRMM based flood indices for Gaziantep, Turkey. *Nat. Hazards* **2017**, *88*, 821–834. [CrossRef]

45. Pombo, S.; De Oliveira, R.P. Evaluation of extreme precipitation estimates from TRMM in Angola. *J. Hydrol.* **2015**, *523*, 663–679. [CrossRef]

46. Tekeli, A.E.; Fouli, H. Reducing False Flood Warnings of TRMM Rain Rates Thresholds over Riyadh City, Saudi Arabia by Utilizing AMSR-E Soil Moisture Information. *Water Resour. Manag.* **2017**, *541*, 471–479. [CrossRef]

47. WMO. Available online: https://gis.pecad.fas.usda.gov/WmoStationExplorer/ (accessed on 15 August 2017).

*water*

MDPI

*Article*

# Automated Extraction of Urban Water Bodies from ZY-3 Multi-Spectral Imagery

Fan Yang [1,†], Jianhua Guo [1,2,*,†], Hai Tan [1,2] and Jingxue Wang [1]

1   School of Geomatics, Liaoning Technical University, Fuxin 123000, China;
    yangfan2008beijing@126.com (F.Y.); tanhai@casm.ac.cn (H.T.); xiaoxue1861@163.com (J.W.)
2   Satellite Surveying and Mapping Application Center, National Administration of Surveying,
    Mapping and Geoinformation, Beijing 100048, China
*   Correspondence: nkszjx@163.com; Tel.: +86-41-8335-1991
†   These authors contributed equally to this work.

Academic Editors: Hongjie Xie and Xianwei Wang
Received: 31 October 2016; Accepted: 14 February 2017; Published: 21 February 2017

**Abstract:** The extraction of urban water bodies from high-resolution remote sensing images, which has been a hotspot in researches, has drawn a lot of attention both domestic and abroad. A challenging issue is to distinguish the shadow of high-rise buildings from water bodies. To tackle this issue, we propose the automatic urban water extraction method (AUWEM) to extract urban water bodies from high-resolution remote sensing images. First, in order to improve the extraction accuracy, we refine the NDWI algorithm. Instead of Band2 in NDWI, we select the first principal component after PCA transformation as well as Band1 for ZY-3 multi-spectral image data to construct two new indices, namely NNDWI1, which is sensitive to turbid water, and NNDWI2, which is sensitive to the water body whose spectral information is interfered by vegetation. We superimpose the image threshold segmentation results generated by applying NNDWI1 and NNDWI2, then detect and remove the shadows in the small areas of the segmentation results using object-oriented shadow detection technology, and finally obtain the results of the urban water extraction. By comparing the Maximum Likelihood Method (MaxLike) and NDWI, we find that the average Kappa coefficients of AUWEM, NDWI and MaxLike in the five experimental areas are about 93%, 86.2% and 88.6%, respectively. AUWEM exhibits lower omission error rates and commission error rates compared with the NDWI and MaxLike. The average total error rates of the three methods are about 11.9%, 18.2%, and 22.1%, respectively. AUWEM not only shows higher water edge detection accuracy, but it also is relatively stable with the change of threshold. Therefore, it can satisfy demands of extracting water bodies from ZY-3 images.

**Keywords:** ZY-3 images; urban water bodies; automatic water extraction; NDWI; PCA transformation; shadow detection

## 1. Introduction

Cities are the crystallization of highly developed civilization. As an important factor in the urban ecosystem, water bodies play a critical role in maintaining stability of the urban ecosystem [1]. Their changes are closely related with people's life. Negative changes may lead to disasters, pollution, water shortage, or even epidemics [2]. Therefore, understanding the distribution and changes of urban water has become the focus of people's attention.

In recent years, with the development and application of remote sensing technology, it has played an increasingly important role in natural resource surveying [3,4], dynamic monitoring [5,6], and natural surface water planning [7,8], thus attracting researchers' attention. Remote sensing images enable us to observe the earth from a totally different perspective and monitor its real-time changes.

Water bodies are common ground object in remote sensing images, the rapid acquisition of their dynamic information is apparently valuable for water resource survey, water conservancy planning and environmental monitoring and protection [9]. Among current water extraction technologies, a mainstream method is using remote sensing data to gather urban water information in a timely and accurate way [10]. Thus far, researchers have proposed many methods to extract water using remote sensing images [10,11]. These models could be roughly divided into four categories: (a) single-band or multiple-band threshold methods [12,13]; (b) water indices [14–16]; (c) linear un-mixing models [17]; and (d) supervised or unsupervised classification methods [18,19]. Other methods that are not as frequent used as the above include water extraction technology based on digital elevation models [20,21], microwave remote sensing imagery [22–24] and object oriented technology [25,26]. In general, the water indices are most commonly used in practice because of their simple, convenient and fairly accurate algorithm models [27].

The water indices are under constant refinement. The first model, Normalized Difference Water Index (NDWI), proposed by McFeeters [16], is based on the principle of Normalized Difference Vegetation Index (NDVI). Its basic idea is to extract water bodies by enhancing water information and suppressing non-water information. Xu Hanqiu [17] found that the NDWI algorithm could not effectively inhibit the impact of buildings and proposed a refined version, in which he used the Shortwave Infrared (SWIR) instead of the NIR in the original NDWI algorithm. The new algorithm was called the Modified Normalized Difference Water Index (MNDWI). It exhibits higher accuracy, but it still could not distinguish shadows. Therefore, Feyisa G L [18] proposed a method called the automated water extraction index (AWEI) to adapt to different environments. Five bands of Landsat5 Thematic Mapper (Band1, Band2, Band4, Band5, and Band7) were used to compute the index to enhance the contrast between water and non-water information which could be used to model the water images with or without shadows.

Most of the algorithms, however, are proposed based on medium- or low-resolution remote sensing images. Because of resolution limitations, smaller water bodies cannot be extracted effectively, especially in urban areas where the size of water bodies varies and there are many small artificial lakes and rivers [28]. Therefore, we should prioritize the use of high-resolution remote sensing images in those areas. The ZY-3 satellite is China's first civil high-resolution stereo mapping satellite launched in 9 December 2012. Equipped with four sets of optical cameras, it includes an orthographic panchromatic time delay and integration charge-coupled device (TDI CCD) camera with the ground resolution of 2.1 m, two front-view and rear-view panchromatic TDI CCD cameras with the ground resolution of 3.6 m, and an orthographic multi-spectral camera with the ground resolution of 5.8 m. The acquired data are mainly used for topographic mapping [29], digital elevation modeling [30] and resource investigation [31]. Therefore, it is an ideal multi-spectral image data source for urban water extraction [31].

Recently, with the increase of image resolution, most of the high-resolution remote sensing images (such as those from WorldView-2, IKONOS, RapidEye and ZY-3 satellites) do not have so many available bands for water extraction compared with those from LandsatTM/ETM+/OLI imagery, rendering the MNDWI and AWEI algorithms useless. After all, most high-resolution remote sensing images only have four bands (blue, green, red and near-infrared), lacking the SWIR necessary to compute the MNDWI/AWEI indices [31]. It is therefore problematic to use the NDWI to extract urban water from high-resolution images. For instance, it is difficult to remove shadows, especially those of high-rise buildings in urban areas. The problem dramatically worsens when analyzing high-resolution images [32], thus it is difficult to distinguish between water bodies and shadows [25,33].

To tackle urban water extraction issue, some scholars have pioneered on this subject and proposed some preliminary solutions such as the object oriented technology to detect shadows by computing their texture features [34]. It can achieve expected results, but is relatively complex and time-consuming in the texture description and computation [35]. Therefore, it is not an optimal model for the shadow detection. Another method based on Support Vector Machine (SVM) feature training can be used to remove the impact of shadows on urban water extraction [31]. However, the SVM training is time-consuming, especially when there are many training samples with high-dimension eigenvectors [36]. Some researchers combine the morphological shadow index (MSI) [37] and the NDWI to extract urban water bodies from WorldView-2high-resolution imagery, in order to increase the detection accuracy [38]. The principle of this method is simple, but since the urban water extraction method is based on the NDWI algorithm, the detection accuracy is not very high, especially when detecting small areas of water surrounded by lush vegetation. In those areas, the spectral features of water will be severely contaminated and extremely unstable [39]. In addition, urban water bodies are typically sediment-laden and algae polluted, and thus exhibit different optical features compared with non-contaminated natural water bodies [31].

Therefore, to remove the limitations of traditional NDWI indices in water extraction and improve the initial classification accuracy, we propose the NNDWI1, which is sensitive to turbid water bodies, and NNDWI2, which is sensitive to water bodies whose spectral information is seriously disturbed by that of vegetation, based on the analysis of water features and shadows. To remove the disturbance of shadows of high-rise buildings to the water extraction results, and to better express the features of shadows and water bodies, we use the Object-Oriented Technology to classify the water bodies and shadows. Meanwhile, if the features expressed by the operators are too complex, it will not be conducive to reduce the computational time. Thus, it is better to use operators that express the spectral rather than textural features of ground objects in the algorithm in order to improve the computational efficiency. To further improve the efficiency, we use thresholds rather than the time-consuming classification algorithm to differentiate water bodies from shadows. The experimental results show that the automatic urban water extraction method (AUWEM) algorithm can better identify shadows and water bodies, and improve the urban water detection accuracy.

## 2. Study Areas and Data

### 2.1. Study Areas

To verify the feasibility of the automatic urban water extraction method (AUWEM) algorithm, we select five images featuring different areas with different environments including lakes and rivers within territory of China for experiments. The selected areas were located in Beijing, Guangzhou, Suzhou and Wuhan. As for Wuhan, the city is an ideal place for experiment because of its large amount of rivers and lakes as well as rich diversity of water bodies, so we select two different coverage areas for experiment. Details of the experimental areas are described in the following Table 1, and the corresponding images from ZY-3 satellite are detailed in Table 2.

**Table 1.** Description of studied areas.

| City's Name and Location | Area Coverage (Pixels) | Water Body Type | Topography | Climate | Color Infrared Composite (4/3/2 Band Combination) |
|---|---|---|---|---|---|
| Beijing (39.9° N, 116.3° E) | 1479 × 1550 (77.1 km²) | Rivers Polluted lakes Clear lake | Plain | Warm temperate semi humid continental monsoon climate | |
| Guangzhou (23° N, 113.6° E) | 2351 × 2644 (209.1 km²) | Rivers Ponds Polluted lakes Clear lake | Basin, plain | Typical monsoon climate in South Asia | |
| Suzhou (31.2° N, 120.5° E) | 2351 × 2644 (209.1 km²) | Rivers Ponds Polluted lakes Clear lake | Basin, plain, hills. | Subtropical humid monsoon climate | |

**Table 1.** *Cont.*

| City's Name and Location | Area Coverage (Pixels) | Water Body Type | Topography | Climate | Color Infrared Composite (4/3/2 Band Combination) |
|---|---|---|---|---|---|
| Wuhan_1 (30.5° N, 114.3° E) | 2245 × 2521 (190.4 km$^2$) | Rivers Ponds Large polluted lakes Large clear lakes | Basin, plain, hills. | Subtropical humid monsoon climate | |
| Wuhan_2 (30.5° N, 114.3° E) | 2894 × 3396 (330.6 km$^2$) | Rivers Ponds Large polluted lakes Large clear lakes | Basin, plain, hills. | Subtropical humid monsoon climate | |

**Table 2.** ZY-3 satellite Parameters.

| Item | Contents |
|---|---|
| Camera model | Panchromatic orthographic; Panchromatic front-view and rear-view; multi-spectral orthographic |
| Resolution | Sub-satellite points full-color: 2.1 m; front- and rear-view 22° full color: 3.6 m; sub-satellite points multi-spectral: 5.8 m |
| Wavelength | Panchromatic: 450 nm–800 nm Multi-spectral: Band1 (450 nm–520 nm); Band2 (520 nm–590 nm) Band3 (630 nm–690 nm); Band4 (770 nm–890 nm) |
| Width | Sub-satellite points Panchromatic: 50 km, single-view 2500 km$^2$; Sub-satellite points multi-spectral: 52 km, single-view 2704 km$^2$ |
| Revisit cycle | 5 days |
| Daily image acquisition | Panchromatic: nearly 1,000,000 km$^2$/day; Fusion: nearly 1,000,000 km$^2$/day |

### 2.2. Experimental ZY3 Imagery and Its Corresponding Reference Imagery

ZY-3 Images used in the experiments can be queried and ordered from http://sjfw.sasmac.cn/product/order/productsearchmap.htm. We use theZY-3 multi-spectral data to extract water. All image data are Level 1A products, which have been adjusted through radiometric and geometric correction. All the images used in the experiments were cloud free.ZY-3 satellite parameters are shown in Table 2. The experimental image information is described in the following Table 3.

**Table 3.** Description of ZY-3 scenes.

| Test Site | ZY-3 Scenes | | |
|---|---|---|---|
| | Acquisition Date | Path | Row |
| Beijing | 28 November 2013 | 002 | 125 |
| Guangzhou | 20 October 2013 | 895 | 167 |
| Suzhou | 17 December 2015 | 882 | 147 |
| Wuhan_1 | 24 July 2016 | 001 | 149 |
| Wuhan_2 | 28 March 2016 | 897 | 148 |

The reference imagery is used to evaluate the urban water classification accuracy. To acquire the corresponding reference imagery, we manually delineate the water edge in high-resolution imagery, which is obtained by fusion of ZY-3's high-resolution Panchromatic and ZY-3's Multispectral Images.

During the experiment, we asked an experienced analyst to manually map out the water bodies. To prevent arbitrariness, all referential images corresponding to five experimental areas were drawn by a single person. It took about 10 days, including eight days of imagery creation and two days of double-checking. Before manually mapping out water bodies and non-water areas, we collected and studied a large amount of related samples so that relevant criteria can be set up to improve the accuracy of water boundary mapping. Figure 1 shows the five referential images that are manually drawn. Here, the water bodies are in blue, and non-water area areas are in black. The relevant criteria for water body delineate are as follows:

1. Delineate precision of the fuzzy boundary of water body is within three pixels while the clear boundary of water body is within one pixels.
2. Less than or equal to one pixels of water body information is not given to delineate.
3. We choose reference of higher resolution Google map image in order to distinguish between water body and building shadow as well as the seemingly water body and non-water body.
4. Urban water system is basically interconnected with each, other except for the river intercepted by bridge.

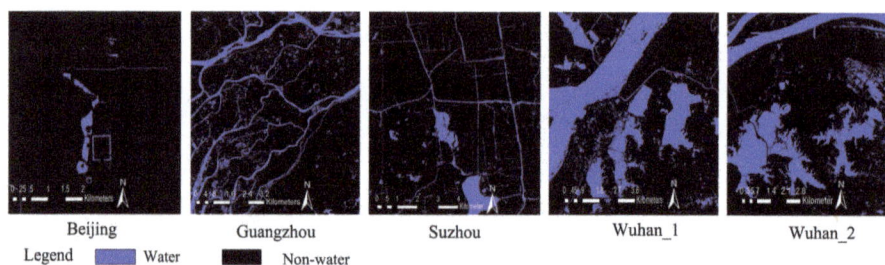

Figure 1. Manually drawn referential imagery.

## 3. Method

### 3.1. Satellite Image Preprocessing

We used in the study the level-1 imagery taken from ZY-3 satellite without Ortho-rectification, therefore we used RPC+30m DEM to process the experimental images and applied Ortho-rectification without control points. We used FLAASH (Fast Line-of-Sight Atmospheric correction model Analysis of Spectral Hypercubus) for atmospheric correction [40]. All of the above steps were completed in ENVI5.2 software.

The radiometric calibration coefficient of ZY-3 FLAASH atmospheric correction can be downloaded from http://www.cresda.com/CN/Downloads/dbcs/index.shtml. The spectral response function could be downloaded from http://www.cresda.com/CN/Downloads/gpxyhs/index.shtml.

Figure 2 depicts the spectral curves of ground objects before and after atmospheric correction. We can see from this figure that there is huge difference between the two spectral curves of pixels. The one after the atmospheric correction is more consistent with the actual features of ground objects.

| | Road | Vegetation | Water | Shadow |
|---|---|---|---|---|
| Ground objects |  |  |  |  |
| The spectral curves before atmospheric correction |  |  |  |  |
| The spectral curves after atmospheric correction |  |  |  |  |

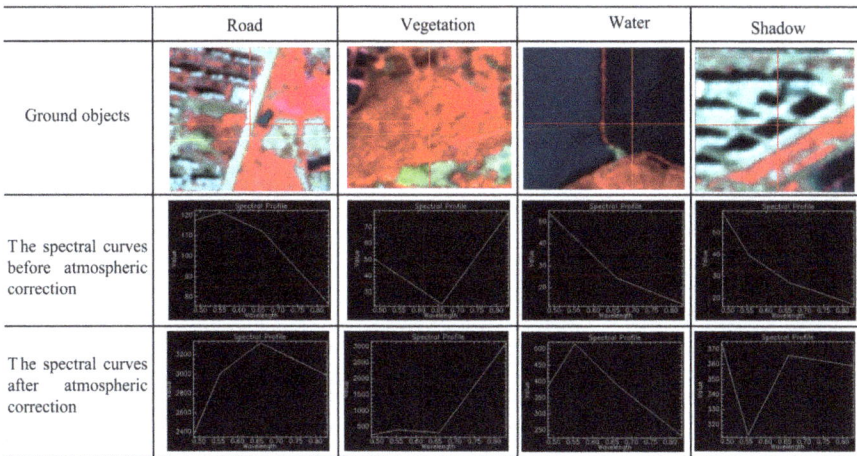

**Figure 2.** Comparison of ground objects' spectral curves before and after the atmospheric correction.

### 3.2. Normalized Difference Water Index (NDWI)

The NDWI was first proposed by McFeeters in 1996 and successfully applied to detect the surface water in multi-spectral imagery from Landsat Multi-spectral Scanner (MSS) [14]. The definition is as follows:

$$NDWI = \frac{(Green - NIR)}{(Green + NIR)} \tag{1}$$

According to this equation and the spectral feature curves of ground objects, the NDWI index value of water surface is greater than 0, the NDWI value of soil and other ground objects with high reflectivity approximately equals 0, while the NDWI value of vegetation is below 0 because the reflectivity of the vegetation on the infrared band is higher than on the green band. As a result, the water can be easily extracted from multi-spectral images.

### 3.3. New Normalized Difference Water Indexes (NNDWI)

In our study, the computation of NNDWI comprises of two steps:

1.  Use the ZY-3 Blue band (Band1) to replace the green band in Equation (1) to obtain NNDWI1, i.e.,

$$NNDWI1 = \frac{(Blue - NIR)}{(Blue + NIR)} \tag{2}$$

2.  Four bands of ZY-3 imagery were processed by the Principal Component Analysis (PCA) transformation [41], use the first principle component after PCA transformation to replace the Green band in Equation (1) to obtain NNDWI2, i.e.,

$$NNDWI2 = \frac{(Component1 - NIR)}{(Component1 + NIR)} \tag{3}$$

where *Component1* is the first principal component after PCA transformation. The PCA transformation reflects the methodology of dimension reduction [41]. From the mathematic perspective, it is to find a set of basis vectors which can most efficiently express the relations among various data. From the geometrical perspective, it is to rotate the original coordinate axis and get an orthogonal one, so that all data points reach the maximum dispersion along the new axis direction. When applied to the image analysis, it is to find as few basis images as possible to preserve the maximum information of the original images, thus achieving the purpose of feature extraction.

In our study, the initial water extraction results are generated by the superimposition of the threshold segmentation results from two water indexes, namely NNDWI1 and NNDW2. Therefore, NNDWI is expressed as follows:

$$NNDWI = (segmentation\_NNDWI1) \cup (segmentation\_NNDWI2) \tag{4}$$

In Equation (4), *segmentation_NNDWI1* and *segmentation_NNDWI2* represent the threshold segmentation results generated by NNDWI1 and NNDWI2 index image, respectively.

The result generated by NNDWI integrates the water extraction results from both algorithms, thus the omission caused by a singular index is avoided. As shown below in Figure 3, NNDWI2 algorithm is not sensitive to turbid water, whereas NNDWI1 is a complement because of its sensitivity to turbid water. Therefore, in practice, these two algorithms can be combined to generate a composite water extraction result instead of two separate ones, thus the subsequent water extraction accuracy can be enhanced.

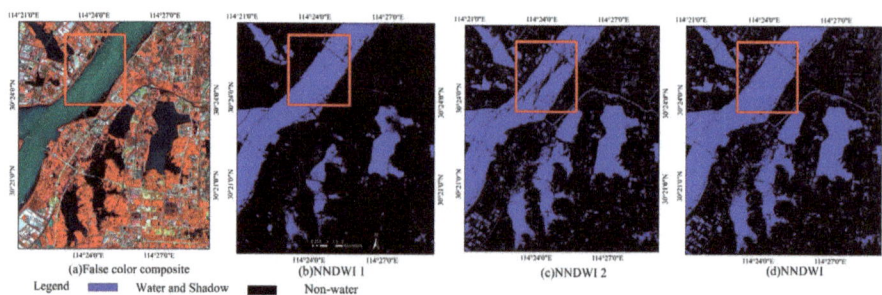

**Figure 3.** Different water extraction results generated by NNDWI1, NNDWI2 and NNDWI, respectively.

*3.4. Shadow Detection Based on Object Oriented Technology*

3.4.1. Shadow Objects

In the initial water extraction results generated by NNDWI, shadows are extracted along with the water bodies. While analyzing the image data extracted using NNDWI, we find that the areas of shadows are generally smaller than those of water bodies, except for some small artificial ponds and lakes in the city. Therefore, in practice, we only need to detect objects that cover small areas. These objects will encompass almost all possible shadows and small area water bodies. The model for acquiring small-area objects can be described as follows:

$$\begin{cases} component = water & if\ area(component) > t\ ,\ component \in NNDWI \\ component = \ shadow\ or\ water & if\ area(component) \leq t, component \in NNDWI \end{cases} \tag{5}$$

where *t* indicates the set segmentation threshold, whose value is the number of pixels that enables the maximum shadow objects; it is a minimum detectable size of water bodies that equals exclusion. The number of pixels of the largest shadow area varies in different images, resulting in different values of *t*, which should be set accordingly. The experimental statistics show that if we set $2000 < t < 5000$, the results will be satisfactory. *component* indicates the discrete objects in the water extraction results generated using NNDWI, including water and shadow areas. *area(component)* indicates the object areas: if *area(component)* > *t*, then it indicates the water objects, while, if *area(component)* ≤ *t*, then it indicates either small area water or shadow objects.

It is impossible to extract all the shadow pixels from the water extraction results generated by using NNDWI. For better application of the Object Oriented Technology, the acquired shadow objects are under morphological dilation [42], so that the dilated objects can better include shadow pixels in the area. Meanwhile, to limit the dilation results in the actual shadow areas, we use the threshold segmentation results on the near infrared band (Band4) of ZY-3 images as the constraint. (Due to relatively low reflectivity of water and shadows on the near infrared band (Band4), the values of water and shadow pixels are relatively small. The water and shadow areas are in dark black on this band. The threshold segmentation can effectively enable the extraction of water and shadow objects. Therefore, the threshold segmentation results of Band4 serve as a constraint.) Specifically, the constraint on the dilation results is set by intersecting the dilated images and those under threshold segmentation on Band4, expressed as below:

$$component2 = (dilate\_component) \cap (segmentation\_Band4) \tag{6}$$

In Equation (6), the *dilate_component* indicates dilation results of *component* (i.e., the objects of water/shadow whose areas are below the threshold); and the *segmentation_Band4* indicates threshold segmentation results on the near-infrared band (Band4). How the dilation results are constrained by way of intersection is shown in Figure 4.

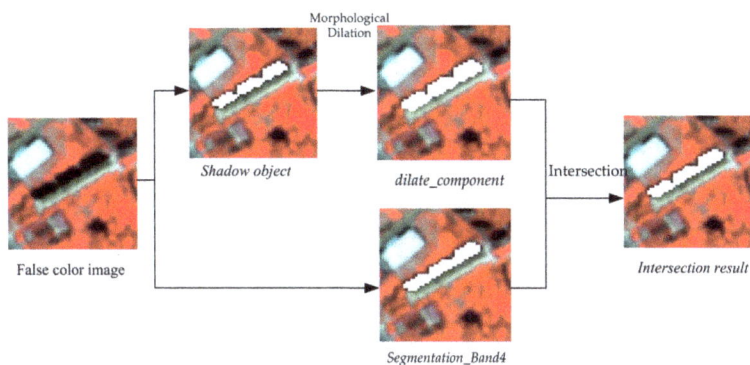

**Figure 4.** Diagram of dilation constraint.

### 3.4.2. The Shadow Objects Description (The Description of Spectral Feature Relations between Water-Body Pixels and Shadow-Area Pixels)

Generally, the water extraction results generated by the NNDWI only cover water and shadow areas. Thus, we only need to analyze their features and find the proper ones. In the study, we find that textural features can be used to effectively describe shadows and water bodies, but those of ground objects (such as Gray Level Co-occurrence Matrix, GLCM) are complex and time-consuming to compute and thus unfit for the classification of water bodies and shadows. As a result, we use the spectral features of ground objects to describe the pixels of water and shadow areas and distinguish between them.

Through an extensive analysis of the spectral feature curves of water bodies and shadows, we find that, in general, the spectral relation of water pixels satisfies the following inequality:

$$Band2 > Band4 \tag{7}$$

The spectral curves of shadow-area pixels are more complicated. When the sunshine is blocked by buildings, there will be shadows. The spectral features of the pixels in the shaded areas typically resemble those of other ground objects, such as vegetation, cement and soil. After analyzing the spectral features of those areas, we summarized five different spectral curve models, as shown in Figure 5.

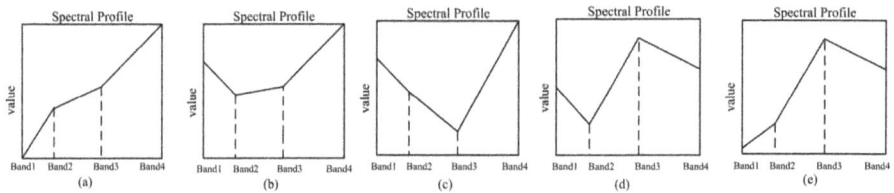

**Figure 5.** The spectral feature curves of the shadow-area pixels: (**a–e**) typical spectral curves of five types of pixels.

Accordingly, we can set up the following model that shows the spectral relations of shadow pixels:

$$
\begin{cases}
Band2 > Band1 \\
Band3 > Band2 \\
Band4 > Band3
\end{cases}
\tag{8}
$$

$$
\begin{cases}
Band1 > Band2 \\
Band4 > Band2 \\
Band4 > Band3
\end{cases}
\tag{9}
$$

$$
\begin{cases}
Band3 > Band2 \\
Band3 > Band4 \\
Band4 > Band2
\end{cases}
\tag{10}
$$

If the spectral curves in the experimental results generated by the NNDWI index correspond with the pixels shown in the above three models, they will be classified as shadow pixels, and vice versa.

### 3.4.3. The Shadow Objects Detection Method

In the experiments, the classification of each small-area discrete object is determined. First, the spectral relation of each pixel of discrete objects is described to judge whether it satisfies the constraint of a shadow pixel. The number of shadow pixels in each object is recorded. According to extensive statistical experiments, we find that if the proportion of shadow pixels exceeds the threshold T, then the object can be classified as a shadow area. Otherwise, the object is classified as a water body. The judgment function can be expressed as:

$$
\begin{cases}
component2 = water & if \frac{m}{n} \leq T \\
component2 = shadow & if \frac{m}{n} > T
\end{cases}
\tag{11}
$$

where n indicates the total number of pixels of an object, and m indicates the number of its shadow pixels. The threshold T is an empirical number optimized through experiments. In a statistical analysis of the shadow pixels of the ZY-3 images, we find that when T equals 0.5, water and shadow objects can be effectively differentiated.

### 3.5. Urban Water Extraction and Its Accuracy Evaluation

Figure 6 depicts the steps of the AUWEM algorithm. First, preprocess the imagery (by using Ortho rectification and atmospheric correction). Second, use the NNDWI described in Section 3.3 to obtain the initial water extraction results. Third, use the shadow detection method of the Object Oriented Technology detailed in Section 3.4 to detect shadow objects. Finally, remove detected shadow objects to obtain the final results of urban water extraction. The overall flow chart of AUWEM is shown in Figure 6. In order to compare image classification accuracy, we use six indicators to describe the extraction accuracy of different algorithms, including producer accuracy, user accuracy, Kappa coefficient, omission error, commission error and total error.

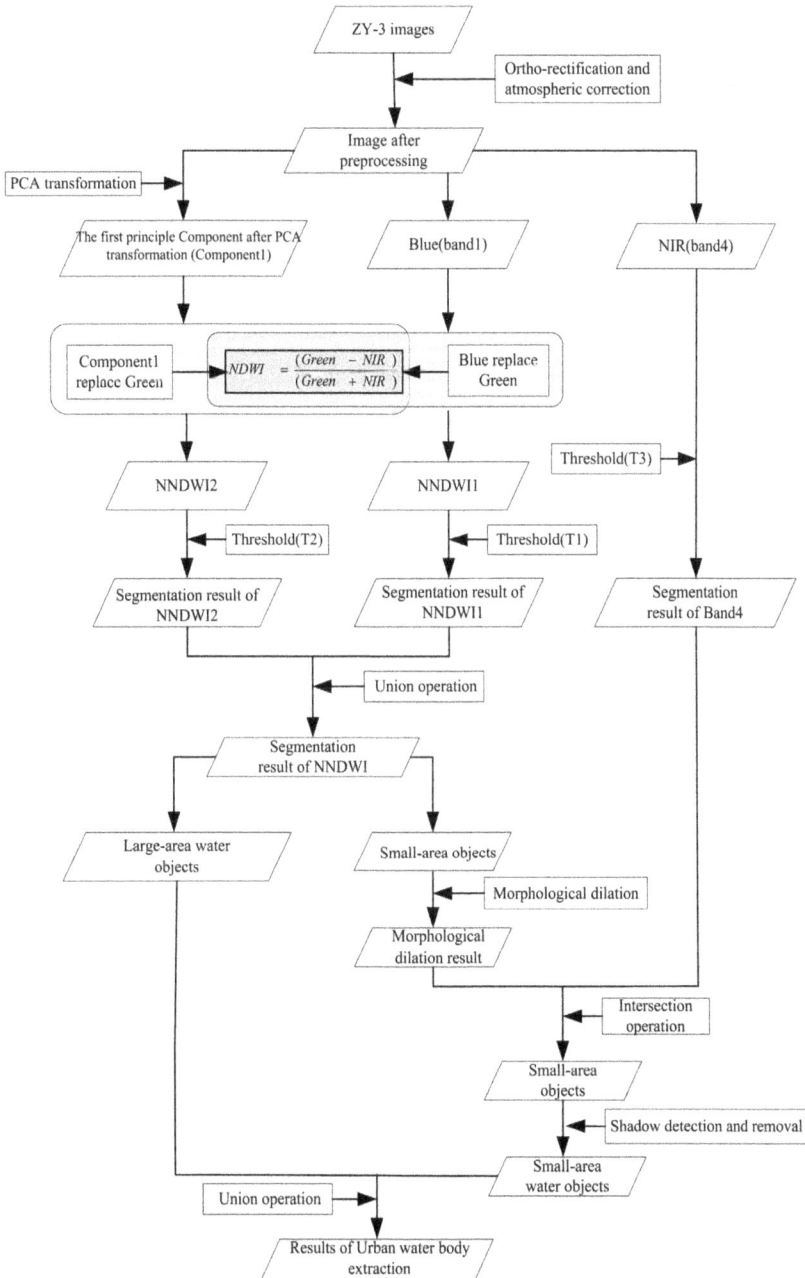

**Figure 6.** The overall flowchart of AUWEM.

## 4. Experimental Results and Analysis

### 4.1. Water Extraction Maps

To demonstrate the feasibility of the algorithm, we compare the water extraction results generated by using NDWI algorithm and the supervised Maximum Likelihood (MaxLike) classifier was also included in our comparison as the latter one is one of the most widely used methods in land cover classification [16]. Table 4 shows the settings of threshold parameters in different algorithms that are used to extract water from each area. To evaluate the accuracy of the three algorithms, high-resolution fusion imageries are used as the accuracy reference data. We obtain the reference imageries by manually delineating the water edge in fusion imagery, whose information is shown in Table 3. We compare reference imageries with the classification results generated by the three algorithms. For visual interpretation and analysis of classification results generated by different algorithm, the correct classification of water pixels is colored in blue, correct classification of non-water pixels in black. If there are erroneous classifications, corresponding pixels will be highlighted in white.

The experimental results are shown in Figure 7. To facilitate the observation and analysis, we select a small area in yellow rectangular frame from the image, and the classification results are shown in Figure 8. According to the results, the classification accuracy of AUWEM was better than that of NDWI and MaxLike. The AUWEM algorithm excels in classifying mixed pixels of the water edge (judging from the water classification results shown in Beijing, Wuhan_1and Wuhan_2), detecting small pond water compared with NDWI and the MaxLike (judging from classification results shown in Suzhou), and removing shadows of buildings (judging from the classification results shown in Suzhou and Wuhan_2). The NNDWI algorithm is excellent in extracting water bodies that are turbid or whose spectral information is seriously disturbed by vegetation. Therefore, it shows better edge classification results compared with the NDWI algorithm. On the other hand, the classification results of the MaxLike depend on selection of water samples. A limited number of samples will result in unsatisfactory results, especially when the edge pixels are seriously affected by the mixed spectrum. Similarly, small rivers in urban areas are usually flanked with trees, so their spectral information will be seriously disturbed by that of the vegetation. Therefore, the NDWI and MaxLike are inadequate to extract water bodies of small rivers. The Object-Oriented Technology is adopted to differentiate shadows from water bodies by expressing their spectral features, in order to eliminate the influence of high-rise urban buildings on water extraction results.

**Table 4.** Threshold setting of the three algorithms in different experimental areas. Among them, T, T1, T2 and T3 are the threshold of NDWI, NNDWI1, NNDWI2 and Band4, respectively.

| Method | Threshold | | | | |
|---|---|---|---|---|---|
| | Beijing | Guangzhou | Suzhou | Wuhan_1 | Wuhan_2 |
| AUWEM | T1 = 0, T2 = 0, T3 = 38 | T1 = 0, T2 = 0, T3 = 20 | T1 = 0, T2 = 0, T3 = 25 | T1 = 0, T2 = 0, T3 = 45 | T1 = 0, T2 = 0, T3 = 65 |
| NDWI | T = −0.04 | T = −0.07 | T = 0.07 | T = 0.08 | T = 0.02 |
| MaxLike | - | - | - | - | - |

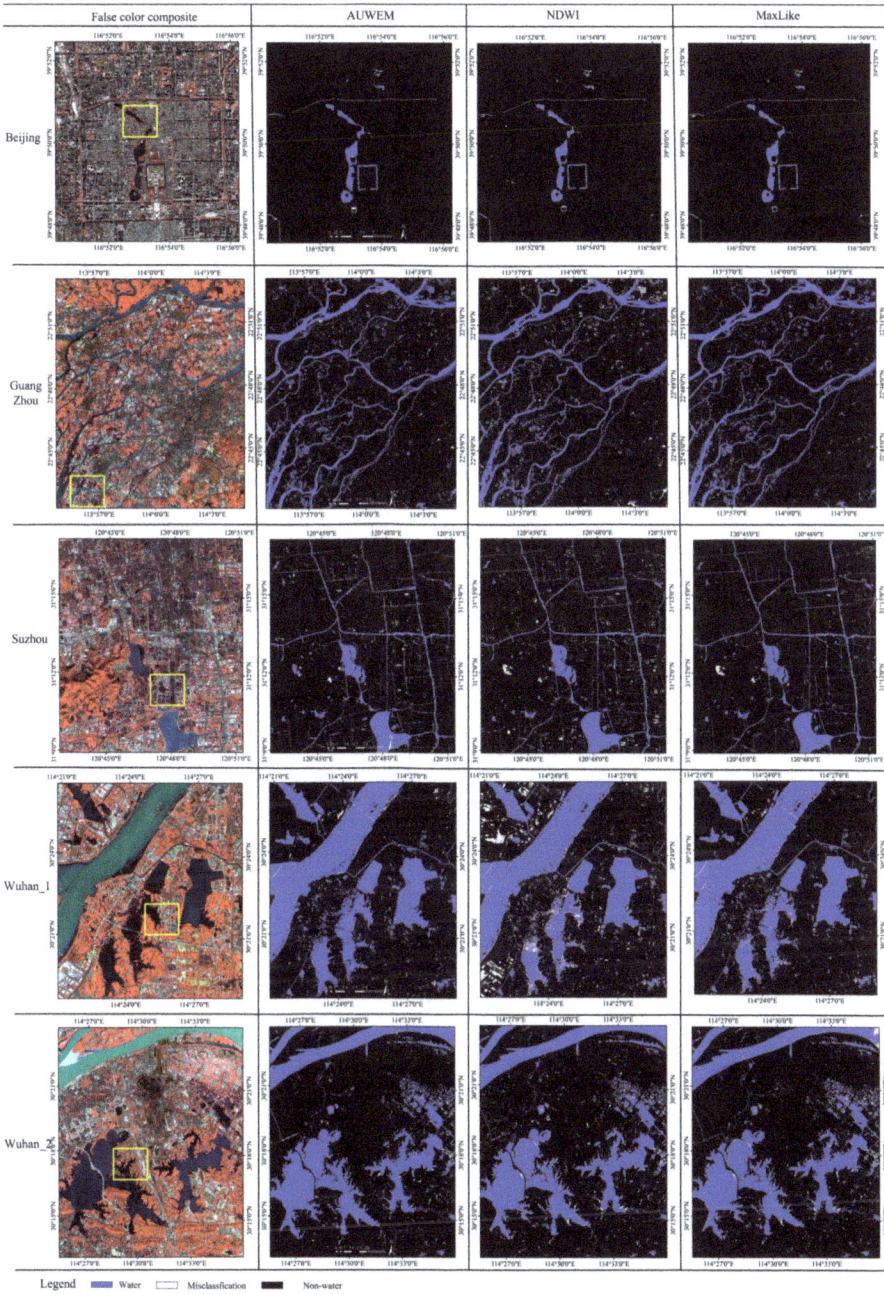

**Figure 7.** Comparison of water extraction results of three algorithms in different experimental areas.

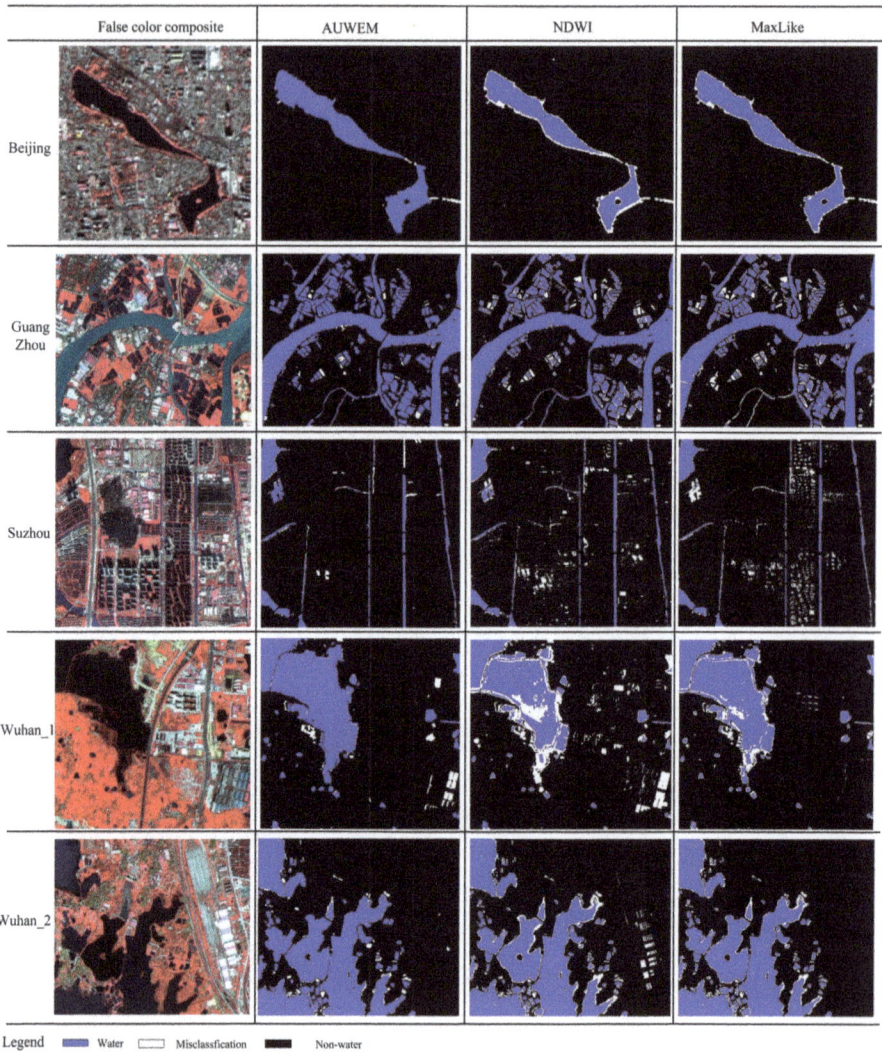

| | False color composite | AUWEM | NDWI | MaxLike |
|---|---|---|---|---|
| Beijing | | | | |
| Guang Zhou | | | | |
| Suzhou | | | | |
| Wuhan_1 | | | | |
| Wuhan_2 | | | | |

Legend  ■ Water  □ Misclassification  ■ Non-water

**Figure 8.** Comparison of water classification results among different algorithms in local areas (a small area in yellow rectangular frame from the image of Figure 8).

## 4.2. Water Extraction Accuracy

Accuracy of water extraction can be evaluated by visual interpretation and one-by-one pixel comparison. The visual interpretation has been discussed in Section 4.1. In this section, we will evaluate the classification accuracy by using some quantitative indicators. Table 5 shows the comparison of water classification accuracy among three algorithms in different experimental areas. A statistical analysis of Table 5 indicates that the classification accuracy of AUWEM is greater than that of NDWI and MaxLike. AUWEM algorithm exhibits the greatest classification accuracy in five experimental areas with the average Kappa coefficient of 93%; the NDWI exhibits the lowest classification accuracy with the average Kappa coefficient of about 84.4%; and the MaxLike falls in between, with the average Kappa coefficient of about 88.6%. As shown in the schedule, we use the detailed statistics of the confusion matrix to describe the classification accuracy of the three algorithms in different experimental areas. Among them, Tables A1–A15 shows the detailed classification accuracy of the three algorithms in different experimental areas.

**Table 5.** The statistics of accuracy of three algorithms in different experimental areas.

| Classification Algorithm | Beijing (1479 × 1550) | Guangzhou (2973 × 3495) | Suzhou (2351 × 2644) | Wuhan_1 (2245 × 2521) | Wuhan_2 (2894 × 3396) |
|---|---|---|---|---|---|
| | Kappa (%) | Kappa (%) | Kappa (%) | Kappa (%) | Kappa (%) |
| AUWEM | 91.6924 | 95.5355 | 87.8783 | 96.3811 | 93.7445 |
| NDWI | 83.0431 | 85.2771 | 78.8652 | 84.6675 | 90.2501 |
| MaxLike | 84.3326 | 91.7285 | 85.6260 | 91.8418 | 90.7601 |

Figure 9 shows the histogram of water classification accuracy of three different algorithms in the five experimental areas. From the histogram, we can find that the water extraction classification accuracy of AUWEM algorithm is higher than that of NDWI and MaxLike. The commission error of AUWEM is below 5% in most experimental area except in Suzhou (9.5%). The omission error rate of AUWEM is significantly lower than that of NDWI and MaxLike in all the five areas. When both the commission and omission error rates are low, the total error rate will be minimal. From the histogram, we can find that the proposed algorithm exhibits the lowest total error rate, followed by the MaxLike and NDWI. The approximate average total error rates of the three algorithms are about 11.9%, 18.2% and 22.1%, respectively.

In terms of the water classification producer accuracy, the AUWEM algorithm ranks first with the average accuracy of about 91.6%, followed by MaxLike with an average of about 84.8% and NDWI with an average of about 82.9%. In terms of the user accuracy, MaxLike ranks first with the average accuracy of about 96.6%, followed by the proposed algorithm with an average of about 96.4% and NDWI with an average of about 91.2%.

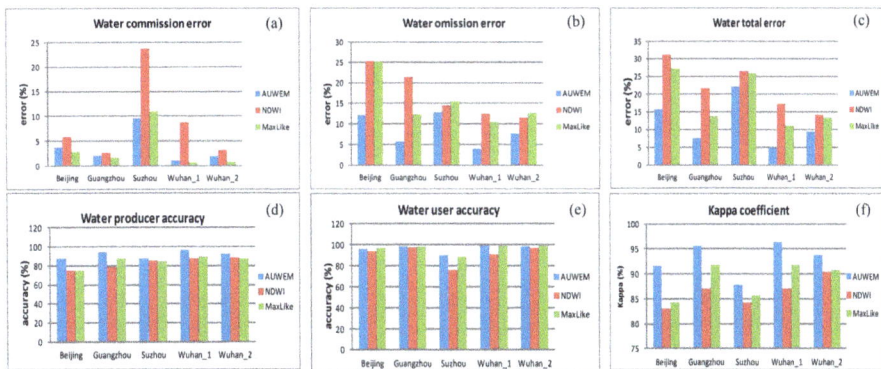

**Figure 9.** A comparison of classification accuracy among different algorithms in five experimental areas. (**a**) water commission error; (**b**) Water omission error; (**c**) Water total error; (**d**) Water producer accuracy; (**e**) Water user accuracy; (**f**) Kappa coefficient.

### 4.3. An Analysis of Water-Edge Pixel Extraction Accuracy

In order to evaluate the edge detection accuracy of the three algorithms more objectively, we design the algorithm below. The steps are as follows:

1. Use the reference image to acquire the water edge by applying the Canny operator.
2. Apply the morphological dilation to the acquired edge to establish a buffer zone centered around the edge with a radius of four pixels.

3. Determine the pixels in the buffer zone. Suppose that the total number of pixels in the buffer zone is $N$, the number of correctly classified pixels is $N_R$, the number of omitted pixels is $N_O$, and the number of commission error is $N_c$, then:

$$A = \frac{N_R}{N} \times 100\% \tag{12}$$

$$E_O = \frac{N_O}{N} \times 100\% \tag{13}$$

$$E_C = \frac{N_C}{N} \times 100\% \tag{14}$$

where $A + E_o + E_c = 100\%$. $A$ indicates the proportion of correctly classified edge pixels (accuracy of edge detection), $E_o$ indicates the proportion of omitted edge pixels (omission error), and $E_c$ is the proportion of commissioned edge pixels (commission error). The edge detection results generated by the approach indicate a comparative rather than absolute conclusion. After all, the reference imageries we use are manually obtained so there will be limitations in visual observations and statistical results are an approximate reflection of the algorithms' edge extraction accuracy. The process of obtaining the algorithm to acquire the water edge area for evaluation is shown below in Figure 10.

Table 6 showed the statistics about the water edge detection accuracy of above methods in the experimental areas. The statistics include the commission error, omission error and the accuracy of edge detection. Comparison in Figure 11 clearly shows that the edge detection accuracy of the AUWEM algorithm exceeds that of NDWI and MaxLike. The maximum and minimum rates of correct classification of water edge pixels by AUWEM algorithm are 93.7691% (shown in Guangzhou) and 79.5798% (shown in Wuhan_2); the maximum and minimum correct rates of NDWI are 84.0917% (shown in Suzhou) and 69.8310% (shown in Beijing); the maximum and minimum correct rates of MaxLike are 85.8149% (shown in Guangzhou) and 69.7974% (shown in Wuhan_2).

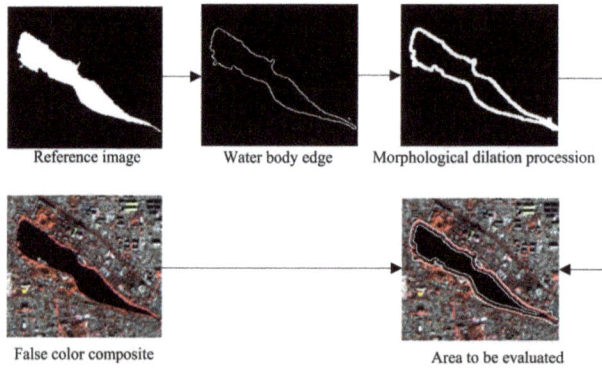

**Figure 10.** Process of acquiring water edge area for evaluation. Edge of the reference images are extracted and processed by morphological dilation to acquire the water edge for evaluation.

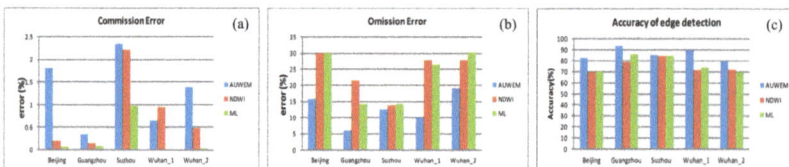

**Figure 11.** Comparison of water edge detection accuracy among different algorithms in five experimental areas. (**a**) Commission Error; (**b**) Omission Error; (**c**) Accuracy of edge detection.

**Table 6.** Statistics about water edge detection accuracy of different algorithms in five experimental areas.

| Site | Method | Commission Error (%) | Omission Error (%) | A (%) |
|------|--------|----------------------|--------------------|-------|
| | AUWEM | 1.8032 | 15.8446 | 82.3522 |
| Beijing | NDWI | 0.2042 | 29.9648 | 69.8310 |
| | MaxLike | 0.0738 | 29.9747 | 69.9515 |
| | AUWEM | 0.3417 | 5.8892 | 93.7691 |
| Guangzhou | NDWI | 0.1438 | 21.4114 | 78.4448 |
| | MaxLike | 0.0833 | 14.1019 | 85.8149 |
| | AUWEM | 2.3455 | 12.5791 | 85.0755 |
| Suzhou | NDWI | 2.2140 | 13.6943 | 84.0917 |
| | MaxLike | 0.9649 | 14.2155 | 84.8196 |
| | AUWEM | 0.6422 | 9.8925 | 89.4653 |
| Wuhan_1 | NDWI | 0.9452 | 27.8494 | 71.2054 |
| | MaxLike | 0.0211 | 26.3919 | 73.5870 |
| | AUWEM | 1.3827 | 19.0375 | 79.5798 |
| Wuhan_2 | NDWI | 0.4743 | 27.8402 | 71.6855 |
| | MaxLike | 0.0335 | 30.1691 | 69.7974 |

## 5. Discussion

### 5.1. Effect of PCA Transformation

By replacing Green in Equation (1) with the first principal component of PCA transformation, we obtain the improved NNDWI2. The NNDWI2 computational result has good resistance to mixed spectral interference, especially when the water bodies are eutrophicated or surrounded by dense vegetation. The pixels of those water bodies exhibit the spectral information of non-water because they are affected and interfered by the spectral information of vegetation like algae, thus their detection will be severely disturbed. According to the classification results shown in Figure 12, the pixels of the water bodies whose spectral information is interfered can be effectively classified in threshold segmentation results of NNDWI2. The number of misclassified pixels generated using the algorithm is less than that generated using NDWI and MaxLike. In addition, the water edge pixels in the images are effectively classified, thus the overall water extraction accuracy is enhanced.

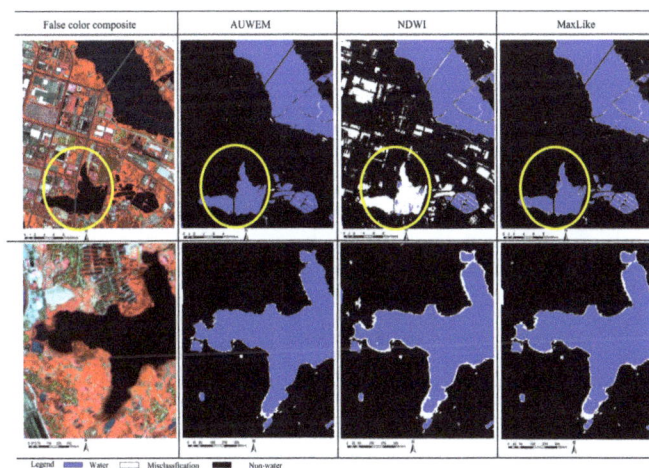

**Figure 12.** Comparison of results of classifying pixels of spectrally contaminated water bodies among different algorithms. The yellow circle indicates an area clearly undetected by the NDWI algorithm. The water body in this area is eutrophicated with a lot of algal vegetation that affects its spectral information, making it hard for the NDWI to detect.

## 5.2. Effect of Intersection

In Section 3.4.1, we set the constraint on dilation results by intersecting the dilated images and those under threshold segmentation on Band4. However, how many pixels are in the result of the segmentation prior to intersection, and how many pixels are there after the intersection? We choose the following four urban areas for the experiment. The results are shown in the Figure 13. The value changes of water body/shadow pixels before and after computing the intersection are shown in the Table 7. The statistics show that the number of pixels increases after the computation in Figure 13a–c where there are many shadows. After zooming in Figure 13a, we find that after the computation, the building shadows correctly represent the shaded areas. However, the number of pixels is reduced in Figure 13d after the computation, indicating that the computation can result in the removal of error detections generated by the NNDWI algorithm. It can be explained by the experimental results in Figure 14.

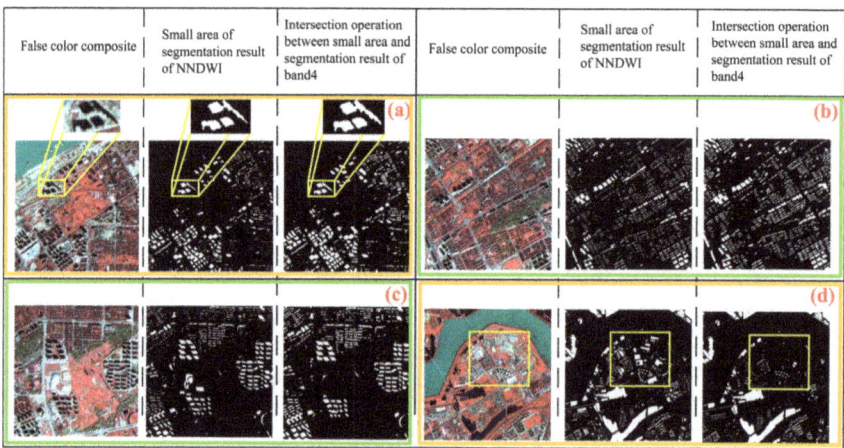

**Figure 13.** Intersection operation results. (**a**) First experimental results; (**b**) Second experimental results; (**c**) Third experimental results; (**d**) Fourth experimental results.

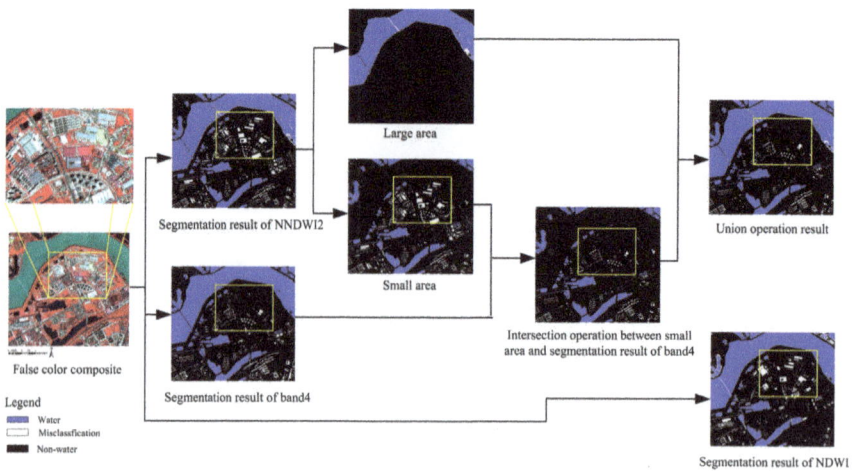

**Figure 14.** Comparison between intersection result and NDWI result.

**Table 7.** Statistics shows the changes of the number of water body/shadow pixels before and after the computation of intersection. Nb represents the number of pixels of water/shadow before the intersection, and Na represents the number of pixels of water/shadow after the intersection.

| Image Name | Image Size | Nb | Na | Nb-Na |
|:---:|:---:|:---:|:---:|:---:|
| a | 361 × 361 | 11,883 | 15,888 | 4005 |
| b | 327 × 335 | 12,336 | 17,630 | 5294 |
| c | 299 × 319 | 9923 | 12,218 | 2295 |
| d | 677 × 762 | 76,932 | 57,389 | −19,543 |

As shown in Figure 14, the ground objects in the yellow rectangle are misclassified as water by both NNDWI and NDWI. In fact, these objects are the roof surface of buildings. On the other hand, the objects in this area can be correctly classified by using threshold segmentation result on Band4. After morphological dilation of small-area objects, intersecting it with the images under threshold segmentation on Band4 enables the correction of pixel classification in this area, thus the subsequent classification accuracy of the water bodies will be enhanced.

### 5.3. Shadow Detection Ability of the Shadow Object Description Method

Since the shadow detection algorithm model is established on the premise of extracting water and shadow, we cannot guarantee a sound result by solely relying on it, as shown in Figure 15. The spectral features of the shadows are similar to those of such ground objects as cement surface, soil, vegetation, etc. In our study, we find that the spectral features of such ground objects are presented in the shaded areas. Therefore, it is not ideal to solely use the spectral relation model to detect shadows. Otherwise, almost all of the objects other than water bodies will be detected as shadows. In that case, the imagery is classified into water and non-water areas. However, when zooming in, we find that the water edge detection accuracy is poor; the pixels in water edges cannot be detected properly. On the other hand, when the shadow detection model is used in the NNDWI extraction results, the effect is quite satisfactory, as shown in Figure 16.

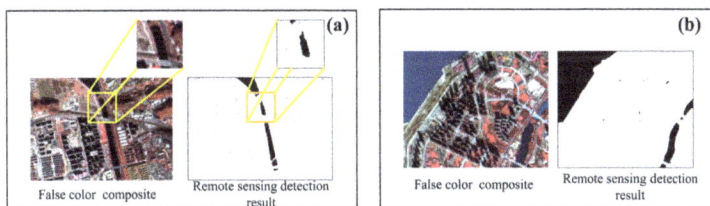

**Figure 15.** Shadow detection results generated when solely applying the model. (**a**) First experimental results; (**b**) Second experimental results.

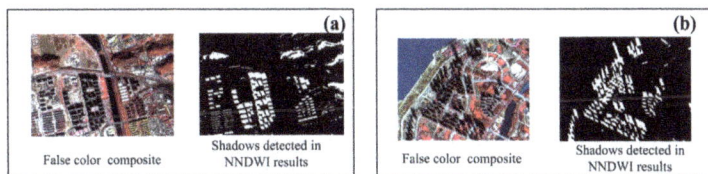

**Figure 16.** The shadow detection results generated when combining the model with the NNDWI extraction results. (**a**) First experimental results; (**b**) Second experimental results.

From the experiment described above, we can conclude that a combination of the model and the NNDWI extraction results will enable us to effectively detect shadows. When applied solely, the model is not competent in detecting shadows, resulting in misclassification.

*5.4. Threshold Setting and Stability of Algorithm in Correlation Computation*

Although there are many problems concerning threshold setting in AUWEM algorithm, it is necessary to set three thresholds, namely NNDWI1, NNDWI2 and Band4 segmentation thresholds. The optimized segmentation threshold value of near-infrared (Band4) is obtained by gray histogram. Before image histogram statistics, we use the Equation (15) to normalize the segmented image pixel value into the range of (0~255). The standardized expression is shown as follows:

$$y = 255 \times \frac{(x - x_{min})}{(x_{max} - x_{min})} \tag{15}$$

where $y$ indicates the standardized value, $x$ indicates all of the pixel values that need to be processed on Band4, $x_{min}$ indicates the minimum value on Band4, and $x_{max}$ indicates the maximum value on Band4.

NNDWI1 is more sensitive to the turbid water. When the threshold value is set to 0, the turbid water will be effectively extracted. As for NNDWI2 threshold setting, we can analyze and discuss in detail the following figures. Figure 17a is the false color image for experimental analysis, and Figure 17b shows the pixel value of the first principal component after the PCA transformation on the four bands of the image. It can be seen from the figure that the pixel values of water areas are below 0 in the first principal component (the maximum pixel value is $-176.333$), while the pixel values of non-water areas are above 0 (the minimum pixel value is 39.8416); in the NNDWI2 calculation results, as shown in Figure 17c, the pixel values of water areas are above 0 (the minimum and maximum pixel values are 2.65607 and 25.17149, respectively), while the pixel values of non-water areas are below 0 (the minimum and maximum pixel values are $-6.90065$ and $-0.44693$, respectively). The difference between the minimum value of water areas and the maximum value of non-water areas is 3.103 (in some parts of the image, the actual difference is even greater). Therefore, the optimal segmentation threshold of the images after the computation of NNDWI2 can be set to 0. This is also verified by other experiments, and zero can be used as the best segmentation threshold of NNWI2 index image.

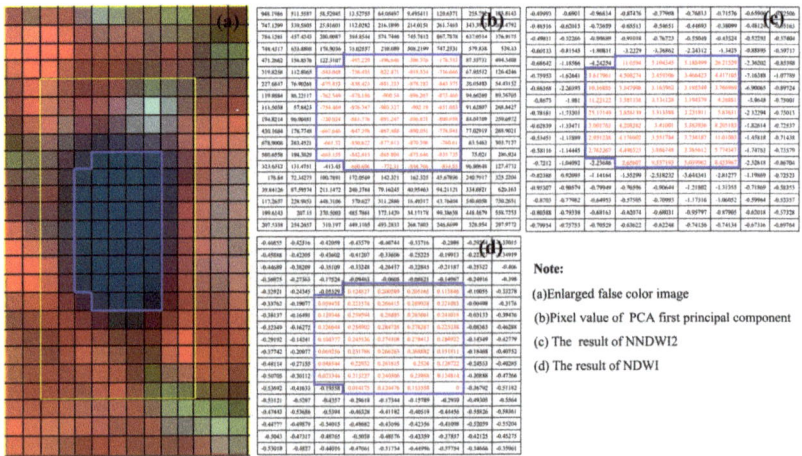

Note:

(a) Enlarged false color image

(b) Pixel value of PCA first principal component

(c) The result of NNDWI2

(d) The result of NDWI

**Figure 17.** The different index of pixels after the PCA transformation, NNDWI2 and NDWI, respectively.

In Figures 18–21, we compare the water extraction accuracy among algorithms when the threshold changes. The statistical results show that AUWEM algorithm will not have an obvious impact on classification accuracy when the threshold is within the range of T $\pm$ $\triangle$T. (T is the selected or optimal threshold. In Figures 18–20, $\triangle$T = 0.05, and in Figure 17, $\triangle$T = 3.) On the other hand, NDWI's accuracy is greatly affected when the threshold changes. By analyzing the accuracy data of NDWI in Figure 18, we can find that the water extraction accuracy changes drastically when the threshold changes, the variance are 0.4639 (Beijing), 0.7902 (Guangzhou), 1.0588 (Suzhou), 0.2651 (Wuhan_1) and 0.4749 (Wuhan_2). Thus, the changes in threshold affect NDWI's accuracy (especially in Guangzhou and Suzhou). It shows that the algorithm is unstable. In Figures 19 and 20, we find that when the threshold changes, the accuracy of NNDWI1 and NNDWI2 is almost unchanged. In Figure 21, we find that the accuracy on Band4 is to some extent influenced by the changes in threshold, but such influence is minimal, and the mean square deviation of the accuracy in the experimental areas corroborates with the observation (variance are 0.0433 (Beijing), 0.0056 (Guangzhou), 0.0013 (Suzhou), 0.0011 (Wuhan_1) and 0.0066 (Wuhan_2)). In summary of the statistical analysis of Figures 18–21, we can conclude that when the thresholds change, the water extraction accuracy of AUWEM algorithm is more stable than that of NDWI. Even though three threshold values need to be set, the setting is quite simple, so there is no need to consider too many influencing factors.

When the threshold changes, the water extraction accuracy on Band4 is to some extent influenced. Through the experimental analysis, we find that it is mainly caused by the way we compute intersection in Section 3.4.1. The computation results in constraints on the dilation. The related analysis is shown in Figure 22. From the figure, we can find that different threshold segmentation results cover different areas, but the area variation is very small, within the range of $T \pm \triangle T$. The threshold will not impact on the water in terms of covering area and detection, thus barely affecting the detection accuracy.

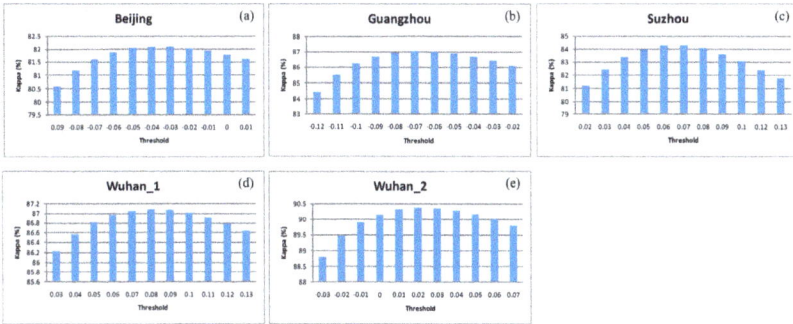

**Figure 18.** A comparison among changes of NDWI's water extraction accuracy when the threshold changes. (**a**) Water extraction accuracy of Beijing; (**b**) Water extraction accuracy of Guangzhou; (**c**) Water extraction accuracy of Suzhou; (**d**) Water extraction accuracy of Wuhan_1; (**e**) Water extraction accuracy of Wuhan_2.

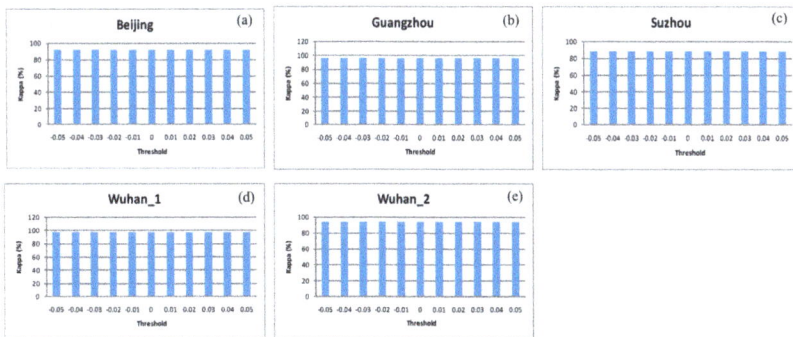

**Figure 19.** The changes in NNDWI1 when the threshold changes and Band4 and NNDWI2 remain unchanged. (**a**) Water extraction accuracy of Beijing; (**b**) Water extraction accuracy of Guangzhou; (**c**) Water extraction accuracy of Suzhou; (**d**) Water extraction accuracy of Wuhan_1; (**e**) Water extraction accuracy of Wuhan_2.

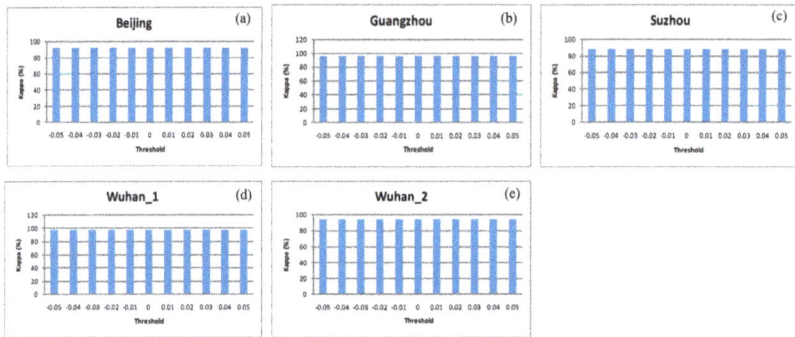

**Figure 20.** The changes in NNDWI2 when the threshold changes and Band4 and NNDWI1 remain unchanged. (**a**) Water extraction accuracy of Beijing; (**b**) Water extraction accuracy of Guangzhou; (**c**) Water extraction accuracy of Suzhou; (**d**) Water extraction accuracy of Wuhan_1; (**e**) Water extraction accuracy of Wuhan_2.

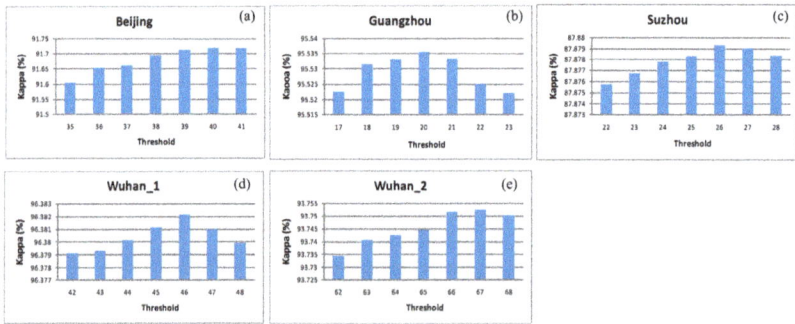

**Figure 21.** The changes in Band4 when the threshold changes and NNDWI1 and NNDWI2 remain unchanged. (**a**) Water extraction accuracy of Beijing; (**b**) Water extraction accuracy of Guangzhou; (**c**) Water extraction accuracy of Suzhou; (**d**) Water extraction accuracy of Wuhan_1; (**e**) Water extraction accuracy of Wuhan_2.

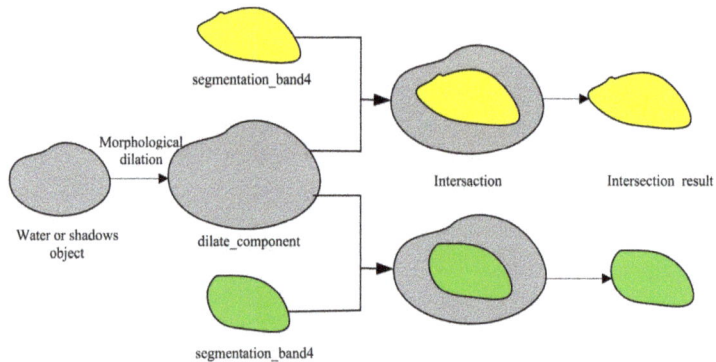

**Figure 22.** Comparison of intersection on Band4 under different thresholds as constraint. The yellow segmentation area is larger than the green one, so according to the results in the figure, after the intersection there will be water body or shadow objects that cover different areas and are to be detected.

## 5.5. Summary

Although results are quite satisfactory in different experimental areas, some issues remain to be considered, such as seasons, the sun's height angle, components of the atmosphere, and the chemical composition of water bodies. All of these factors have an impact on the reflection features. Different atmospheric correction for subsequent image segmentation threshold may be different, thus affecting the subsequent water detection accuracy; the same atmospheric correction method will exhibit different atmospheric correction accuracy under different weather conditions, especially when there is heavy haze. Heavy haze has been a serious issue in Chinese urban areas during wintertime in recent years. The current atmospheric correction model may not necessarily work well when correcting atmospheric haze. In some areas of the imagery, shadows and water bodies are adjacent. If water body area is large enough, the whole area will be classified as water. Our algorithm is proposed for ZY-3 image data, so whether it has a wider applicability or not needs to be validated by image data from other sources and in different areas. These issues are worth of our follow-up study and verification.

## 6. Conclusions

We propose a new method for urban water extraction from high-resolution remote sensing images. In order to improve the accuracy of water extraction, we improve the NDWI algorithm and propose two new water indices, namely the NNDWI1 which is sensitive to turbid water, andNNDWI2 which is sensitive to water bodies whose spectral information is interfered by that of vegetation. We superimpose NNDWI1 and NNDWI2 image segmentation results, and then use Object-Oriented Technology to detect and remove shadows in the small areas, in order to obtain the final results of urban water extraction. Our experiments test the accuracy of algorithms in five urban areas. According to the results, the AUWEM algorithm has greater water extraction accuracy compared with NDWI and the MaxLike, with an average Kappa coefficient of 93% and an average total error rate of about 11.9%. In contrast, the average Kappa coefficient and error rate of the MaxLike are about 88.6% and 18.2%, respectively; the average Kappa coefficient and error rate of NDWI is about 86.2% and 22.1%, respectively. In addition, AUWEM exhibits greater accuracy when detecting water edge and small rivers. It can effectively distinguish shadows of high buildings from water bodies to improve the overall accuracy. More importantly, AUWEM has more stable detection accuracy than NDWI has when the threshold changes. It can also be applicable for other water features extraction, and can be applied to monitor and study the changes in water bodies in other places.

**Acknowledgments:** The authors wish to thank the editors, reviewers, and Yueming Peng's help. This work was supported by outstanding postgraduate development schemes of School of Geomatics, Liaoning Technical University (YS201503), Key Laboratory Fund Item of Liaoning Provincial Department of Education (LJZS001), Scientific research project of Liaoning Provincial Department of Education (LJYB023) and the National key research and development program of China (2016YFB0501403).

**Author Contributions:** Jianhua Guo was responsible for the research design, experiment and analysis, and drafting of the manuscript. Fan Yang reviewed the manuscript and was responsible for the analysis of data and main technical guidance. Hai Tan and Jingxue Wang were responsible for drafting and revising the manuscript and took part in the discussion of experiment design. All authors read and approved the manuscript.

**Conflicts of Interest:** The authors declare no conflict of interest.

## Appendix A

**Table A1.** Statistic results of image water extraction based on maximum likelihood method in Beijing area.

| Class | Ground Truth (Pixels) | | | | |
|---|---|---|---|---|---|
| | Water | No_Water | Total | Produc Accuracy (%) | Omission Error (%) |
| Water | 34,961 | 11,657 | 46,618 | 74.9946 | 25.0054 |
| No_water | 1061 | 2,244,771 | 2,245,832 | 99.9528 | 0.0472 |
| Total | 36,022 | 2,256,428 | 2,292,450 | | |
| User Accuracy (%) | 97.0546 | 99.4834 | | | |
| Commission Error (%) | 2.9454 | 0.5166 | | | |
| Overall Accuracy = 99.4452%; Kappa Coefficient = 84.3326% | | | | | |

**Table A2.** Statistic results of image water extraction based on the NDWI index in Beijing.

| Class | Water | No_Water | Total | Produc Accuracy (%) | Omission Error (%) |
|---|---|---|---|---|---|
| | | | Ground Truth (Pixels) | | |
| Water | 34,827 | 11,791 | 46,618 | 74.7072 | 25.2928 |
| No_water | 2125 | 2,243,707 | 2,245,832 | 99.9054 | 0.0946 |
| Total | 36,952 | 2,255,498 | 2,292,450 | | |
| User Accuracy (%) | 94.2493 | 99.4772 | | | |
| Commission Error (%) | 5.7507 | 0.5228 | | | |
| | | Overall Accuracy = 99.3930%; Kappa Coefficient = 83.0431% | | | |

**Table A3.** Statistic results of image water extraction based on AUWEM in Beijing.

| Class | Water | No_Water | Total | Produc Accuracy (%) | Omission Error (%) |
|---|---|---|---|---|---|
| | | | Ground Truth (Pixels) | | |
| Water | 40,929 | 5689 | 46,618 | 87.7966 | 12.2034 |
| No_water | 1571 | 2,244,261 | 2,245,832 | 99.9300 | 0.0700 |
| Total | 42,500 | 2,249,950 | 2,292,450 | | |
| User Accuracy (%) | 96.3035 | 99.7471 | | | |
| Commission Error (%) | 3.6965 | 0.2529 | | | |
| | | Overall accuracy = 99.6833%; Kappa Coefficient = 91.6924% | | | |

## Appendix B

**Table A4.** Statistic results of image water extraction based on maximum likelihood method in Guangzhou.

| Class | Water | No_Water | Total | Produc Accuracy (%) | Omission Error (%) |
|---|---|---|---|---|---|
| | | | Ground Truth (Pixels) | | |
| Water | 1,212,617 | 169,976 | 1,382,593 | 87.7060 | 12.2940 |
| No_water | 19,157 | 8,988,885 | 9,008,042 | 99.7873 | 0.2127 |
| Total | 1,231,774 | 9,158,861 | 10,390,635 | | |
| User Accuracy (%) | 98.4448 | 98.1441 | | | |
| Commission Error (%) | 1.5552 | 1.8559 | | | |
| | | Overall accuracy = 98.1798%; Kappa Coefficient = 91.7285% | | | |

**Table A5.** Statistic results of image water extraction based on the NDWI index in Guangzhou.

| Class | Water | No_Water | Total | Produc Accuracy (%) | Omission Error (%) |
|---|---|---|---|---|---|
| | | | Ground Truth (Pixels) | | |
| Water | 1,087,494 | 295,099 | 1,382,593 | 78.6561 | 21.3439 |
| No_water | 29,105 | 8,978,937 | 9,008,042 | 99.6769 | 0.3231 |
| Total | 1,116,599 | 9,274,036 | 10,390,635 | | |
| User Accuracy (%) | 97.3934 | 96.8180 | | | |
| Commission Error (%) | 2.6066 | 3.1820 | | | |
| | | Overall accuracy = 96.8798%; Kappa Coefficient = 85.2771% | | | |

**Table A6.** Statistic results of image water extraction based on AUWEM in Guangzhou.

| Class | Water | No_Water | Total | Produc Accuracy (%) | Omission Error (%) |
|---|---|---|---|---|---|
| | | | Ground Truth (Pixels) | | |
| Water | 1,304,001 | 78,592 | 1,382,593 | 94.3156 | 5.6844 |
| No_water | 26,733 | 8,981,309 | 9,008,042 | 99.7032 | 0.2968 |
| Total | 1,330,734 | 9,059,901 | 10,390,635 | | |
| User Accuracy (%) | 97.9911 | 99.1325 | | | |
| Commission Error (%) | 2.0089 | 0.8675 | | | |
| | | Overall accuracy = 98.9863%; Kappa Coefficient = 95.5355% | | | |

## Appendix C

**Table A7.** Statistic results of image water extraction based on maximum likelihood method in Suzhou.

| Class | Ground Truth (Pixels) | | | | |
| --- | --- | --- | --- | --- | --- |
| | Water | No_Water | Total | Produc Accuracy (%) | Omission Error (%) |
| Water | 415,717 | 76,225 | 491,942 | 84.5053 | 15.4947 |
| No_water | 50,948 | 5,673,154 | 5,724,102 | 99.1099 | 0.8901 |
| Total | 466,665 | 5,749,379 | 6,216,044 | | |
| User Accuracy (%) | 89.0825 | 98.6742 | | | |
| Commission Error (%) | 10.9175 | 1.3258 | | | |
| | | Overall accuracy = 97.9541%; Kappa Coefficient = 85.6260% | | | |

**Table A8.** Statistic results of image water extraction based on the NDWI index in Suzhou.

| Class | Ground Truth (Pixels) | | | | |
| --- | --- | --- | --- | --- | --- |
| | Water | No_Water | Total | Produc Accuracy (%) | Omission Error (%) |
| Water | 420,726 | 71,216 | 491,942 | 85.5235 | 14.4765 |
| No_water | 130,884 | 5,593,218 | 5,724,102 | 97.7135 | 2.2865 |
| Total | 551,610 | 5,664,434 | 6,216,044 | | |
| User Accuracy (%) | 76.2724 | 98.7428 | | | |
| Commission Error (%) | 23.7276 | 1.2572 | | | |
| | | Overall accuracy = 96.7487%; Kappa Coefficient = 78.8652% | | | |

**Table A9.** Statistic results of image water extraction based on AUWEM in Suzhou.

| Class | Ground Truth (Pixels) | | | | |
| --- | --- | --- | --- | --- | --- |
| | Water | No_Water | Total | Produc Accuracy (%) | Omission Error (%) |
| Water | 429,101 | 62,841 | 491,942 | 87.2259 | 12.7741 |
| No_water | 45,182 | 5,678,920 | 5,724,102 | 99.2107 | 0.7893 |
| Total | 474,283 | 5,741,761 | 6,216,044 | | |
| User Accuracy (%) | 90.4736 | 98.9055 | | | |
| Commission Error (%) | 9.5264 | 1.0945 | | | |
| | | Overall accuracy = 98.2622%Kappa Coefficient = 87.8783% | | | |

## Appendix D

**Table A10.** Statistic results of image water extraction based on maximum likelihood method in Wuhan_2.

| Class | Ground Truth (Pixels) | | | | |
| --- | --- | --- | --- | --- | --- |
| | Water | No_Water | Total | Produc Accuracy (%) | Omission Error (%) |
| Water | 1,562,974 | 182,267 | 1,745,241 | 89.5563 | 10.4437 |
| No_water | 9274 | 3,905,130 | 3,914,404 | 99.7631 | 0.2369 |
| Total | 1,572,248 | 4,087,397 | 5,659,645 | | |
| User Accuracy (%) | 99.4101 | 95.5408 | | | |
| Commission Error (%) | 0.5899 | 4.4592 | | | |
| | | Overall accuracy = 96.6157%; Kappa Coefficient = 91.8418% | | | |

**Table A11.** Statistic results of image water extraction based on the NDWI index in Wuhan_2.

| Class | Ground Truth (Pixels) | | | | |
| --- | --- | --- | --- | --- | --- |
| | Water | No_Water | Total | Produc Accuracy (%) | Omission Error (%) |
| Water | 1,526,202 | 219,039 | 1,745,241 | 87.4494 | 12.5506 |
| No_water | 146,867 | 3,767,537 | 3,914,404 | 96.2480 | 5.4944 |
| Total | 1,673,069 | 3,986,576 | 5,659,645 | | |
| User Accuracy (%) | 91.2217 | 94.5056 | | | |
| Commission Error (%) | 8.7783 | 3.7520 | | | |
| | | Overall accuracy = 93.5348%; Kappa Coefficient = 84.6675% | | | |

**Table A12.** Statistic results of image water extraction based on AUWEM in Wuhan_2.

| Class | Ground Truth (Pixels) | | | | |
|---|---|---|---|---|---|
| | Water | No_Water | Total | Produc Accuracy (%) | Omission Error (%) |
| Water | 1,676,387 | 68,854 | 1,745,241 | 96.0548 | 3.9452 |
| No_water | 17,803 | 3,896,601 | 3,914,404 | 99.5452 | 0.4548 |
| Total | 1,694,190 | 3,965,455 | 5,659,645 | | |
| User Accuracy (%) | 98.9492 | 98.2637 | | | |
| Commission Error (%) | 1.0508 | 1.7363 | | | |
| | Overall accuracy = 98.4689%; Kappa Coefficient = 96.3811% | | | | |

## Appendix E

**Table A13.** Statistic results of image water extraction based on maximum likelihood method in Wuhan_3.

| Class | Ground Truth (Pixels) | | | | |
|---|---|---|---|---|---|
| | Water | No_Water | Total | Produc Accuracy (%) | Omission Error (%) |
| Water | 2,084,870 | 303,303 | 2,388,173 | 87.2998 | 12.7002 |
| No_water | 17,198 | 7,422,653 | 7,439,851 | 99.7688 | 0.2312 |
| Total | 2,102,068 | 7,725,956 | 9,828,024 | | |
| User Accuracy (%) | 99.1819 | 96.0742 | | | |
| Commission Error (%) | 0.7201 | 3.9258 | | | |
| | Overall accuracy = 96.7389%; Kappa Coefficient = 90.7601% | | | | |

**Table A14.** Statistic results of image water extraction based on the NDWI index inWuhan_3.

| Class | Ground Truth (Pixels) | | | | |
|---|---|---|---|---|---|
| | Water | No_Water | Total | Produc Accuracy (%) | Omission Error (%) |
| Water | 2,114,412 | 273,761 | 2,388,173 | 88.5368 | 11.4632 |
| No_water | 68,478 | 7,371,373 | 7,439,851 | 99.0796 | 0.9204 |
| Total | 2,182,890 | 7,645,134 | 9,828,024 | | |
| User Accuracy (%) | 96.8630 | 96.4191 | | | |
| Commission Error (%) | 3.1370 | 3.5809 | | | |
| | Overall accuracy = 96.5177%; Kappa Coefficient = 90.2501% | | | | |

**Table A15.** Statistic results of image water extraction based on AUWEM in Wuhan_3.

| Class | Ground Truth (Pixels) | | | | |
|---|---|---|---|---|---|
| | Water | No_Water | Total | Produc Accuracy (%) | Omission Error (%) |
| Water | 2,207,784 | 180,389 | 2,388,173 | 92.4466 | 7.5534 |
| No_water | 41,320 | 7,398,531 | 7,439,851 | 99.4446 | 0.5554 |
| Total | 2,249,104 | 7,578,920 | 9,828,024 | | |
| User Accuracy (%) | 98.1628 | 97.6199 | | | |
| Commission Error (%) | 1.8372 | 2.3801 | | | |
| | Overall accuracy = 97.7441%; Kappa Coefficient = 93.7445% | | | | |

## References

1. Stabler, L.B. Management regimes affect woody plant productivity and water use efficiency in an urbandesert ecosystem. *Urban Ecosyst.* **2008**, *11*, 197–211. [CrossRef]
2. Mcfeeters, S.K. Using the Normalized Difference Water Index (NDWI) within a Geographic Information System to Detect Swimming Pools for Mosquito Abatement: A Practical Approach. *Remote Sens.* **2013**, *5*, 3544–3561. [CrossRef]
3. Zhai, W.; Huang, C. Fast building damage mapping using a single post-earthquake PolSAR image: A case study of the 2010 Yushu earthquake. *Earth Planets Space* **2016**, *68*, 1–12. [CrossRef]
4. Raju, P.L.N.; Sarma, K.K.; Barman, D.; Handique, B.K.; Chutia, D.; Kundu, S.S.; Das, R.; Chakraborty, K.; Das, R.; Goswami, J.; et al. Operational remote sensing services in north eastern region of India for natural resources management, early warning for disaster risk reduction and dissemination of information and services. *ISPRS—Int. Arch. Photogramm. Remote Sens. Spat. Inform. Sci.* **2016**, *XLI-B4*, 767–775. [CrossRef]
5. Muriithi, F.K. Land use and land cover (LULC) changes in semi-arid sub-watersheds of Laikipia and Athi River basins, Kenya, as influenced by expanding intensive commercial horticulture. *Remote Sens. Appl. Soc. Environ.* **2016**, *3*, 73–88. [CrossRef]

6.  Byun, Y.; Han, Y.; Chae, T. Image fusion-based change detection for flood extent extraction using bi-temporal very high-resolution satellite images. *Remote Sens.* **2015**, *7*, 10347–10363. [CrossRef]

7.  Kang, L.; Zhang, S.; Ding, Y.; He, X. Extraction and preference ordering of multireservoir water supply rules in dry years. *Water* **2016**, *8*, 28. [CrossRef]

8.  Katz, D. Undermining demand management with supply management: Moral hazard in Israeli water policies. *Water* **2016**, *8*, 159. [CrossRef]

9.  Deus, D.; Gloaguen, R. Remote sensing analysis of lake dynamics in semi-arid regions: Implication for water resource management. Lake Manyara, east African Rift, northern Tanzania. *Water* **2013**, *5*, 698–727. [CrossRef]

10. Zhai, K.; Xiaoqing, W.U.; Qin, Y.; Du, P. Comparison of surface water extraction performances of different classic water indices using OLI and TM imageries in different situations. *Geo-Spat. Inform. Sci.* **2015**, *18*, 32–42. [CrossRef]

11. Gautam, V.K.; Gaurav, P.K.; Murugan, P.; Annadurai, M. Assessment of surface water dynamics in Bangalore using WRI, NDWI, MNDWI, supervised classification and K-T transformation. *Aquat. Procedia* **2015**, *4*, 739–746. [CrossRef]

12. Bryant, R.G.; Rainey, M.P. Investigation of flood inundation on playas within the Zone of Chotts, using a time-series of AVHRR. *Remote Sens. Environ.* **2002**, *82*, 360–375. [CrossRef]

13. Sun, F.; Sun, W.; Chen, J.; Gong, P. Comparison and improvement of methods for identifying water bodies in remotely sensed imagery. *Int. J. Remote Sens.* **2012**, *33*, 6854–6875. [CrossRef]

14. Mcfeeters, S.K. The use of the Normalized Difference Water Index (NDWI) in the delineation of open water features. *Int. J. Remote Sens.* **1996**, *17*, 1425–1432. [CrossRef]

15. Xu, H. Modification of normalized difference water index (NDWI) to enhance open water features in remotely sensed imagery. *Int. J. Remote Sens.* **2006**, *27*, 3025–3033. [CrossRef]

16. Feyisa, G.L.; Meilby, H.; Fensholt, R.; Proud, S.R. Automated Water Extraction Index: A new technique for surface water mapping using Landsat imagery. *Remote Sens. Environ.* **2014**, *140*, 23–35. [CrossRef]

17. Rogers, A.S.; Kearney, M.S. Reducing signature variability in unmixing coastal marsh Thematic Mapper scenes using spectral indices. *Int. J. Remote Sens.* **2004**, *25*, 2317–2335. [CrossRef]

18. Lira, J. Segmentation and morphology of open water bodies from multispectral images. *Int. J. Remote Sens.* **2006**, *27*, 4015–4038. [CrossRef]

19. Lv, W.; Yu, Q.; Yu, W. Water Extraction in SAR Images Using GLCM and Support Vector Machine. In Proceedings of the IEEE 10th International Conference on Signal Processing Proceedings, Beijing, China, 24–28 October 2010.

20. Wendleder, A.; Breunig, M.; Martin, K.; Wessel, B.; Roth, A. Water body detection from TanDEM-X data: Concept and first evaluation of an accurate water indication mask. *Soc. Sci. Electron. Publ.* **2011**, *25*, 3779–3782.

21. Wendleder, A.; Wessel, B.; Roth, A.; Breunig, M.; Martin, K.; Wagenbrenner, S. TanDEM-X water indication mask: Generation and first evaluation results. *IEEE J. Sel. Top. Appl. Earth Obs. Remote Sens.* **2012**, *6*, 1–9. [CrossRef]

22. Wang, Y.; Ruan, R.; She, Y.; Yan, M. Extraction of water information based on RADARSAT SAR and Landsat ETM+. *Procedia Environ. Sci.* **2011**, *10*, 2301–2306. [CrossRef]

23. Wang, K.; Trinder, J.C. Applied Watershed Segmentation Algorithm for Water Body Extraction in Airborne SAR Image. In Proceedings of the European Conference on Synthetic Aperture Radar, Aachen, Germany, 2–4 June 2014.

24. Ahtonen, P.; Hallikainen, M. Automatic Detection of Water Bodies from Spaceborne SAR Images. In Proceedings of the 2005 IEEE International on Geoscience and Remote Sensing Symposium, IGARSS '05, Seoul, Korea, 25–29 July 2005.

25. Li, B.; Zhang, H.; Xu, F. Water Extraction in High Resolution Remote Sensing Image Based on Hierarchical Spectrum and Shape Features. In Proceedings of the 35th International Symposium on Remote Sensing of Environment (ISRSE35), Beijing, China, 22–26 April 2013.

26. Deng, Y.; Zhang, H.; Wang, C.; Liu, M. Object-Oriented Water Extraction of PolSAR Image Based on Target Decomposition. In Proceedings of the 2015 IEEE 5th Asia-Pacific Conference on Synthetic Aperture Radar (APSAR), Singapore, 1–4 September 2015.

27. Jiang, H.; Feng, M.; Zhu, Y.; Lu, N.; Huang, J.; Xiao, T. An automated method for extracting rivers and lakes from Landsat imagery. *Remote Sens.* **2014**, *6*, 5067–5089. [CrossRef]

28. Steele, M.K.; Heffernan, J.B. Morphological characteristics of urban water bodies: Mechanisms of change and implications for ecosystem function. *Ecol. Appl.* **2014**, *24*, 1070–1084. [CrossRef] [PubMed]

29. Yu, X.; Li, B.; Shao, J.; Zhou, J.; Duan, H. Land Cover Change Detection of Ezhou Huarong District Based on Multi-Temporal ZY-3 Satellite Images. In Proceedings of the 2015 IEEE 23rd International Conference on Geoinformatics, Wuhan, China, 19–21 June 2015.

30. Dong, Y.; Chen, W.; Chang, H.; Zhang, Y.; Feng, R.; Meng, L. Assessment of orthoimage and DEM derived from ZY-3 stereo image in Northeastern China. *Surv. Rev.* **2015**, *48*, 247–257. [CrossRef]

31. Yao, F.; Wang, C.; Dong, D.; Luo, J.; Shen, Z.; Yang, K. High-resolution mapping of urban surface water using ZY-3 multi-spectral imagery. *Remote Sens.* **2015**, *7*, 12336–12355. [CrossRef]

32. Su, N.; Zhang, Y.; Tian, S.; Yan, Y.; Miao, X. Shadow detection and removal for occluded object information recovery in urban high-resolution panchromatic satellite images. *IEEE J. Sel. Top. Appl. Earth Obs. Remote Sens.* **2016**, *9*, 1–15. [CrossRef]

33. Li, Y.; Gong, P.; Sasagawa, T. Integrated shadow removal based on photogrammetry and image analysis. *Int. J. Remote Sens.* **2005**, *26*, 3911–3929. [CrossRef]

34. Zhou, W.; Huang, G.; Troy, A.; Cadenasso, M.L. Object-based land cover classification of shaded areas in high spatial resolution imagery of urban areas: A comparison study. *Remote Sens. Environ.* **2009**, *113*, 1769–1777. [CrossRef]

35. Xie, J.; Zhang, L.; You, J.; Shiu, S. Effective texture classification by texton encoding induced statistical features. *Pattern Recognit.* **2014**, *48*, 447–457. [CrossRef]

36. Lizarazo, I. SVM-based segmentation and classification of remotely sensed data. *Int. J. Remote Sens.* **2008**, *29*, 7277–7283. [CrossRef]

37. Huang, X.; Zhang, L. Morphological building/shadow index for building extraction from high-resolution imagery over urban areas. *IEEE J. Sel. Top. Appl. Earth Obs. Remote Sens.* **2012**, *5*, 161–172. [CrossRef]

38. Xie, C.; Huang, X.; Zeng, W.; Fang, X. A novel water index for urban high-resolution eight-band WorldView-2 imagery. *Int. J. Digit. Earth* **2016**, *9*, 925–941. [CrossRef]

39. Sivanpilla, R.; Miller, S.N. Improvements in mapping water bodies using ASTER data. *Ecol. Inform.* **2010**, *5*, 73–78. [CrossRef]

40. Nazeer, M.; Nichol, J.E.; Yung, Y.K. Evaluation of atmospheric correction models and Landsat surface reflectance product in an urban coastal environment. *Int. J. Remote Sens.* **2014**, *35*, 6271–6291. [CrossRef]

41. Pechenizkiy, M.; Tsymbal, A.; Puuronen, S. PCA-based feature transformation for classification: Issues in medical diagnostics. *Proc. IEEE Symp. Comput. Based Med. Syst.* **2004**, *1*, 535–540.

42. Heijmans, H.J.A.M.; Ronse, C. The algebraic basis of mathematical morphology I. Dilations and erosions. *Comput. Vis. Graph. Image Process.* **1990**, *50*, 245–295. [CrossRef]

*Article*

# Monitoring of the Spatio-Temporal Dynamics of the Floods in the Guayas Watershed (Ecuadorian Pacific Coast) Using Global Monitoring ENVISAT ASAR Images and Rainfall Data

Frédéric Frappart [1,2,*], Luc Bourrel [1,*], Nicolas Brodu [3], Ximena Riofrío Salazar [1,4], Frédéric Baup [5], José Darrozes [1] and Rodrigo Pombosa [6]

1   Géosciences Environnement Toulouse (GET), UMR 5563, CNRS/IRD/UPS, Observatoire Midi-Pyrénées (OMP), 14 Avenue Edouard Belin, 31400 Toulouse, France; ximena.riofrio@gmail.com (X.R.S.); jose.darrozes@get.omp.eu (J.D.)
2   Laboratoire d'Etudes en Géophysique et Océanographie Spatiales (LEGOS), UMR 5566, CNRS/IRD/UPS, Observatoire Midi-Pyrénées (OMP), 14 Avenue Edouard Belin, 31400 Toulouse, France
3   Inria Bordeaux Sud-Ouest, Géostat, 200 Avenue de la Vieille Tour, 33405 Talence, France; nicolas.brodu@inria.fr
4   Secretaría de Educación Superior, Ciencia y Tecnología (SENESCYT), Whymper E7-37 y Alpallana, 170516 Quito, Ecuador
5   Centre d'Etudes Spatiales de la BIOsphère (CESBIO), UMR5126, UPS/CNRS/IRD/CNES, Observatoire Midi-Pyrénées (OMP), 14 Avenue Edouard Belin, 31400 Toulouse, France; frederic.baup@cesbio.cnes.fr
6   Instituto Nacional de Meteorología e Hidrología (INAMHI), Iñaquito N36-14 y Corea, 160310 Quito, Ecuador; rpombosa@inamhi.gob.ec
*   Correspondence: frederic.frappart@legos.obs-mip.fr (F.F.); luc.bourrel@ird.fr (L.B.)

Academic Editor: Hongjie Xie
Received: 13 November 2016; Accepted: 21 December 2016; Published: 1 January 2017

**Abstract:** The floods are an annual phenomenon on the Pacific Coast of Ecuador and can become devastating during El Niño years, especially in the Guayas watershed (32,300 km$^2$), the largest drainage basin of the South American western side of the Andes. As limited information on flood extent in this basin is available, this study presents a monitoring of the spatio-temporal dynamics of floods in the Guayas Basin, between 2005 and 2008, using a change detection method applied to ENVISAT ASAR Global Monitoring SAR images acquired at a spatial resolution of 1 km. The method is composed of three steps. First, a supervised classification was performed to identify pixels of open water present in the Guayas Basin. Then, the separability of their radar signature from signatures of other classes was determined during the four dry seasons from 2005 to 2008. In the end, standardized anomalies of backscattering coefficient were computed during the four wet seasons of the study period to detect changes between dry and wet seasons. Different thresholds were tested to identify the flooded areas in the watershed using external information from the Dartmouth Flood Observatory. A value of $-2.30 \pm 0.05$ was found suitable to estimate the number of inundated pixels and limit the number of false detection (below 10%). Using this threshold, monthly maps of inundation were estimated during the wet season (December to May) from 2004 to 2008. The most frequently inundated areas were found to be located along the Babahoyo River, a tributary in the east of the basin. Large interannual variability in the flood extent is observed at the flood peak (from 50 to 580 km$^2$), consistent with the rainfall in the Guayas watershed during the study period.

**Keywords:** flood; SAR; ENVISAT ASAR; rainfall; Guayas; Ecuadorian Pacific slope

## 1. Introduction

El Niño Southern Oscillation (ENSO) has a strong influence on rainfall patterns on the Pacific Coast of South America, from Ecuador to Chile, at an interannual time scale e.g., [1–4], affecting the likelihood of droughts and floods [5]. Positive anomalies of sea surface temperatures along the coast of Peru and Ecuador occurring during ENSO events induced torrential rains that cause high discharge and large flood events [6]. In the Guayas Basin, located in the southwest of Ecuador, in the Pacific slope of the Andes Cordillera, large flood events that occurred during the major El Niño episodes of 1965–1966, 1972–1973, 1982–1983, and especially 1997–1998, were responsible for many casualties and had important socio-economic impacts on housing, agriculture, and fisheries [7].

Data acquired by in situ hydrometeorological networks and remotely sensed images are commonly used for monitoring the spatial extent of the floods and even for forecasting and warning [8,9]. In spite of the good spatial and temporal resolutions of multispectral images, their use is limited in tropical areas as they are vulnerable to cloud cover and unable to detect water under dense vegetation cover. In the Guayas Basin, the almost permanent cloud cover during the rainy season does not allow the use of multispectral images, even the high frequency daily or the 8-day composite MODIS reflectance products, to monitor the temporal variations of the flood extent. Synthetic Aperture Radar (SAR) images, that provide useful information under all-weather conditions, including cloud cover, have been widely used for flood mapping and wetland delineation, especially at L-band as this microwave frequency band is able to penetrate dense vegetation cover e.g., [10–13]. Nevertheless, L-band images acquired by sensors such as JERS-1, PALSAR onboard ALOS (repeat-period of 46 days), did not have a sufficient temporal resolution to be used for flood monitoring. Images from PALSAR-2 on-board ALOS-2, available since May 2014, were not considered as they only cover two rainy seasons in the Guayas watershed. Several studies showed that SAR images acquired at the higher C-band frequency can also be used for detecting floods in less vegetated areas e.g., [14–17]. SAR images acquired in Wide Swath (WS) mode by the Advanced SAR (ASAR) onboard ENVISAT, at a spatial resolution of 150 m, were widely used for the monitoring of flood dynamics from local to regional scales [16–19]. Only 29 ASAR images were acquired in WS mode between November 2002 and March 2012 and just 14 during the rainy seasons due to acquisitions in other modes. No images were acquired during the 2004, 2005 and 2009 rainy seasons, and only two or three the other years, for a rainy season that last 6 months. The low number of available ASAR WS images is insufficient to allow the monitoring of the time variations of the flood in the Guayas watershed. Images acquired by ASAR in Scan SAR Global Monitoring (GM) mode at coarse spatial resolution (1 km), for a temporal resolution generally better than one month, offered the opportunity to monitor inundation extents from regional to global scales [20–22]. Before the launch of Sentinel-1 that has been providing high resolution images (20 m of spatial resolution) with a temporal sampling of a few days since April 2014, ASAR images acquired in GM mode had the advantages to provide frequent (twice or third a month) observations of the same scene and to provide a complete observation of medium-size watershed in one acquisition.

In this study, we used ENVISAT ASAR images acquired in GM mode to monitor the floods in the Guayas watershed between 2005 and 2008. We first present the study area and the datasets. Then, we describe the change detection method used for determining the inundation extent in each image. The last part presents a sensitivity analysis of the change detection method, the spatio-temporal variations of the flood extent and the relationship between the time-series of rainfall and inundation extent during the observation period.

## 2. Study Area

The Guayas River Basin, located in the southwestern part of Ecuador, is the largest tropical agricultural watershed and estuarine system of the Pacific slope of South America. It extends between latitudes 0° S and 3° S, and longitudes 81° W and 78° W and covers an area of 32,300 km$^2$ (Figure 1). Its watershed represents 13% of the total area of Ecuador where 40% of the Ecuadorian population lives [23]. It corresponds to the most productive region of Ecuador for agriculture and aquaculture [24].

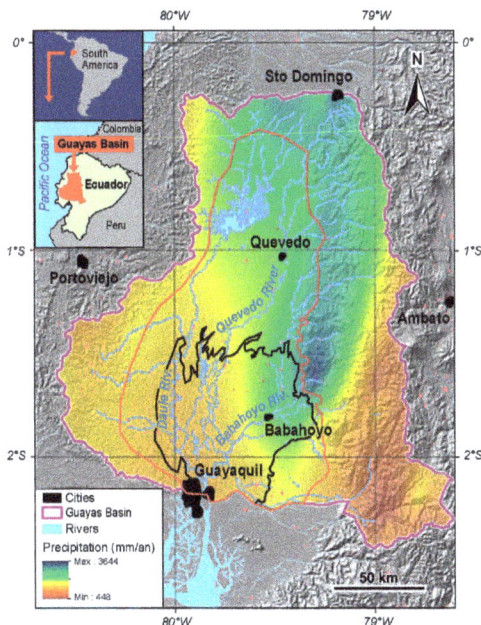

**Figure 1.** The Guayas Basin (boundary in purple) is located in the southwest Pacific slope of Ecuadorian Andes. The lowland (boundary in red) occupies 15,000 km$^2$ of the 32,300 km$^2$ of the watershed area, and the floodplain 5000 km$^2$. The mean annual rainfall isohyets map was elaborated for the 1963–2009 period using cokriging method.

Its climate is characterized by the occurrence of the rainy season from December to May. The average precipitation in the Guayas basin for the 1963–2009 period is 1849 mm during the hydrological year and 1130 mm during the rainy season. This value increased respectively to 4769 mm, 2412 mm and 6786 mm for the El Niño episodes of 82–83, 91–92 and 97–98, corresponding respectively to ratios El Niño/Normal events of 2.6, 1.3 and 3.7 [25].

The altitude of this drainage basin ranges from sea level to 6310 m at Mount Chimborazo in the Andes Cordillera. The lower plain of the Guayas Basin is the region corresponding to altitudes lower than 200 m. Its area is ~15,000 km$^2$ [23], i.e., ~46% of the surface of the watershed. In this lower plain, the region likely to be affected by floods corresponds to almost flat areas (slopes ≤ 5%). They represent 30% of the lower plain or ~5000 km$^2$.

The Daule, Babahoyo and Quevedo Rivers are the largest tributaries of western and eastern parts that merge downstream to form the Guayas River and subsequently the Guayas Estuary that flow into the Gulf of Guayaquil (Figure 1). The mean discharge to the sea of the Guayas River varies from 200 m$^3$/s during the dry season to 1600 m$^3$/s at the peak of flow [26,27]. It can reach up to 5000 m$^3$/s during strong ENSO events as 1982/1983 or 1997/1998 [26].

## 3. Datasets

### 3.1. ENVISAT ASAR GM Mode Images

ENVISAT mission was launched on 1 March 2002 by the European Space Agency (ESA). It orbits at an average altitude of 790 km, with an inclination of 98.54°, on a sun-synchronous orbit with a 35-day repeat cycle. It carried 10 scientific instruments including ASAR operating at a central frequency of 5.331 GHz (C-band). This sensor offered multiple acquisition modes for SAR images at various spatial and temporal resolutions and alternating polarizations. Among them, the ScanSAR GM mode acquired images in a swath of 405 km at a spatial resolution of 1 km and temporal resolution between four and seven days when no other acquisition in a different mode was ordered [28]. This mode is very useful for the monitoring of dynamic processes, such as soil moisture or floods, from regional to global scales e.g., [20–22,29,30]. In this study, we used 92 ASAR images acquired in GM mode, with HH polarization and 1 km spatial resolution, from December 2004 to September 2008, that encompass the whole Guayas Basin (Figure 2). HH polarization is the more adequate polarization for flood mapping on ASAR [31]. During the rainy season, there is generally one acquisition every two weeks at least that covers the whole basin.

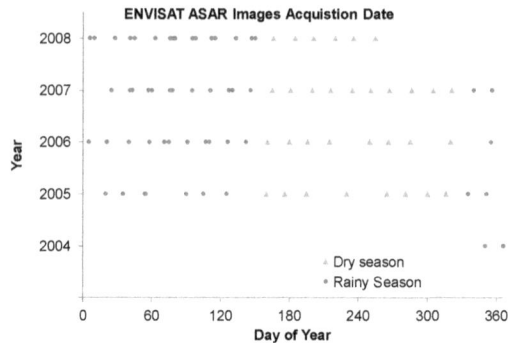

**Figure 2.** Acquisition dates of ENVISAT ASAR images in GM mode over the Guayas Basin between December 2004 and September 2008 for dry (orange squares) and rainy (blue dots) seasons. These images were made available by ESA through the EOLi (Earth Observation Link) Earth Observation Catalogue and Ordering Services [32].

*3.2. Land Cover Map of Ecuador*

A land cover map of the Ecuador was produced by the Ecuadorian Ministry of Agriculture (Ministerio de Agricultura, Ganadería, Acuacultura y Pesca—MAGAP) in 2002 over the whole country at the spatial scale of 1:250,000 using Landsat TM multispectral images acquired between 1999 and 2001 validated with a ground control assessment performed in 2001–2002. The main land cover types present in the Guayas watershed are the following:

(i)   forests (native and cultivated) occupying an area of 9206 $km^2$ (29% of the watershed area), permanent crops including banana, sugar cane, fruit trees, plantain, African palm, cacao, and coffee occupying an area of 6087 $km^2$ (20% of the watershed area),

(ii)  annual crops including corn, rice, soybeans, and vegetables occupying an area of 9181 $km^2$ (29% of the watershed area),

(iii) pastures occupying an area of 4130 $km^2$ (13% of the watershed area), water bodies occupying an area of 487 $km^2$ (2% of the watershed area), urban areas occupying an area of 223 $km^2$ (1% of the watershed area).

More details about intra-class land cover repartition can be found in [33].

*3.3. Surface Water Record from the Dartmouth Flood Observatory*

The Dartmouth Flood Observatory (DFO) provides a unique source of information on floods due to its global coverage [34]. It comprises the Surface Water Record (SWR), a comprehensive record of satellite-observed changes in the Earth's inland surface waters. This dataset compiles the observed history of flooding, starting in the year 2000. Extent of surface water is mostly derived from NASA's MODIS Terra and Aqua sensors with, in some cases, additional information from Radarsat, ASTER, or other higher spatial resolution data [35]. Water areas are accumulated over 10 days to minimize the effect of cloud cover. Inundation maps are made available at a spatial resolution of 250 m on $10° \times 10°$ tiles. A color code indicates maximum flood extent each year. It is important to note, that using this representation mode, it is impossible to know how many times a pixel was inundated during the observation period. Besides, this dataset does not allow the monitoring of the variations of flood extent during the hydrological cycle. A specific color indicates the reference water status (February, 2000, from the Shuttle Radar Topography Mission Water Body—SWBD data). In the case of the Guayas Basin, MODIS images during the rainy season cannot be used due to the cloud cover. So, this dataset is limited to SWBD in this specific area completed with the inundation extent observed during the two large flood events of 1998 and 2002. In our study, it is used as static information on the maximum flood extent in the Guayas Basin. The flood extent image encompassing the eastern part of the basin (east of 80° W of longitude) was downloaded at the Flood Observatory [36]. No data are available for the western part that corresponds to the western tributaries of the Daule River and a part of its stream.

### 3.4. Rainfall

Monthly rainfall records from 310 meteorological stations located over the Pacific slope of the Ecuadorian Andes (5° S–1.5° N and 77.5° W–81° W) were provided by INAMHI (National Institute of Meteorology and Hydrology of Ecuador) from 1963 to 2009. A careful quality check of this data was performed using the regional vector method similarly to what was performed in [3,4]. This dataset was interpolated at a spatial resolution of 1 km using a co-kriging method as in [4].

### 4. Methods

The methodology used to process the time series of SAR images is composed of the following three main steps: the preprocessing of the SAR images, the classification of the SAR images acquired during the dry season for land cover purposes and assessment of the characterization of the radar backscattering response between open water and non-inundated surfaces during the dry season, and the flood detection during the dry season (Figure 3).

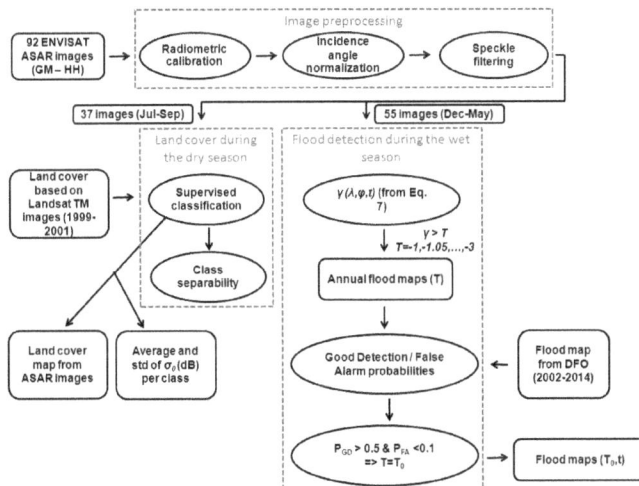

**Figure 3.** Schematic view of the processing of the ENVISAT ASAR images acquired in GM mode between December 2004 and September 2008. The method is composed of three steps: a pre-processing of the images, a supervised classification to determine the land cover using the images acquired during the dry season (July–September), the change detection method applied to monitor the flood extent during the wet season (December–May).

*4.1. ENVISAT ASAR GM Mode Images Preprocessing*

The ASAR GM images were preprocessed with respect to the following steps: radiometric calibration, incidence angle normalization, speckle reduction and geometrically correction using the Next ESA SAR Toolbox (NEST) [37]. Images of radar backscattering coefficients ($\sigma_0$ or sigma naught) at the incidence angle $\theta$ of acquisition of each image were derived from the brightness amplitudes expressed in Digital Numbers (*DN*) in the ASAR GM products using:

$$\sigma_0(\theta) = \frac{DN^2}{K} \sin\theta \tag{1}$$

where *K* is the absolute calibration constant [38].

The effect of the incidence angle on the surface backscattering was taken into account through a normalization procedure [39]. The normalized backscatter coefficient is given by:

$$\sigma_0^{norm} = \frac{\sigma_0(\theta)}{F(\theta)} \tag{2}$$

where $F(\theta)$ is a peculiar function to the target environmental conditions that has the following form [39,40]:

$$F(\theta) = \cos^\alpha \theta \tag{3}$$

where $\alpha$ is a coefficient depending on the predominant scattering mechanism and the sensor condition [41].

In the following, $\sigma_0^{norm}$ is noted $\sigma_0$ for simplification.

Finally, the radar images were spatially filtered to reduce speckle effects using a classical Lee filter [42] with a window size of 5 by 5 pixels. Examples of resulting images are presented in Figure 4a,b for dry and rainy seasons, respectively.

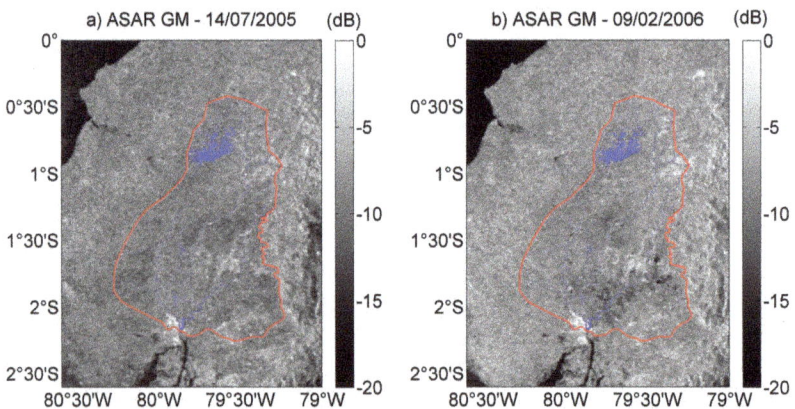

**Figure 4.** Images of backscattering obtained after the pre-processing of ENVISAT ASAR images in GM mode acquired during the (**a**) dry (14 July 2005); and (**b**) rainy (9 February 2006) seasons in the Guayas Basin lowland. The boundary of the lowland is represented using a red line and open water appears in blue.

*4.2. Land Cover during the Dry Season Using a Supervised Classification*

The change detection method used to identify flooded pixels during the wet season requires a well-defined initial state corresponding to the backscattering during the dry season. It is necessary to check the separability of the class corresponding to water from the other classes. Eleven supervised classifications were performed on 37 ASAR images acquired during the dry seasons, from July to September, over 2005–2008 and not only the maximum of likelihood technique as done previously in [43].

The feature space consists of all dry season backscattering coefficients for each given pixel. Classifiers thus work in a large dimensional vector space and discriminate seasonal patterns of each of the five classes present in the Guayas watershed according to the Landsat TM-based land use map: water bodies (lakes, rivers), permanent crops (palm trees, cocoa, bananas, coffee), seasonal crops (rice, corn, soy), grazing fields, cities. The training data were chosen using the land use map defined in 2002 using Landsat TM images. The selected training sites (2292 pixels randomly selected in the land use map) correspond to the five classes present in the Guayas watershed according to the Landsat TM-based land use map. A 10-fold cross-validation was used to set the meta-parameters (e.g., $K$ in k-nearest neighbors) and report the corresponding accuracy as the performance metric for each classifier in Section 5 (Results). As the results of the classification during the dry season are used in the followings for detecting changes in backscattering related to floods, it is necessary to choose the method that provides the best results in terms of accuracy. The separability between the classes was assessed using:

(i) the M-statistics, originally introduced to discriminate burned and unburned pixels in multi-spectral images [44], was also applied to SAR images in land use applications [45]. It is computed as follows:

$$M = \frac{\mu_i - \mu_j}{\sigma_i - \sigma_j} \tag{4}$$

where $\sigma_k$ and $\mu_k$ are the mean and standard deviation of the $k$th class respectively. $M$ values greater than 1 indicate a reasonable separability increasing with the value of $M$.

(ii) the Jeffries-Matusita distance defined as follows [46]:

$$d_{J-M}(i,j) = \sqrt{2(1 - e^{-\alpha})} \tag{5}$$

with

$$\alpha = \frac{1}{2}(\mu_i - \mu_j)^T \left(\frac{C_i + C_j}{2}\right)^{-1}(\mu_i - \mu_j) + \frac{1}{2}\ln\left(\frac{\frac{1}{2}|C_i + C_j|}{\sqrt{|C_i||C_j|}}\right) \tag{6}$$

where $C_k$ is the covariance matrix of the $k$th class. It varies from 0 (no separation) to $\sqrt{2}$–1.41 (complete separation).

## 4.3. Flood Detection

A change detection method was applied to the ASAR images acquired during the rainy seasons between 2005 and 2008. For each pixel of coordinates $(\lambda, \varphi)$ of any 55 ASAR images acquired during the wet season at time $t$, we computed the following normalized anomaly of $\sigma_0$ ($\gamma$), defined as follows:

$$\gamma(\lambda, \varphi, t) = \frac{\sigma_0(\lambda, \varphi, t) - \langle\sigma_0(\lambda, \varphi, t)\rangle_{dry}}{std(\sigma_0(\lambda, \varphi, t))_{dry}} \tag{7}$$

where $<\sigma_0(\lambda,\varphi,t)>_{dry}$ and $std(\sigma_0(\lambda,\varphi,t))_{dry}$ are the average and the standard deviation of $\sigma_0$ during the 2005–2008 dry seasons, respectively.

Flooded pixels present lower backscattering than bare soil or vegetation covered ones as the radar electromagnetic wave is specularly reflected by water surfaces. As a consequence, the lower $\gamma$ is, the more inundated the pixel is. As the vegetation cover in the lowland of the Guayas Basin is not composed of forests, but of pastures, seasonal and permanent crops, the presence of water under vegetation will cause a decrease in radar backscattering e.g., [14]. The risk to have $std(\sigma_0(\lambda,\varphi,t))_{dry}$ close to zero is limited at the spatial resolution of 1 km. The only available external source of information on the flood in the Guayas watershed is the SWR from the DFO. Most of the flooded areas present in the SWR since 2000 have their extent included in the previous large flood events of 2002, and especially 1998. This dataset is commonly used for estimating flood extent limits when processing other remotely sensed observations e.g., [47,48]. Our goal is to determine a threshold to discriminate changes caused by the floods from changes in land cover, due to the vegetation growth, or due to the presence of soil moisture. We determined the number of inundated and falsely inundated pixels in this watershed using the surface water extent 2000–2014 from the DFO SWR for each value of $\gamma$ varying from $-1$ to $-3$ with a step of 0.05 as follows:

(i)     for each ASAR image, a binary image representing the flood extent at time $t$ is obtained: all ASAR pixels with values lower or equal to $\gamma$ are considered flooded and pixels with values greater than $\gamma$ not flooded.

(ii)    annual maps of maximum flood extent are then obtained. In these maps, a pixel is considered to be well identified as flooded if it is also inundated in the SWR whereas it is badly identified as if it is not also identified as inundated in the SWR.

(iii)   annual ratios of good and false detections are determined as the numbers of well and falsely detected as inundated pixels divided by the numbers of pixels identified as inundated and non-inundated in the DFO SWR, respectively.

Thresholds on good and false detections allow the determination of the value of the threshold on $\gamma$. For every value of $\gamma$, the inundated area $S$ is then estimated as follows:

$$S(t) = R_e^2 \sum_{i \in A} \delta(\lambda_i, \varphi_i, t) \cos(\varphi_i) \Delta\lambda\Delta\varphi \tag{8}$$

where $R_e$ the radius of the Earth equals 6378 km, $\lambda_i$ and $\varphi_i$ are the longitude and latitude of the $i$th pixel inside the Guayas watershed of area $A$, $\delta(\lambda_i,\varphi_i,t)$ equals one if the pixel is inundated and 0 if not, $\Delta\lambda$ and $\Delta\varphi$ are the grid steps in longitude and latitude, respectively, that are equal to $0.0045°$.

*4.4. Relative Frequency of Inundation*

A relative frequency of inundation (RFI) map, based on the time-series of inundation map derived from the ASAR images acquired between December 2004 and June 2008, was estimated for the Guayas Basin floodplain. Following a method similar to the one proposed by [49], the RFI, expressed in %, of a pixel of geographical coordinates $(\lambda;\varphi)$ is defined as follows:

$$RFI(\lambda;\varphi) = \frac{\sum_{n=1}^{N} \delta(\lambda, \varphi, n)}{N} \times 100 \tag{9}$$

where $N$ is the number of images acquired during the rainy season over the whole study period and $\delta(\lambda,\varphi,n)$ equals one if the pixel is inundated and 0 if not.

## 5. Results

*5.1. Land Cover from ASAR GM Images during the Dry Season*

The performances of the 11 different techniques (i.e., k-nearest neighbors, linear and Gaussian SVM, random forest, extremely random trees, adaboost, naive Bayes, logit, linear and quadratic discriminants, ridge regression) were assessed in terms of multi-class accuracy (5 classes). Among them, the k-nearest neighbor classification appears to be the most efficient and the most robust for discriminating the classes based on their temporal patterns in the dry season with an accuracy of 98.2% $\pm$ 0.7%. In the followings, all the results are presented for this supervised classification. The resulting land cover map derived from the k-nearest neighbor method is presented in Figure 5.

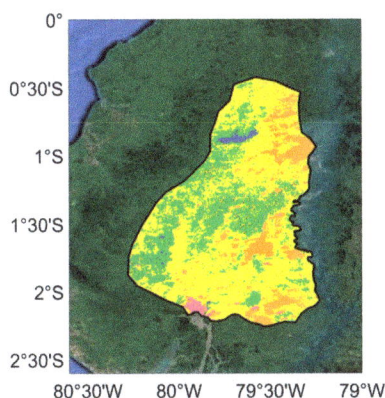

**Figure 5.** Land use from ENVISAT ASAR GM images acquired during the dry seasons (June to November) between 2005 and 2008, as determined by k-nearest neighbors supervised classification. The resulting classes are the followings: open water (blue), permanent crops (green), seasonal crops (orange), grazing fields (yellow), cities (pink). The boundary of the lowland is represented using a black line. Background image is from Google Maps.

The five classes correspond to open water (lakes and reservoirs) as the Guayas river stream and its tributaries cannot be identified at a spatial resolution of 1 km, permanent crops composed of tropical arboriculture (banana, oil palm, coffee and cacao trees), seasonal crops (mostly rice, corn, soya, and shrubby vegetation), mostly pastures and sugar cane, towns, respectively.

The results of the class separability are presented in Table 1. It appears clearly that the different classes are well separated from one another using both criteria, especially using the Jeffries-Matusita distance. The Jeffries-Matusita distance between each pair of classes is higher than 1.40 except between permanent crops and seasonal crops. In this case, its value is 1.34 which corresponds to a good separation between the two classes as this value is quite close to the maximum value of the Jeffries-Matusita distance ($\sqrt{2}$). Very good separability is also observed using the M-statistics (M > 1), except for the separability with cities class, due to the large standard deviation (see Table 2) observed in this class, a consequence of the low spatial resolution of the ENVISAT ASAR GM images (1 km). However, even for this class, the M-statistics present values close to 1 or a little bit higher (Table 1). In all cases, a good separability is observed between open water and the other classes.

**Table 1.** Results of the separability between classes obtained from a k-nearest neighbors supervised classification performed on ENVISAT ASAR GM images acquired during the dry season (July–September) between 2005 and 2008 using the M-statistics and the Jeffries-Matusita distance.

| Separability between Classes | | Open Water | Permanent Crops | Seasonal Crops | Grazing Fields | Cities |
|---|---|---|---|---|---|---|
| Open water | M-statistics | - | 6.67 | 2.14 | 4.37 | 1.08 |
| | J-M | - | 1.41 | 1.41 | 1.41 | 1.41 |
| Permanent crops | M-statistics | - | - | 1.55 | 2.91 | 0.98 |
| | J-M | - | - | 1.34 | 1.41 | 1.41 |
| Seasonal crops | M-statistics | - | - | - | 1.20 | 0.90 |
| | J-M | - | - | - | 1.41 | 1.41 |
| Grazing fields | M-statistics | - | - | - | - | 0.93 |
| | J-M | - | - | - | - | 1.41 |
| Cities | M-statistics | - | - | - | - | - |
| | J-M | - | - | - | - | - |

**Table 2.** Results of the k-nearest neighbors supervised classification performed on ENVISAT ASAR GM images acquired during the dry season (July–September) between 2005 and 2008. For each class, average backscattering (dB), associated standard deviation (dB), and percentage of the watershed area (%) are given.

| Dry Season Classification Class Name | Average Backscattering (dB) | Standard Deviation of Backscattering (dB) | Percentage of the Watershed Area (%) |
|---|---|---|---|
| Open water | −11.30 | 1.68 | 0.9 |
| Permanent crops | −8.63 | 1.29 | 15.0 |
| Seasonal crops | −6.19 | 1.70 | 54.0 |
| Grazing fields | −7.65 | 1.29 | 29.0 |
| Cities | −1.87 | 2.74 | 1.1 |

Using the results from the k-nearest neighbors supervised classification, land cover in the Guayas watershed is dominated by seasonal crops (54%), characterized by $\sigma_0 = -6.19 \pm 1.70$ dB, followed by grazing fields (29%), with $\sigma_0 = -7.65 \pm 1.29$ dB, permanent crops (15%), with $\sigma_0 = -8.63 \pm 1.29$ dB, cities (1.1%), with $\sigma_0 = -1.87 \pm 2.74$ dB and open water (0.9%), with $\sigma_0 = -11.30 \pm 1.68$ dB (Table 2 and Figure 6).

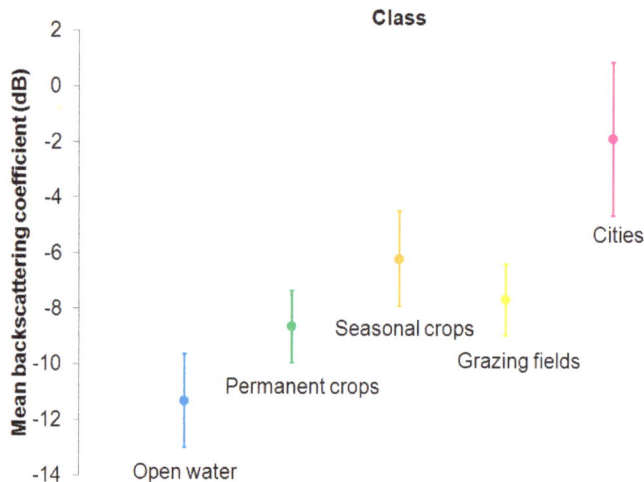

**Figure 6.** Average backscattering coefficient (dB) and associated standard deviation (dB) for each class resulting from a k-nearest neighbors supervised classification performed during the dry season (July–September) between 2005 and 2008: open water (blue), permanent crops (green), seasonal crops (orange), grazing fields (yellow), cities (pink).

### 5.2. Flood Detection during the Rainy Season

Normalized anomalies of backscattering $\gamma$ were computed for the 55 ASAR images acquired during the rainy season using (7). We determined the number of pixels whose value is beyond the threshold for each rainy season from 2005 to 2008 for threshold values varying from −1 to −3 with a step of −0.05. These pixels are then considered as flooded and can be used to estimate annual maps of inundation when stacking the binary images of floods during each year's rainy season. The range of the threshold values corresponds to intervals of confidence varying from 68.3% to 99.7% compared to its value during the dry season, according to the three-sigma rule of thumb assuming a normal distribution of the data.

Examples of resulting maps for different threshold values are presented in Figure 7 for 2008 that was the wetter year of the observation period. Compared to the SWR from DFO (Figure 7a), lower threshold values led to flooded areas in excess (for instance, thresholds of −1.0 and −1.5 in Figure 7b,c), whereas flood maps obtained using higher threshold values present very limited inundated areas (for instance, thresholds of −2.5 and −3.0 in Figure 7e,f). The map obtained using $\gamma$ equals to −2.0 exhibits a similar pattern as the SWR (Figure 7a,d, respectively). To more accurately determine the thresholds value, we compared the SWR and $\gamma$ maps pixel by pixel. A pixel is considered to be well identified as flooded if its value is below the threshold and is inundated in the SWR whereas it is poorly identified as flooded if its value is lower or equal to the threshold but it is not identified as inundated in the SWR. The results of the change detection method limited to the floodplain area are presented on Figure 8. We fixed the following thresholds: as no large ENSO was recorded between 2005 and 2008, we considered that the minimum threshold value for inundated pixels should correspond to a number of pixels higher than 50% of the number of inundated pixels of the SWR in the Guayas floodplain during the wettest year of the study period. According to Figure 8a, it corresponds to a minimum value of $\gamma$ higher than −2.35. The corresponding ratios are 10.8%, 30.9%, 12.1% and 50.2% for 2005, 2006, 2007, and 2008, respectively. Similarly, the number of pixels falsely detected as inundated should be lower than 10% of the number of inundated pixels of the SWR in the Guayas watershed during the wetter year, considering the possible missed detection of flooded areas in this dataset. According to Figure 8b, it corresponds to a maximum value of $\gamma$ lower than −2.25. The corresponding ratios are 3.7%, 7.1%, 2.2% and 9.5% for 2005, 2006, 2007, and 2008, respectively. Using these criteria, $\gamma$ varies between −2.25 and −2.35. Converted (7) into dB, it means that the difference between (i) the change in backscattering between the dry season and the time *t* during the rainy season; and (ii) the standard deviation of the backscattering during the dry season is greater than 3.52, 3.63 and 3.71 dB for threshold values of −2.25, −2.30, and −2.35, respectively. According to Table 2, these values are larger than the standard deviation of any of the five classes considered in this study. As these classes were found be well separated, it is reasonable to consider that these thresholds can be used for separating inundated and non-inundated pixels. Maximum annual flood extent obtained using thresholds of −2.25, −2.30, −2.35 between 2005 and 2008 are presented in Figure 9. Very similar patterns are observed with the flood extent decreasing with $\gamma$. The pattern of inundated areas is very consistent with the hydrological characteristics of the Guayas floodplain. Areas detected as flooded are present along the major tributaries to the Guayas (i.e., Babahoyo, Quevedo, Daule) or in the wetlands such as the RAMSAR site Abras de Mantequilla wetland, located between the Quevedo and Daule rivers, which plays a role in flood attenuation and streamflow regulation [50]. Large interannual variability can also be observed during the 2005–2008 time-span. A more detailed analysis of the results is provided in Section 5.3.

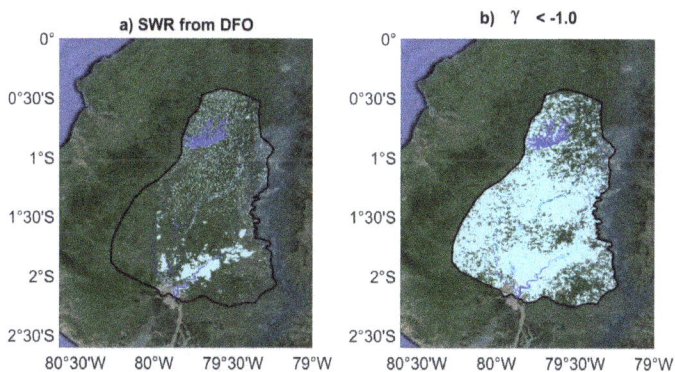

**Figure 7.** *Cont.*

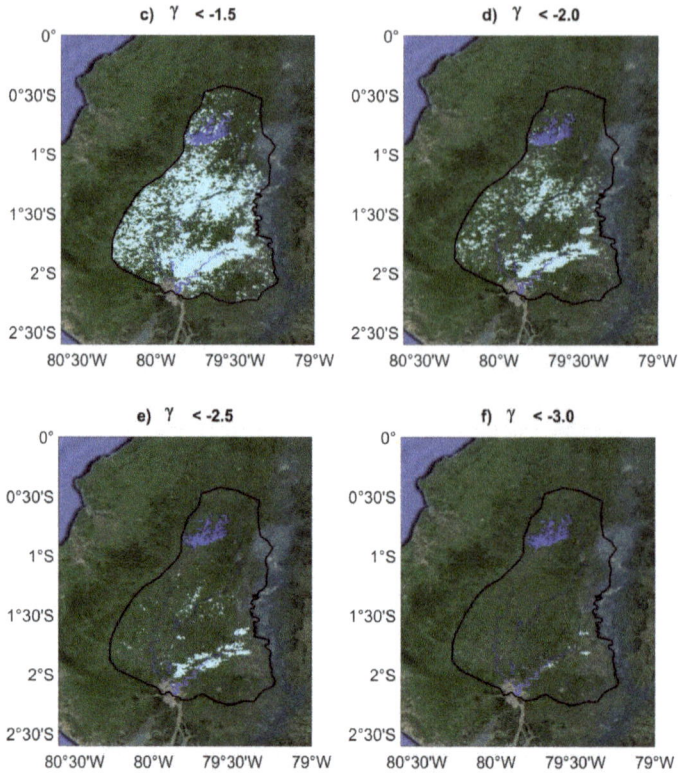

**Figure 7.** Inundation in the Guayas watershed using the SWR from DFO (**a**). Normalized anomalies of backscattering coefficients ($\gamma$) lower than a given threshold appear in light blue. The threshold values equal $-1.0$ (**b**); $-1.5$ (**c**); $-2.0$ (**d**); $-2.5$ (**e**); and $-3.0$ (**f**) for 2008. Background image is from Google Maps. Permanent water bodies appear in blue.

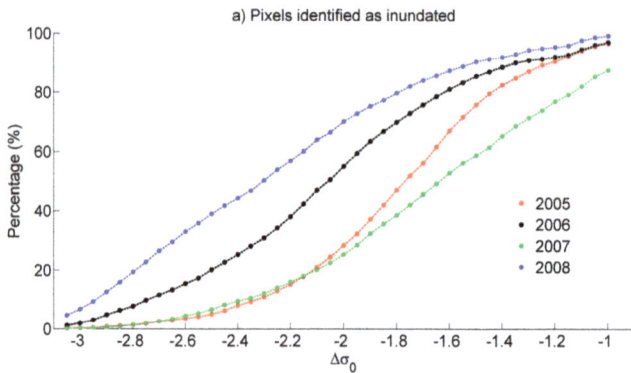

**Figure 8.** *Cont.*

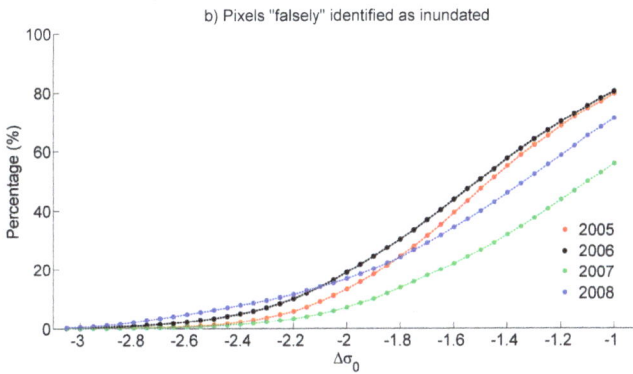

b) Pixels "falsely" identified as inundated

**Figure 8.** Percentage of pixels detected as inundated (**a**) and falsely identified as inundated during the rainy seasons (December–June) from 2005 to 2008 as a function of the normalized anomaly of backscattering though comparison with the SWR from DFO in the Guayas floodplain (**b**).

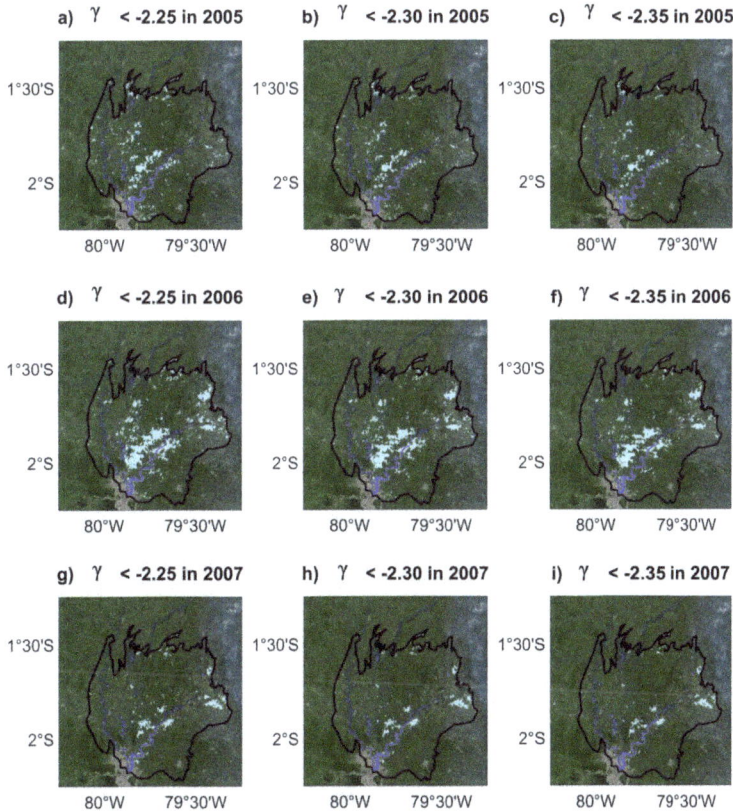

**Figure 9.** *Cont.*

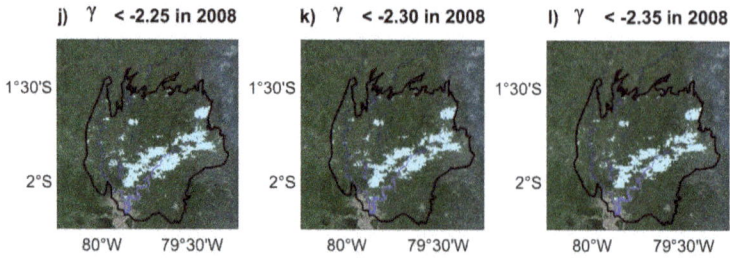

**Figure 9.** Maps of annual normalized anomalies of backscattering coefficients lower than a given threshold (light blue) in the Guayas floodplain (black line). The threshold values $\gamma$ equals to $-2.25$ in 2005 (**a**); 2006 (**d**); 2007 (**g**); and 2008 (**j**); $-2.30$ in 2005 (**b**); 2006 (**e**); 2007 (**h**); and 2008 (**k**); $-2.35$ in 2005 (**c**); 2006 (**f**); 2007 (**i**); and 2008 (**l**). Background image is from Google Maps. Permanent water bodies appear in blue.

### 5.3. Time Variations of Inundated Areas

Using the thresholds determined in Section 5.2, monthly variations of inundation extent were estimated in the Guayas Basin. These monthly variations were obtained as the cumulated extent of the inundated areas observed each month using ASAR images. Time variations of inundated areas during the wet seasons (December–May) from 2005 to 2008 are presented in Figures 10 and 11 for $\gamma = -2.30$. During the observation period, small and disseminated flooded areas were present in December (Figure 10a,g and Figure 11a,g) and January (Figure 10b,h and Figure 11b,h). Largest flood area extents are observed between February and May (Figures 10c–f and 11i–l) with a large interannual variability. Floods generally started in February in the upstream part of the Daule River and the central part of the Quevedo River (Figure 10c,i and Figure 11c,i). Then, floods were mainly located along the central and downstream parts of the Quevedo and Babahoyo Rivers in March and April (Figure 10d,e,j,k and Figure 11d,e,j,k). Floods of smaller extent were detected in May along the Quevedo and Daule Rivers (Figure 10f,l and Figure 11f,l). Inundated areas are most frequently observed along the Babahoyo and Quevedo Rivers and in the Abras de Mantequilla wetland (225 km$^2$ at 79°45′ W and 1°30′ S). In 2006 and 2008, flooded areas were also detected in the upstream part of the Babahoyo River and along the Daule River (Figures 10g–l and 11g–l). In 2008, very extensive floods were detected in March and April along the Quevedo and Babahoyo river streams.

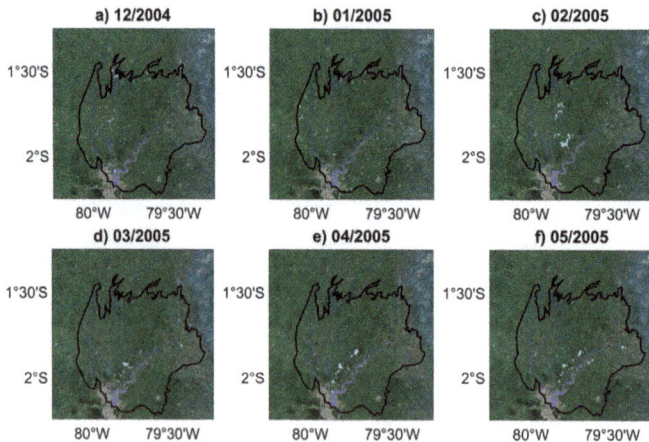

**Figure 10.** *Cont.*

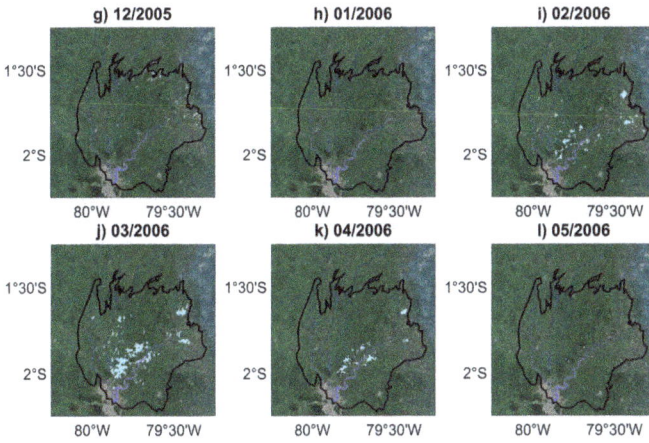

**Figure 10.** Monthly maps of inundation for normalized anomalies of backscattering coefficients lower than −2.30 (light blue) in the Guayas floodplain (black line) during the wet seasons of 2005 and 2006 (**a–f**) from December 2004 to May 2005; and (**g–l**) from December 2005 to May 2006. Background image is from Google Maps. Permanent water bodies appear in blue.

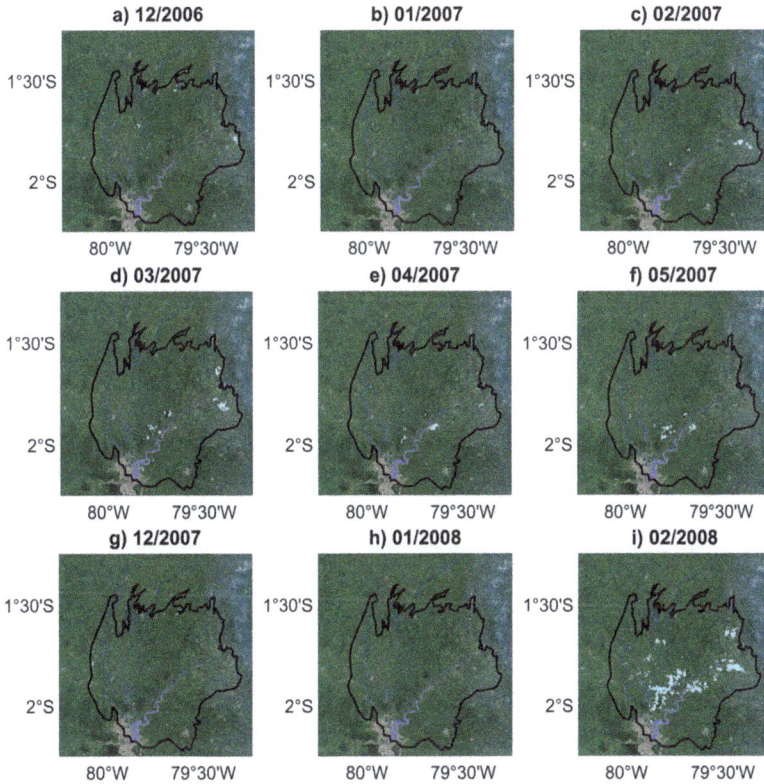

**Figure 11.** *Cont.*

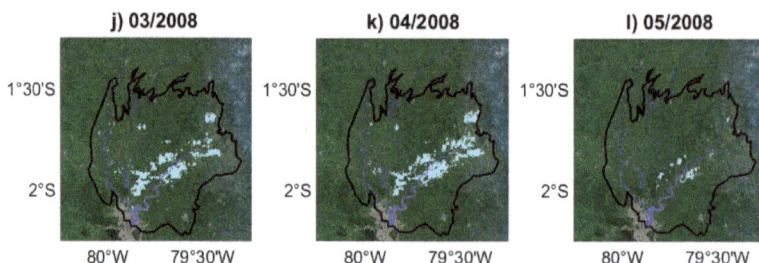

**j) 03/2008**  **k) 04/2008**  **l) 05/2008**

**Figure 11.** Monthly maps of inundation for normalized anomalies of backscattering coefficients lower than −2.30 (light blue) in the Guayas floodplain (black line) during the wet seasons of 2005 and 2006 (**a–f**) from December 2006 to May 2007; and (**g–l**) from December 2007 to May 2008. Background image is from Google Maps. Permanent water bodies appear in blue.

RFI maps of the Guayas floodplain were estimated using ENVISAT ASAR GM images acquired during the rainy seasons between December 2004 and June 2008 for the three threshold value used earlier. As they exhibit very similar patterns, only the one obtained using the −2.30 threshold value is presented in Figure 12. Four main areas of large flood occurrence (>25%): the junction between the Babahoyo and Quevedo Rivers, in the upstream north east of Guayaquil, the north and south banks of the Babahoyo River close to the city of Babahoyo, and the upstream eastern part of the floodplain. Secondary maxima (15%) can also be observed along the Daule River.

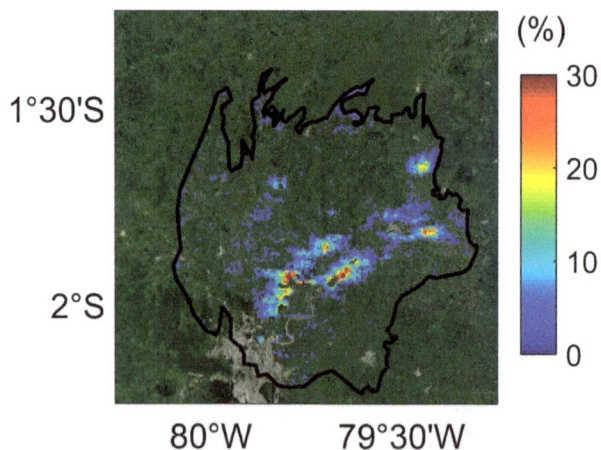

**Figure 12.** RFI map derived from the ASAR-based inundation extent map for the −2.30 threshold values applied to the normalized anomalies of backscattering coefficients ($\gamma$).

Figure 13 shows the time variations of flood extent and rainfall between 2005 and 2008. The hydrological years 2006 and 2008 are characterized by larger flood events (318 and 585 km$^2$ occurring in March 2006 and April 2008, respectively, as detected using ASAR images) than 2005 and 2007 (53 and 59 km$^2$ occurring in January 2005 and March 2007, respectively). The total of rainfall for the wet season (December–May) over the whole Guayas Basin for the same time-period were for 757, 1121, 950 and 1366 mm 2005, 2006, 2007 and 2008 respectively. In other terms, these totals of rainfall are 33%, 1% and 16% below the 1963–2009 average rainfall during the rainy season for 2005, 2006 and 2007, respectively, and 20% above it in 2008, meaning that 2005 and 2007 were drier than usual wet seasons, 2006 was a normal wet season and 2008 was a wetter than usual rainy season. When comparing the inundated areas obtained using the change detection method to the total rainfall during the wet season, a good consistency between these two hydrological variables can be noticed. A correlation of 0.75 was found between these two variables at monthly time-scale, with the total amount of rainfall preceding the flood of one month (inundation extent was set to zero between June and November of each year).

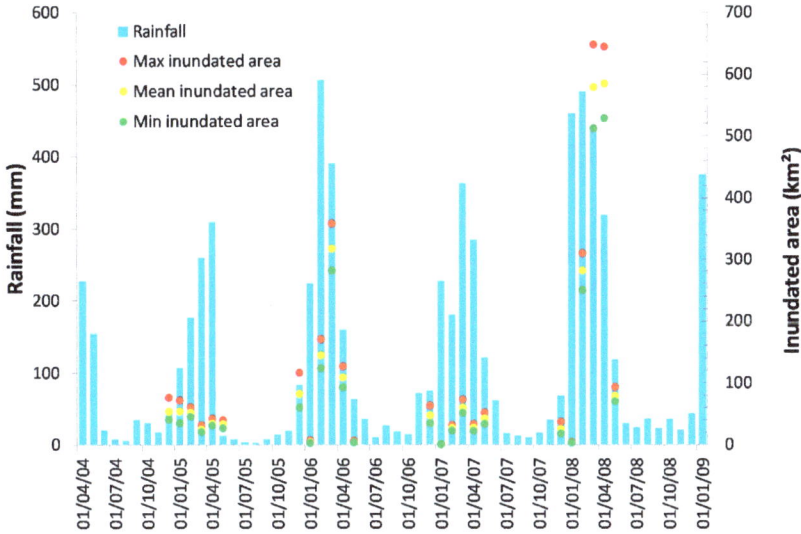

**Figure 13.** Time series of monthly rainfall (mm—blue bars) and inundation extent (km$^2$) for three threshold values applied to the normalized anomalies of backscattering coefficients ($\sigma_0$) and corresponding to a minimal (−2.35—green dots), a mean (−2.30—yellow dots) and a maximal (−2.25—red dots) inundation area (km$^2$).

## 6. Conclusions

ENVISAT ASAR GM images were used to monitor the spatio-temporal dynamics of the flood in the Guayas watershed between 2005 and 2008 applying a change detection method based on the difference in backscattering at C-band between the wet and the dry seasons. First, a supervised classification method was applied to determine the possibility to separate open water from the other classes. Among the different supervised classification techniques applied to determine the land cover in the Guayas Basin using ASAR images acquired during the dry season, the k-nearest neighbors technique provided the better results with an accuracy of 98.2% ± 0.7%. Open water, with a mean backscattering of −11.30 ± 1.68 dB, was clearly separated from the other classes for both the M-statistics and the Jeffries-Matusita distance criteria. This allows the use of the change detection method. In spite of the coarse spatial resolution of these SAR images, inundated areas detection provides consistent results with the sparse information currently available on the distribution of the floodplain in this watershed. Flood most frequently occurred in the eastern part of the floodplain (Andean part), along the Babahoyo river stream, but also on the western part (coastal reliefs) during normal and wetter than usual years. The large interannual variability in the inundation extent, with larger inundated areas detected in 2006 and 2008, was also found to be consistent with the temporal distribution of the rainfall during the wet season.

This study is the first step toward understanding the spatio-temporal (at seasonal and interannual time scales) dynamics of the flood in the Guayas Basin using multi-sensors and multi-resolution SAR images and their relationship with ENSO on a longer time-scale. With the recent launches of Sentinel-1A in April 2014, Sentinel-2A in June 2015 and Sentinel-1B in April 2016, satellite SAR (C-band)

and multispectral images are now globally available at an unprecedented spatio-temporal resolution of a couple of tens of meters every few days. These new datasets will allow better monitoring of land cover and flood locations in watersheds such as the Guayas Basin.

**Acknowledgments:** This work was supported by SENESCYT under X.R.S. scholarship. The authors thank ESA for providing the ENVISAT ASAR images through the "Monitoring floods in the Guayas basin (Ecuador) using ERS-2 SAR and ENVISAT ASAR images" (ID 13844) proposal.

**Author Contributions:** Frédéric Frappart and Luc Bourrel conceived, designed the study and processed the ASAR images. Nicolas Brodu and Ximena Riofrío Salazar processed the ASAR images. Frédéric Baup, José Darrozes and Rodrigo Pombosa were involved in the analysis of the results. All authors contributed to the writing of the manuscript.

**Conflicts of Interest:** The authors declare no conflict of interest.

## References

1. Horel, J.D.; Cornejo Garrido, A.G. Convection along the coast of Northern Peru during 1983—Spatial and temporal variation of clouds and rainfall. *Mon. Weather Rev.* **1986**, *114*, 2091–2105. [CrossRef]
2. Bendix, A.; Bendix, J. Heavy rainfall episodes in Ecuador during El Niño events and associated regional atmospheric circulation and SST patterns. *Adv. Geosci.* **2006**, *6*, 43–49. [CrossRef]
3. Bourrel, L.; Rau, P.; Dewitte, B.; Labat, D.; Lavado, W.; Coutaud, A.; Vera, A.; Alvarado, A.; Ordoñez, J. Low-frequency modulation and trend of the relationship between ENSO and precipitation along the Northern to Center Peruvian Pacific coast. *Hydrol. Proc.* **2015**, *29*, 1252–1266. [CrossRef]
4. Rau, P.; Bourrel, L.; Labat, D.; Melo, P.; Dewitte, B.; Frappart, F.; Lavado, W.; Felipe, O. Regionalization of rainfall over the Peruvian Pacific slope and coast. *Int. J. Clim.* **2016**. [CrossRef]
5. Lyon, B.; Barnston, A.G. ENSO and the spatial extent of interannual precipitation extremes in tropical land areas. *J. Clim.* **2005**, *18*, 5095–5109. [CrossRef]
6. Vuille, M.; Raymond, S.B.; Keimig, F. Climate Variability in the Andes of Ecuador and its relation to tropical Pacific and Atlantic Sea Surface Temperature anomalies. *J. Clim.* **2000**, *13*, 2520–2535. [CrossRef]
7. Arteaga, K.; Tutasi, P.; Jiménez, R. Climatic variability related to El Niño in Ecuador—A historical background. *Adv. Geosci.* **2006**, *6*, 237–241. [CrossRef]
8. Carsel, K.M.; Pingel, N.D.; Ford, D.T. Quantifying the benefit of a flood warning system. *Nat. Hazards Rev.* **2004**, *5*, 131–140. [CrossRef]
9. Werner, M.; Reggiani, P.; De Roo, A.; Bates, P.; Sprokkereef, E. Flood forecasting and warning at the river basin and at the European scale. *Nat. Hazards* **2005**, *36*, 25–42. [CrossRef]
10. Rosenqvist, A.; Birkett, C.M. Evaluation of JERS-1 SAR mosaics for hydrological applications in the Congo river basin. *Int. J. Remote Sens.* **2002**, *23*, 1283–1302. [CrossRef]
11. Hess, L.L.; Melack, J.M.; Novo, E.M.L.M.; Barbosa, C.C.F.; Gastil, M. Dual-season mapping of wetland inundation and vegetation for the Central Amazon region. *Remote Sens. Environ.* **2003**, *87*, 404–428. [CrossRef]
12. Frappart, F.; Seyler, F.; Martinez, J.-M.; León, J.G.; Cazenave, A. Floodplain water storage in the Negro River basin estimated from microwave remote sensing of inundation area and water levels. *Remote Sens. Environ.* **2005**, *99*, 387–399. [CrossRef]
13. Betbeder, J.; Gond, V.; Frappart, F.; Baghdadi, N.; Briant, G.; Bartholomé, E. Mapping of Central Africa forested wetlands using remote sensing. *IEEE J. Sel. Top. Appl. Earth Obs. Remote Sens.* **2014**, *7*, 531–542. [CrossRef]
14. Kasischke, E.S.; Smith, K.B.; Bourgeau-Chavez, L.L.; Romanowicz, E.A.; Brunzell, S.; Richardson, C.J. Effects of seasonal hydrologic patterns in south Florida wetlands on radar backscatter measured from ERS-2 SAR imagery. *Remote Sens. Environ.* **2003**, *88*, 423–441. [CrossRef]
15. Bourrel, L.; Phillips, L.; Moreau, S. The dynamics of floods in the Bolivian Amazon Basin. *Hydrol. Proc.* **2009**, *23*, 3161–3167. [CrossRef]
16. Kuenzer, C.; Guo, H.; Huth, J.; Leinenkugel, P.; Li, X.; Dech, S. Flood mapping and flood dynamics of the Mekong delta: ENVISAT-ASAR-WSM based time series analyses. *Remote Sens.* **2013**, *5*, 687–715. [CrossRef]

17. Westerhoff, R.S.; Kleuskens, M.P.H.; Winsemius, H.C.; Huizinga, H.J.; Brakenridge, G.R.; Bishop, C. Automated global water mapping based on wide-swath orbital synthetic-aperture radar. *Hydrol. Earth Syst. Sci.* **2013**, *17*, 651–663. [CrossRef]

18. Schumann, G.; Di Baldassarre, G.; Bates, P.D. The utility of spaceborne radar to render flood inundation maps based on multialgorithm ensembles. *IEEE Trans. Geosci. Remote Sens.* **2009**, *47*, 2801–2807. [CrossRef]

19. Schlaffer, S.; Matgen, P.; Hollaus, M.; Wagner, W. Flood detection from multi-temporal SAR data using harmonic analysis and change detection. *Int. J. Appl. Earth Obs. Geoinf.* **2015**, *38*, 15–24. [CrossRef]

20. Bartsch, A.; Doubkova, M.; Path, C.; Sabel, D.; Wagner, W.; Wolski, P. River flow and wetland monitoring with ENVISAT ASAR global mode in the Okavango Basin and Delta. In Proceedings of the Second IASTED Africa WRM Conference, Gaborone, Botswana, 8–10 September 2008; Acta Press: Anaheim, CA, USA, 2008; Volume 602, pp. 152–156.

21. Bartsch, A.; Wagner, W.; Scipal, K.; Pathe, C.; Sabel, D.; Wolski, P. Global monitoring of wetlands—The value of ENVISAT ASAR Global mode. *J. Environ. Manag.* **2009**, *90*, 2226–2233. [CrossRef] [PubMed]

22. O'Grady, D.; Leblanc, M.; Gillieson, D. Use of ENVISAT ASAR Global Monitoring Mode to complement optical data in the mapping of rapid broad-scale flooding in Pakistan. *Hydrol. Earth Syst. Sci.* **2011**, *15*, 3475–3494. [CrossRef]

23. Rossel, F.; Cassier, E.; Gómez, G. Las inundaciones en la zona costera ecuatoriana: Causas—Obras de protección existentes y previstas. *Bulletin de l'Institut Français d'Études Andines* **1996**, *25*, 399–420.

24. Borbor-Cordova, M.; Boyer, E.; McDowell, W.; Hall, C. Nitrogen and phosphorus budgets for a tropical watershed impacted by agricultural land use: Guayas, Ecuador. *Biogeochemistry* **2006**, *79*, 135–161. [CrossRef]

25. Bourrel, L.; Melo, P.; Vera, A.; Pombosa, R.; Guyot, J.-L. Study of the erosion risks of the Ecuadorian Pacific coast under the influence of ENSO phenomenom: Case of the Esmeraldas and Guayas basins. In Proceedings of the International Conference on the Status and Future of the World's Large Rivers, Vienna, Austria, 11–14 April 2011.

26. Waite, P.J. Competition for water resources of the Rio Guayas, Ecuador. In *Optimal Allocation of Water Resources, Proceedings of the Exeter Symposium, Exeter, UK, 19–30 July 1982*; Lowing, M.J., Ed.; IAHS Publications No. 135; IAHS Redbooks: Exeter, UK, 1982; pp. 79–88.

27. Twilley, R.R.; Cárdenas, W.; Rivera-Monroy, V.H.; Espinoza, J.; Suescum, R.; Armijos, M.M.; Solórzano, L. The Gulf of Guayaquil and the Guayas river estuary, Ecuador. In *Coastal Marine Ecosystems of Latin America. Ecological Studies*; Seeliger, U., Kjerfve, B., Eds.; Springer: Berlin, Germany, 2001; pp. 245–263.

28. Desnos, Y.-L.; Buck, C.; Guijarro, J.; Suchail, J.-L.; Torres, R.; Attema, E. ASAR—Envisat's Advanced Synthetic Aperture Radar. Building on ERS achievements towards future watch missions. *ESA Bull.* **2000**, *102*, 91–100.

29. Pathe, C.; Wagner, W.; Sabel, D.; Doubkova, M.; Basara, J.B. Using ENVISAT ASAR Global Mode Data for Surface Soil Moisture Retrieval Over Oklahoma, USA. *IEEE Trans. Geosci. Remote Sens.* **2009**, *47*, 468–480. [CrossRef]

30. Baup, F.; Mougin, E.; de Rosnay, P.; Hiernaux, P.; Frappart, F.; Frison, P.-L.; Zribi, M.; Viarre, J. Mapping surface soil moisture over the Gourma mesoscale site (Mali) by using ENVISAT ASAR data. *Hydrol. Earth Syst. Sci.* **2011**, *15*, 603–616. [CrossRef]

31. Henry, J.B.; Chastanet, P.; Fellah, K.; Desnos, Y.-L. Envisat multi-polarized ASAR data for flood mapping. *Int. J. Remote Sens.* **2006**, *27*, 1921–1929. [CrossRef]

32. EOLI (Earth Observation Link). The European Space Agency's Client for Earth Observation Catalogue and Ordering Services. Available online: https://earth.esa.int/web/guest/eoli (accessed on 1 April 2013).

33. Tapia Aldas, J.C. Modelización Hidrológica de un Area Experimental en la Cuenca del Rio Guayas en la Producción de Caudales y Sedimentos. Ph.D. Thesis, Universidad Nacional de la Plata, La Plata, Argentina, 2012.

34. Brakenridge, G.R. *Global Active Archive of Large Flood Events, Dartmouth Flood Observatory*; University of Colorado: Boulder, CO, USA, 2012.

35. Brakenridge, G.R.; Anderson, E. MODIS-based flood detection, mapping, and measurement: The potential for operational hydrological applications. In *Transboundary Floods: Reducing the Risks through Flood Management*; Marsalek, J., Stancalie, G., Balint, G., Eds.; Springer: Dordrecht, The Netherlands, 2006; p. 16.

36. *Flood Observatory Database*; University of Colorado: Boulder, CO, USA, 2015. Available online: http://floodobservatory.colorado.edu (accessed on 21 September 2015).

37. Rosich, B.; Meadows, P. *Absolute Calibration of ASAR Level 1 Products*; ESA/ESRIN, ENVI-CLVL-EOPG-TN-03-0010; European Space Agency: Noordwijk, The Netherland, 2004. Available online: https://earth.esa.int/web/guest/-/absolute-calibration-of-asar-level-1-products-generated-with-pf-asar-4503 (accessed on 21 March 2013).

38. European Space Agency (ESA). *Absolute Calibration of ASAR Level 1 Products Generated with PF-ASAR*, 15th ed.; European Space Agency: Rome, Italy, 2004.

39. Baghdadi, N.; Bernier, M.; Gauthier, R.; Neeson, I. Evaluation of C-band SAR data for wetlands mapping. *Int. J. Remote Sens.* **2001**, *22*, 71–88. [CrossRef]

40. Ulaby, F.T.; Moore, R.K.; Fung, A.K. *Microwave Remote Sensing, Active and Passive, 2, Radar Remote Sensing and Surface Scattering and Emission Theory*; Addison-Wesley: Reading, MA, USA, 1982.

41. Shi, J.; Dozier, J.; Rott, H. Snow mapping in Alpine regions with Synthetic Aperture Radar. *IEEE Trans. Geosci. Remote Sens.* **1994**, *32*, 152–158.

42. Lee, J.S. Digital image smoothing and the sigma filter. *Comput. Vis. Graph. Image Process.* **1983**, *24*, 255–269. [CrossRef]

43. Frappart, F.; Bourrel, L.; Riofrio Salazar, X.; Baup, F.; Darrozes, J.; Pombosa, R. Spatiotemporal dynamics of the floods in the Guayas watershed (Ecuadorian Pacific Coast) using ASAR images. In Proceedings of the 2015 IEEE International Geoscience and Remote Sensing Symposium (IGARSS), Milan, Italy, 26–31 July 2015; pp. 2515–2518.

44. Kaufman, Y.J.; Remer, L.A. Detection of forests using mid-IR reflectance: An application for aerosol studies. *IEEE Trans. Geosci. Remote Sens.* **1994**, *32*, 672–683. [CrossRef]

45. Miettinen, J.; Liew, S.C. Separability of insular South-east Asian woody plantation species in the 50 m resolution ALOS PALSAR mosaic product. *Remote Sens. Lett.* **2011**, *2*, 299–307. [CrossRef]

46. Richards, J.A. *Remote Sensing Digital Image Analysis*; Springer: Berlin, Germany, 1999.

47. Gianinetto, M.; Villa, P.; Lechi, G. Postflood damage evaluation using Landsat TM and ETM+ data integrated with DEM. *IEEE Trans. Geosci. Remote Sens.* **2006**, *44*, 236–243. [CrossRef]

48. Papa, F.; Frappart, F.; Malbéteau, Y.; Shamsudduha, M.; Venugopal, V.; Sekhar, M.; Ramillien, G.; Prigent, C.; Aires, F.; Pandey, R.K.; et al. Satellite-derived Surface and Sub-surface Water Storage in the Ganges-Brahmaputra River Basin. *J. Hydrol. Reg. Stud.* **2015**, *4*, 15–35. [CrossRef]

49. Skakun, S.; Kussul, N.; Shelestov, A.; Kussul, O. Flood hazard and flood risk assessment using a time series of satellite images: A case study in Namibia. *Risk Anal.* **2014**, *34*, 1521–1537. [CrossRef] [PubMed]

50. Arias-Hidalgo, M.; Gonzalo, V.-C.; van Griensven, A.; Solórzano, G.; Villa-Cox, R.; Mynett, A.E.; Debels, P. A decision framework for wetland management in a river basin context: The "Abras de Mantequilla" case study in the Guayas River Basin, Ecuador. *Environ. Sci. Policy* **2013**, *34*, 103–114. [CrossRef]

MDPI

St. Alban-Anlage 66

4052 Basel, Switzerland

Tel. +41 61 683 77 34

Fax +41 61 302 89 18

http://www.mdpi.com

*Water* Editorial Office

E-mail: water@mdpi.com

http://www.mdpi.com/journal/water

www.ingramcontent.com/pod-product-compliance
Lightning Source LLC
Chambersburg PA
CBHW051839210326

41597CB00033B/5705